Grazing Communities

Studies in Environmental Anthropology and Ethnobiology

General Editor: **Roy Ellen**, FBA
Emeritus Professor of Anthropology and Human Ecology, University of Kent at Canterbury

Interest in environmental anthropology and ethnobiological knowledge has grown steadily in recent years, reflecting national and international concern about the environment and developing research priorities. "Studies in Environmental Anthropology and Ethnobiology" is an international series based at the University of Kent at Canterbury. It is a vehicle for publishing up-to-date monographs and edited works on particular issues, themes, places, or peoples that focus on the interrelationship between society, culture, and the environment.

Recent volumes:

Volume 29
Grazing Communities
Pastoralism on the Move and Biocultural Heritage Frictions
Edited by Letizia Bindi

Volume 28
Delta Life
Exploring Dynamic Environments Where Rivers Meet the Sea
Edited by Franz Krause and Mark Harris

Volume 27
Nature Wars
Essays around a Contested Concept
Roy Ellen

Volume 26
Ecological Nostalgias
Memory, Affect and Creativity in Times of Ecological Upheavals
Edited by Olivia Angé and David Berliner

Volume 25
Birds of Passage
Hunting and Conservation in Malta
Mark-Anthony Falzon

Volume 24
At Home on the Waves
Human Habitation of the Sea from the Mesolithic to Today
Edited by Tanya J. King and Gary Robinson

Volume 23
Edges, Fringes, Frontiers
Integral Ecology, Indigenous Knowledge and Sustainability in Guyana
Thomas Henfrey

Volume 22
Indigeneity and the Sacred
Indigenous Revival and the Conservation of Sacred Natural Sites in the Americas
Edited by Fausto Sarmiento and Sarah Hitchner

Volume 21
Trees, Knots, and Outriggers
Environmental Knowledge in the Northeast Kula Ring
Frederick H. Damon

Volume 20
Beyond the Lens of Conservation
Malagasy and Swiss Imaginations of One Another
Eva Keller

For a full volume listing, please see the series page on our website:
http://berghahnbooks.com/series/environmental-anthropology-and-ethnobiology

Grazing Communities

Pastoralism on the Move and
Biocultural Heritage Frictions

Edited by Letizia Bindi

berghahn
NEW YORK · OXFORD
www.berghahnbooks.com

First published in 2022 by
Berghahn Books
www.berghahnbooks.com

© 2022, 2024 Letizia Bindi
First paperback edition published in 2024

All rights reserved. Except for the quotation of short passages for the purposes of criticism and review, no part of this book may be reproduced in any form or by any means, electronic or mechanical, including photocopying, recording, or any information storage and retrieval system now known or to be invented, without written permission of the publisher.

Library of Congress Cataloging-in-Publication Data
A C.I.P. cataloging record is available from the Library of Congress
Library of Congress Cataloging in Publication Control Number: 2022014356

British Library Cataloguing in Publication Data
A catalogue record for this book is available from the British Library

ISBN 978-1-80073-475-3 hardback
ISBN 978-1-80539-333-7 paperback
ISBN 978-1-80073-476-0 epub
ISBN 978-1-80073-667-2 web pdf

https://doi.org/10.3167/9781800734753

The electronic open access publication of *Grazing Communities: Pastoralism on the Move and Biocultural Heritage Frictions* has been made available under a CC BY-NC-ND 4.0 license as a part of the Berghahn Open Migration and Development Studies initiative.

This work is published subject to a Creative Commons Attribution Noncommercial No Derivatives 4.0 License. The terms of the license can be found at http://creativecommons.org/licenses/by-nc-nd/4.0/. For uses beyond those covered in the license contact Berghahn Books.

Contents

List of Figures vii

Foreword ix
  Tim Ingold

Introduction 1
  Letizia Bindi

Part I. Pastoralism as a Biocultural Heritage?

Chapter 1. Transhumance in Greece: Multifunctionality as an Asset for Sustainable Development 23
  Athanasios Ragkos

Chapter 2. The Conflict of Itinerant Pastoralism in the Piedmont Po Plain (Collina Po Biosphere Reserve, Italy) 44
  Dino Genovese, Ippolito Ostellino, and Luca Maria Battaglini

Chapter 3. Between Two Different Worlds: Pastoralism and Protected Natural Areas in Provence-Alpes-Côte d'Azur 61
  Jean-Claude Duclos and Patrick Fabre

Chapter 4. Reintroducing Bears and Restoring Shepherding Practices: The Production of a Wild Heritage Landscape in the Central Pyrenees 81
  Lluís Ferrer and Ferran Pons-Raga

Chapter 5. Transhumance in Kelmend, Northern Albania: Traditions, Contemporary Challenges, and Sustainable Development 102
  Martine Wolff

Chapter 6. Revisiting Transhumance from Stilfs, South Tyrol, Italy: The Everyday Diverse Economy of a Forgotten Alternative Food Network 121
  Annalisa Colombino and Jeffrey John Powers

Part II. Discontinuities and Transformations

Chapter 7. Transhumance Is the New Black: Fragile Rangelands and Local Regeneration — 149
Letizia Bindi

Chapter 8. Continuities and Disruptions in Transhumance Practices in the Silesian Beskids (Poland): The Case of Koniaków Village — 174
Katarzyna Marcol and Maciej Kurcz

Chapter 9. Contemporary Transformation of the Pastoral System in the Romanian Carpathian: A Case Study from Maramures Region — 203
Cosmin Marius Ivașcu and Anamaria Iuga

Chapter 10. Mountain Pasture in Friuli (Italy): Past and Present — 222
Špela Ledinek Lozej

Chapter 11. From Nomadism to Ranching Economy: Reindeer Transhumance among the Finnish Sámi — 241
Nuccio Mazzullo and Hannah Strauss-Mazzullo

Chapter 12. Wandering Shepherds: New and Old Transhumances in Sardinia and Sicily — 259
Sebastiano Mannia

Chapter 13. The Coexistence of Transhumance Shepherding Practices and Tourism on Bjelašnica Mountain in Bosnia and Herzegovina — 280
Manca Filak and Žiga Gorišek

Afterword. Desire for Transhumance — 300
Cyril Isnart

Index — 308

Figures

2.1.	Itinerant flock in a corn field after cultivation. Collina Po Biosphere Reserve, 2019. © Dino Genovese	47
3.1.	National park of Mercantour, 2019. © Patrick Fabre	69
3.2.	Natural regional park of Verdon, 2019. © Lionel Roux	72
4.1.	A single large flock guarded by a professional shepherd grazing in a mountain pasture, Sentein, 2019. © Lluís Ferrer	94
6.1.	The Stilfser Alm from the Upper Alm, 2019. © Annalisa Colombino	129
6.2.	Laura, the *Sennerin* of Stilfer Alm, making cheese in the summer, 2016. © Jeffrey John Powers	130
7.1.	Lu vic p'dent, Bojano, 2019. © Rossella De Rosa	154
7.2.	The route of transhumance, Amatrice, 2019. © Letizia Bindi	159
8.1.	The Shepherd's Centre, Koniaków, 2019. © Katarzyna Marcol	177
8.2.	An exhibition of shepherding utensils in the Shepherd's Centre, Koniaków, 2016. © Katarzyna Marcol	178
8.3.	The ritual of mixing of the sheep, Koniaków, 2017. © Katarzyna Marcol	193
9.1.	Spring grazing in the hay meadows situated in close proximity to the village, Șurdești, 2012. © Anamaria Iuga	209
9.2.	Before the sheep leave for grazing, the shepherds' leader plays the natural trumpet, Ieud, 2016. © Anamaria Iuga	211
9.3.	*Casă în câmp* (house in the fields) with *șopru* for hay storage and stables for cows. Sheep are kept under open sky and are moved on the terrain to improve the vegetation, Ieud, 2015. © Cosmin Marius Ivascu	213
10.1.	Herder and cheesemaker Angelo Tessin, Pian Mazzega alp (Malga Pian Mazzega), Piancavallo, 2016. © Špela Ledinek Lozej	230
10.2.	Fleons di sotto alp (Malga Fleons di sotto), 2017. © Špela Ledinek Lozej	233

12.1.	Transumanza in Sardegna, 1959. © János Reismann	266
12.2.	Transumanza, 2020. © Anna Piroddi	273
13.1.	Lukomir village, Bjelašnica mountain, 2014. © Žiga Gorišek	281
13.2.	Ismet while shepherding in Bare, with snowy Bjelašnica in the background, Hadžići, 2014. © Manca Filak	285
13.3.	New location of Lijetna Bašta at the main entrance of Lukomir, 2017. © Žiga Gorišek	288

Foreword

Tim Ingold

"Men make their own history," wrote Karl Marx in the *Eighteenth Brumaire* of 1852, "but they do not make it as they please; they do not make it under circumstances chosen by themselves, but under circumstances existing already, given and transmitted from the past." And not just men, we would nowadays acknowledge, but women and children too! Human beings all, we find ourselves thrown into the world at times, and in places, over which we had no choice, and fated to carry on our lives from there. Yet in the things we do, purposively and often with an eye for what we imagine as the future, we lay down the conditions that generations coming after us will have to deal with in their turn. The combined efforts of historians, sociologists, and anthropologists, over the past century and a half, have scarcely improved on Marx's original insight. Yet it remains limited. For never in history has any human community existed that has not involved, in its collective life and in the reach of its relationships, living beings of various nonhuman kinds. There have always been animals around, not to mention the plant life on which all animals—humans included—depend. Thus, every community, when it comes to its species composition, is necessarily hybrid. But if that is so, then what role do nonhumans play in history? And if they are history-makers too, then whose history is it? It cannot belong exclusively to humans. Hybrid communities can surely fashion only hybrid histories.

I came to think of this as I was reading through the chapters of the remarkable book that you now have before you. Maybe you thought that its subject matter is of marginal significance in the history of the world. Pastoral herdsmen and their flocks, tucked away in borderland enclaves, are they not the peripheral cast-offs of a history that has marched onwards without them? In most contemporary societies, the migratory movements that herdsmen undertake with their animals are treated as the vestiges of a vanishing past, while countryside woven by centuries of grazing has been converted into tapestry, hanging from the walls of an architecture that has turned landscape into scenery. We tend to think of history as a uniquely human achievement, progressively built up against the backdrop of a recalcitrant nature. It is the story of the rise of civilization. And while this story presents herding as an advance over hunting—with the animals, now domesticated, brought under a measure of human control—it was but a small step. The land remained untamed, people still wandered, and those ties to the soil that laid the foundations for civilization remained weak. Progress depends on people settling down. And with that, pastoralism, as a way of life, is bound to disappear. If it survives at all, it will be as an object of conservation, as part of a common human heritage, to be preserved—like the ruins of ancient empires—for the instruction of the young and the enchantment of tourists.

Yet this doctrine of progress is very recent in the larger scheme of things. Originating in Europe, it has been exported in a history of colonialism that has taken it around the globe. These four or five centuries of colonial expansion, however, are a mere blip in the history of the world. And if we take a longer view, we find that for millennium after millennium, people have been living alongside herds of animals, sharing their lands and lives with them, to varying degrees of closeness or intimacy, and moving from place to place—as the animals do—in order to make the most of what a varied environment affords. Indeed, what looks marginal to us today has been the mainstream of history. Empires have risen and fallen, armies have trampled all over the earth, epidemics have swept through populations, philosophies have come and gone, but always there have been people and herds, seeking out ways to cohabit the earth. They are the one enduring constant of history. And here we are today, in a world ravaged by pandemic disease, overrun by the apparatus of war, living out our days in the shell of an imperial order that is collapsing all around us, our dreams of progress up in smoke, and looking to science for salvation. What next? Turning to the *longue durée* of history, the answer is plain. There will be people and animals carrying on their lives together, bringing forth a history that belongs equally to both.

The painful truth, now dawning on us, is that the doctrine of progress—with its corollaries of technological liberation, economic growth, rising living standards and ever-increasing longevity—is incompatible with the sustainability of life on earth. We cannot have both. This realization drives our present obsession with the idea of the Anthropocene, which pits utopian visions of an all-time geo-technical fix against apocalyptic prophecies of human extinction. Our anxiety, however, must bring relief to the animals. For animals don't do progress. Land, for them, is to be inhabited, not colonized; to be grazed, not built upon. Animals don't take chunks out of the earth to seal it against the sky, but they nibble to meet their needs, leaving the rest untouched. To be sure, theirs is neither an easy nor a comfortable life. There are illnesses to suffer, predators to evade, inclement weather to endure. That's been true for humans too. We thought we could escape life's vicissitudes, but now find to our cost that we cannot. No more than any other creatures do we humans have a foregone right to exist. Yet as Marx realized, there's no turning back the clock on history. We are where we are, and have to go on from here. But going on means joining with other lives in a spirit of coexistence, not supplanting them in a race to the top. It means relearning, from the animals and from those who herd them, how to become grazers ourselves. Our collective future depends on it.

Aberdeen, August 2020

Tim Ingold is Professor Emeritus of Social Anthropology at the University of Aberdeen. He has carried out fieldwork among Saami and Finnish people in Lapland, and has written on environment, technology, and social organization in the circumpolar North, on animals in human society, and on human ecology and evolutionary theory. His more recent work explores environmental perception and skilled practice. Ingold's current interests lie on the interface between anthropology, archaeology, art, and architecture. His recent books include *The Perception of the Environment* (2000), *Lines* (2007), *Being Alive* (2011), *Making* (2013), *The Life of Lines* (2015), *Anthropology and/as Education* (2018), and *Anthropology: Why It Matters* (2018). He is also the editor, inter alia, of *Companion Encyclopedia of Anthropology* (1994).

Introduction

Letizia Bindi

Pastoralism as Biocultural Heritage

Pastoralism is one of the most widespread and ancient forms of human subsistence, and one of the most studied by anthropologists. Indeed, much research has been dedicated to this practice even after the economic and political debate increasingly shifted to peasantry and agriculture as the pivotal rural activity from the end of the nineteenth century onwards (Herskovits 1926; Evans-Pritchard 1940; Barth 1961; Campbell 1964; Cole and Wolf 1974; Digard 1981; Ingold 1980; Herzfeld 1985, 1990; Angioni 1989; Galaty and Douglas 1990; Lewis 1961). Transhumance, in turn, is a particular form of husbandry and a knowledge-practice system essentially based on the seasonal movement of shepherds and herders together with their animals in search of grasslands, moving from the mountain to the plain as well as from inland and even mountainous regions to the large pasturages next to the coasts and back. This particular way of raising animals simultaneously defines a form of land use and a way of knowing/defining spaces and landscapes in many different parts of the world (European and non-European) and involving many different species animals (sheep, cows, horses, reindeer, camelids, and so on). This practice provides not only food and other products derived from animals but also provides a range of ecosystem services and common goods such as: a profound maintenance of local areas, a regeneration of the biodiversity of the land and of the animal lines being raised, and the continuation of specific forms of social organization and environmental resource management that are frequently held up today as an alternative to the unsustainability of industrial farming and breeding systems. In many European regions, scholars have documented the presence of transhumant populations since pre-Roman times: these populations are responsible for having profoundly shaped Europe's agrarian landscape and generating a network of cross-regional and cross-border mobility that also underlies the very first exchanges among European populations and cultures (Aime, Allovio, and

Viazzo 2001; Costello and Svensson 2018). At the same time, this practice has given rise to a powerful grammar of spaces, with its own logic, rules, timing, and interactions in which "footprints (are) akin to words or to punctuation" (Ingold and Vergunst 2008: 9; Palladino 2017; Bindi 2020).

Although intensive and sedentary livestock farming has clearly become dominant in the last few decades in Europe, as in many other rural regions of the world, we find many groups of shepherds and herders still engaged in this kind of activity with their specific rules, ways of life, and systems of beliefs. Given that a significant component of traditional shepherds and herders practice more or less extensive forms of transhumance, several classical studies have associated pastoral communities with nomadism/semi-nomadism. This framing sometimes ends up casting ethnographies of these groups as research on nomadic knowledge-practice systems more than studies of a specific way of livestock breeding. It is true that pastoralist communities usually move with their livestock from drylands or cold mountainous regions to more temperate and fertile ones, following the availability of grasslands and more favorable climactic conditions. Nonetheless, the focus of their "life world" should not be considered the nomadic experience, but rather their deep knowledge of territories and routes, their expert management of animals rooted in centuries-old traditions, and the consistent social organization and division of labor that this movement entails.

Transhumance as a whole encompasses biodiversity conservation and enhancement, capillary maintenance of the lands, the protection of ancient forms of settlement often connected with wise and sustainable uses of resources (water, soil, pastures) and traditional forms of cooperation and economic circularity that could today be reconsidered and updated in a profitable way.

More recently, sheep, cattle, and other livestock breeding has been transformed or influenced by processes of modernization, mechanization, and intensive milk/meat/wool production (Arhem 1984; Aronson 1980; Asad 1970; Chatty 1986; Ingold 1980; Nori and Scoones 2019; Salzman and Galaty 1990; Schlee 1989; Scoones 1995; Viazzo 1989). This has generated aspects of uncertainty, discontinuity, and change in practices, a shift in knowledge transmission and a vast socioeconomic transformation. Nonetheless, in many European countries the practice of transhumance still exists as an efficient form of extensive farming that profoundly influences the landscape, biodiversity conservation, raw-material processing, particular uses of vernacular architecture, traditional social structures, systems of knowledge, and practices at large. It is probably in this sense that Tim Ingold opportunely indicates in the Foreword to this volume the "spirit of coexistence" as a possible perspective central to every current of revitalization of extensive and traditional pastoralism in Europe and invites us to "relearn from the animals and from those who herd them, how to become grazers ourselves."

As a traditional and extensive form of livestock farming, transhumance is particularly relevant for inner, mountainous, insular, and fragile areas that play a concrete role of monitoring and safeguarding local areas, combatting the risk of increasing abandonment and environmental degradation. Over the centuries, sheep farming has been known and appreciated above all for its products. In addition to wool, which has lost much of its economic relevance, this system also provides important products from a nutritional point of view, the result of organic production strategies that meet high standards of animal welfare and health. Frequently, pastoral products also represent real sites for preserving local traditions, as demonstrated by the increase in PDO (Protected Designation of Origin) and PGI (Protected Geographical Indication) products, particularly in Europe. Meanwhile, traditional and extensive pastoralism is considered to be more sustainable, healthy, and respectful of the environment and animals and people than other forms of animal rearing, especially since the COVID-19 pandemic has brought to our attention the greatly enhanced risks of viral contagion through contact with highly polluted areas exploited by intensive industrial farming (May, Romberger, and Poole 2019).

This focus on health and sustainability is currently encompassed in the framework of the "One Health" approach, a sort of radical shift in the concept of healthcare developed to respond to contemporary global challenges. This approach promotes "the integration of human, environmental, and animal health through transdisciplinary cooperation and communication and [it seeks] to understand the complex disease interactions between microbes, domesticated animals and wildlife, humans, and their environments as brought about by ongoing globalized networking processes" (Rock et al. 2009). The aim of this approach is to design and implement programs, policies, laws, and research focused on achieving more effective outcomes in terms of public health (food safety, control of zoonoses, combatting antibiotic resistance, and so on). Such a shift represents a significant challenge for traditional pastoralists and transhumant populations in particular as they are obliged to deal with new risks and dramatic environmental and societal changes. In this sense, the "one health" approach is also connected on the one hand to a profound reconceptualization of the contemporary human-animal relationship (Aisher and Damoradaran 2016) and, on the other hand, to a radical critique of both the post-capitalist exploitation of livestock and the "pet-ization" and reification of animals in the urban framework and global market (Tsing Lowenhaupt 2000; Wolf 2015).

Meanwhile, the sustainability of extensive pastoralism is also threatened by the structural distinction between protected areas and pastoral areas (Chapter 3), a situation characterized by intense and significant frictions and an almost ideological as well as rhetorical opposition between environmentalists and pastoralists. In particular, herders face a growing risk of damage from predation connected to the greater proximity and growth of protected areas and parks, spaces in which efforts are underway to repopulate big carnivores (wolves and bears). Such repopulation policies have led to increased attacks on flocks and herds, causing conspicuous losses for the pastoralists who have chosen to continue breeding, rendering their livelihoods less and less certain and sustainable. In reality, pastoralism is by definition a system of meta-biodiversity because, given that this cultural practice is an important form of diversity, it thrives on the biodiversity of the environment in which it operates. A shepherd cannot produce in a degraded environment. This is why many natural parks and protected areas are established in or next to pastoral areas. Today, protected areas and grazing activities alike perform similar functions and meet common objectives: they offer ecosystem services and contribute to protecting and regenerating mountain environments and biodiversity. They even contribute to enhancing the tourist opportunities of certain areas, although tourists are sometimes kept at a distance from grazing flocks and herds due to the risks associated with the presence of predators, thereby impeding the kind of healthy relationships and exchanges that would be typical of this type of production.

Nonetheless, it has been widely documented that shepherds and herders are returning to various European regions (Battaglini et al. 2017; Fabre 2017; Brisebarre, Fabre, and Lebaudy 2009). Those engaged in maintaining and revitalizing the practice of transhumance as well as extensive breeding are therefore creating a potentially beneficial financial resource for depopulated internal and rural areas of Europe and, as such, a powerful tool for enhancing community resilience in the face of abandonment (Adger 2000; Folke 2006; Norris et al. 2008; Wilson 2012; Steiner and Markantoni 2013; Nori and Scoones 2019). Today's pastoral routes and communities thus offer an opportunity for close-grained ethnographies of local development and challenging opportunities for monitoring cultural landscapes (Bender 2001; Müller, Sutter, Wohlgemuth 2019; Müller 2021) as well as vast transformations in this knowledge-practice system. In particular, the practice of transhumance presents an ancient and traditional "life world" deeply rooted in ancient traditions and Indigenous cultures, such as in many peripheral and mountainous inland areas of Europe (Chapters 5, 6, and 12), often coupled with an increasing idealization of ancestralism, exotism, and essentialism in representing pastoral communities (Chapter 7). At the same time, it must be recognized that this knowledge-practice system has been able to endure, despite many uncertainties and difficulties, the passage of time and the influence of late modern, post-capitalist economic trends by rediscovering itself in the light of contemporary ecological, animal-rights, and community-oriented concerns and as a potential tourist attraction.

Pastoralism, and particularly transhumance, in the past represented a traditional and efficient way of responding to hostile environmental conditions; today, these forms of livestock farming seem to address and suggest new directions for adaptation to contemporary changes. The heritage turn thus seems to offer an antidote to the devaluation of the knowledge forms and practices connected with pastoralism that occurred in past decades.

Discontinuities and Transformations

In the last few years, the inclusion of transhumance in the UNESCO ICH List as "the seasonal droving of livestock across migratory routes in the Mediterranean and the Alps" has brought about a dramatic shift in ways of breeding and a deep transformation in traditional forms of pastoralism, now framed as a new global heritage item (Bendix, Eggert, and Peselman 2012). In Europe, the debate on forms of synergy between natural and cultural heritage is especially focused on habitat and landscape conservation (Magnaghi 2010). Research in this area has found transhumance to constitute a biocultural heritage element at the convergence of traditional knowledge and values systems, cultural and environmental landscapes and biodiversity, and an associated customary legal code encompassing a resilient way of earning a livelihood. In European and Mediterranean regions historically involved in transhumance, for example, this practice has deeply influenced both the social structures and ways of life of many people, including their kinship relations, symbolic representations, and settlements (Delavigne and Roy 2004). Moreover, this research on transhumant pathways is connected to the relatively recent debate in the social sciences, landscape design and planning, rural economy and environmental studies on inner areas and their revitalization and sustainable development (De Rossi 2019) that recognizes such spaces as "systemic margins" (Sassen 2014: 238) in which people are able to experiment with new forms of local economy, new ways of belonging and a potential new fundamental economy (Yuval-Davis 2006; Mee and Wright 2009; Wright 2014; Barbera et al. 2016).

This collective volume attempts to make sense of a multi-situated ethnography of transhumance heritagization processes with particular reference to the European regions involved in the recent (December 2019) inclusion of transhumance in the UNESCO ICH List and considering the ongoing move to extend UNESCO recognition to France, Spain, Albania, Croatia, and Romania as well. All the chapters presented in this volume are based on specific ethnographic research including interviews, participant observation, and the ethnographer/anthropologist's involvement in planning regeneration processes and sustainable development, as well as "ecosystem resilience" (Chapters 1, 5, and 6) initiatives, in the local areas under investigation. At the same time, these chapters display a more conceptual and critical approach to ways of representing and "packaging" transhumance, an approach based on a revived articulation between past and present (Chapters 8 and 12).

Some of these cases focus on the recent revitalization of transhumance and pastoralism as a cultural/tourist issue (Chapters 6 and 13), the ambivalent recognition of this form of biocultural heritage within the framework of ICH Lists as part of the "mise en forme" of cultural practices and landscapes (Chapters 3 and 7), and as a matter of communities' participation in the heritagization process. Indeed, transhumance is increasingly considered a tourist attraction more than a real agropastoral practice. With this "heritage-turn," there has been a growth of slow tourism in pastoral areas (Carnegie and McCabe 2008; Melotti 2013; Debarbieux et al. 2014; Monlor and Soy 2015) while pastoral landscapes and their relative products have been commodified (Korf, Hagman, and Emmenegger 2015; Kilburn 2018). Most of the discussions and projects launched in local communities around this practice are also grappling with the influence of national or even supranational levels of government and, more generally, are unfolding in the framework of development processes and the top-down exploitation of local areas (Maffi 2007; Rapport 2007; Bindi 2013). Driven by politics of acknowledgment and recognition, there has been more attention granted to communities' land ownership claims and demands to participate in the management of resources, particularly in the inner and more peripheral regions of various countries.

The inclusion of transhumance in the UNESCO ICH List and associated projects aimed at reviving pastoralism are currently being promoted and discussed in the framework of the participative governance of development processes. This arena includes organizations such as the FAO Pastoralist Knowledge Hub and transnational UNESCO group aimed at implementing the Safeguarding Plan for Transhumance as an Intangible Cultural Heritage (to date including only the European Regional Steering Group, but with the idea of extending membership to non-European countries as well) as well as discursive spaces such as discussions about CAP (Common Agriculture Policy) and the founding definition of the new Rural Development Programs being drafted in each EU country. These frameworks and policies impact pastoralism and transhumance in very different ways: by empowering individuals and informal groups, at times; by building capacity and enhancing community initiatives; and by stratifying the various levels of governance involved in local development processes. It is thus impossible not to include discussion of policy among the multiple aspects of an anthropological analysis of this practice. Today, the arena of transhumance is a crowded space involving many actors as well recurring conflict and frictions between conservationist and development-oriented currents; at times, this arena is characterized by increasingly and almost exclusively heritagized interpretations and narratives of the practice, framed as a tourist attraction to be appreciated as a networks of walkways (not even accompanied by animals, in some cases), a staging that evokes only a life world represented through the tones of nostalgic, folkloric storytelling.

The main aim of this volume is to present a range of ethnographic cases of different transhumant communities around Europe. These cases represent a powerful repertoire of local-level adjustments, local/supralocal policy mediation, and cogent accounts of highly local forms of interactions: between farmers and pastoralists; between lifestyles based on mobility and the sedentariness of late-modernity; between a deeply rooted notion of cultural landscape and the use of the environment as a simple resource provider; between a circular, cooperative and shared perception of family agriculture and husbandry and the "extractivist" logics of standardization and maximization typical of agri-food production; and, last but not least, between a human-animal bond based on coexistence (Haraway 2008; Davis, Maurstad, and Cowles 2013) and cooperation and an objectivizing lens that reifies notions of domestication and animal welfare.

A Multi-Situated and Pluri-Disciplinary Outlook

The anthropological approach to pastoralism has always been focused on practices, the transmission of knowledge and skills, and pastoralists' dynamic, productive but also harmonious relationship with the surrounding environment and their cooperative and mutually sustaining interaction with farmed animals as well as their deep knowledge of the biodiversity characterizing their surroundings. At the same time, it has become clear that understanding all these aspects requires a more holistic, broad-based approach capable of simultaneously considering multiple elements: the landscapes of pastoralism, the different breeds being raised, various techniques for transforming raw materials, and the broader historical, environmental, and cultural value of this biocultural heritage.

In this volume, therefore, we have tried to provide a panorama of different environmental contexts in Europe from a multidisciplinary point of observation. The authors present critical reconsiderations of this practice that range in focus from pastoralism in the Central Pyrenees (Chapter 4) to the Maison of Transhumance in France (Chapter 3); from Sami reindeer herding communities in the Finnish Arctic region (Chapter 11) to the pastoral communities of the Italian islands of Sardinia and Sicilia (Chapter 12) as well as several other cases of transhumance and extensive farming in Italy (Chapters 2, 6, and 7) and along the border with Slovenia (Chapter 10); chapters also address other European mountainous regions such as Romania (Chapter 9), Poland (Chapter 8), Albania (Chapter 5), Greece (Chapter 1), and Bosnia Herzegovina (Chapter 13).

The first section of the volume is essentially centered on formulating a multifocal definition of different forms of pastoralism in Europe as a biocultural heritage issue. In some cases, the chapters are not necessarily focused on using a specifically anthropological gaze; rather, the authors deliberately engage issues relating to the sustainability of the sector and the interaction between ecosystem components and the overall environmental value of pastoral practice. One example of this is the chapter dedicated to the structure of sheep and goat transhumance and its multifunctional value in contemporary Greece. A keen analysis of animal husbandry modernization and continuity is provided by Athanasios Ragkos (Chapter 1) in his observation of the practice of alternating between more intensive, mechanized livestock breeding methods in winter and more extensive, ancient pastoral behaviors in summer, albeit while maintaining traditional forms of social and economic labor organization. Working on the three main concepts of land, labor, and capital, Ragkos analyzes transhumance's modernization process and its ambivalent relationship to market conditions and traditional elements, aspects which are linked, for example, to alternative marketing as in Colombino and Powers's chapter (Chapter 6). Ragkos's attention to rangelands management and related conflicts is a comparative and close look at the increasing bureaucratization and commodification of pastoral routes and the illegal granting of permits on CAP-established grazing areas, a tendency also seen in Italy (Chapter 7).

In the case of the Piemonte Region of Italy, Dino Genovese, Ippolito Ostellino, and Luca Maria Battaglini (Chapter 2) describe vagrant pastoralism in its dynamic points of contact with the protected area of Collina Po, a site where traditional farming activities coexist with a renewed and multifunctional way of inhabiting the land. This coexistence of different lifestyles generates more or less latent conflicts, at times revealing a mutual incomprehension between forms of subsistence and mixed land use as practiced by farmers and shepherds, for example, as well as divergent representations of the landscape. Moreover, there is a persistent inability to recognize the ecosystem services provided by herders and shepherds and their contribution to the maintenance and valorization of landscapes (Chapters 1, 4, and 8).

A similar ambivalent coexistence between pastoralism and protected areas is addressed in the chapter by Jean-Claude Duclos and Patrick Fabre (Chapter 3) dedicated to the Maison of Transhumance and its efforts to safeguard pastoral culture activities in an area that nearly overlaps with the Crau Natural Reserve. Multiple methods of protection and land use come up against each other in this area, raising debates and giving rise to different processes of valorization in a multi-actor and multilevel context of local-area governance and heritagizing practices. This region that historically hosts transhumance is currently governed by layers of conservation and valorization frameworks (a national park, a regional natural park, and a natural reserve), and even the practice itself is the object of a huge heritagization effort aimed at transmitting and preserving traditional pastoralism and recognizing "the symbiotic relationship between the soil and the herd (that) has played a major part in the organization and management of the protection of the dry Crau" (Chapter 3 p. 9). In this sense, investing in transhumance conservation and valorization entails going beyond the "sanctuary model" hitherto characterizing the late-modern logic of protecting natural areas as generators of autonomous economies.

Historical changes in the way human activities shape the natural environment are likewise investigated in Lluís Ferrer and Ferran Pons-Raga's chapter on the reintroduction of bears and the restoration of local shepherding practices in the Central Pyrenees (Chapter 4). In this case as well, the return of bears as a symbol of wilderness in a particular area is considered the environmental hallmark of biodiversity conservation coupled with the recuperation of traditional pastoral practices and an essentially re-wilded/re-naturalized landscape. At the scale of local actors, however, such projects appear quite top-down and imposed from above, with shepherds excluded from any real participation in the governance of local sustainable development. The restoration project thus takes the form of a reinterpretation of the past in an environmentalist key, played out within a powerful heritagizing framework and local-area regeneration logics that prove essentially hegemonic and not at all participatory. The ambivalent coexistence of pastures and protected areas with the consequent increase in medium-large predators is one of the most controversial issues in the management of pastoral areas, a point of conflict that generates a great deal of uncertainty as also outlined in other chapters in this volume (Chapters 1, 3, and 9).

Martine Wolff's chapter on traditional pastoralism in the Northern Albanian region of Kelmend focuses on shepherds' harmonious, balanced relationships with animals, the environment and the landscape as well as the prevailing "mythopoesis" of the shepherd as a heritage-keeper and key stabilizer for economically and socially marginal parts of the Albanian mountains. Part of this ongoing ethnographic effort involves preparing a dossier to submit Albanian transhumance for consideration as part of the process of extending UNESCO ICH List recognition. Shepherds consider this an empowering process as well as an opportunity to think about the permanence and sustainability of a vagrant pastoral practice in the new, ambivalent heritage framework.

Extensive pastoralism is also explored as an alternative food network in the chapter by Annalisa Colombino and Jeoffrey John Powers based on research among the Alms of South Tyrol/Alto Adige, at the borders of Austria, Switzerland, and Italy. Traditional vertical transhumance and everyday life in Alms are conceptualized as a small, local production system based on a complex set of sustainable agriculture and biodiversity conservation practices as well as high-quality cheese-making. The authors find that the Alms's "diverse economy" is developing new networks and circuits of distribution and building a resilient practice of local heritage. Such resilient practices are also revealed to be valuable in other local examples, such as the Polish village of Koniaków (Chapter 8).

The chapters comprising the second section of the volume are focused on the changes and challenges faced by various European pastoral communities in the face of the heritage turn, as represented by recent discourses on mountainous, peripheral, and inner regions with highly cultivated and bred biodiversity as a form of ecologically and socioculturally sustainable agri-food production. At the same time, the chapters in this section adopt different points of view to reflect on the crucial valorization of the specific pastoral practice of transhumance as a UNESCO intangible heritage and their "discontinuities and transformations."

Chapter 7 outlines the sociocultural and economic transformations underlying the progressive definition and interpretation of transhumance as biocultural heritage in the European cultural and touristic scenario and the multilevel governance questions posed by its inclusion in the UNESCO ICH List, but also by the plan for safeguarding sheep tracks and traditional pastoral landscapes with particular reference to several Italian cases. In recent years, transhumance has been deeply heritagized through many different forms of "touristization" and the commodification of typical products of herding practices (milk, meat, and wool). At the same time, this increasingly diffused storytelling is accompanied by controversial uses of pastoral routes and pastures as well as the proliferation of illegal permits on pastures, a trend made possible in some ways by the intricacies of the CAP itself (Calandra 2019; Mencini 2021). This trend is posing challenges to efforts to support pastoral activities, an area that has never really been resolved in European agricultural policy and least of all in Italian policies for this sector.

An extremely interesting example of recent pastoral revitalization is provided by Katarzyna Marcol and Maciej Kurcz in Chapter 8. Typical Carpathian comanaged farming, grazing, and milk production (*salasz*) in which several shepherds' flocks are brought together under the supervision of a *baca* (chief shepherd) is interrogated through a sharp ethnographic investigation of the Koniaków village in the Beskid Mountains and specifically one family's entrepreneurial project (Maria and Piotr Kohut). The authors outline the points of alternating continuity and disruption in the practice and transmission of traditional form of animal husbandry and cheese-production as well as ambitious efforts to transform transhumance into a powerful driver of local-area valorization and tourist development. Moreover, the revival of traditional local forms of pastoralism empowers local actors to define and reconsider their common cultural identity while restoring and rewriting local collective memory. This takes place by rebuilding an embedded, shared past, which includes grappling with the strong, unique effects of the pandemic with its profound repercussions on cooperation and exchange among shepherds and collectively acknowledging the value of their common cultural heritage.

The chapter by Cosmin Marius Ivașcu and Anamaria Iuga presents the specific case of the Maramureș region in Romania. After outlining the different types of pastoral activities in this area and historical systems of familiar and labor organization, the authors describe the pastoral calendar in Maramureș as a way of understanding resource and environmental management, embedded local customs and ceremonial events, and the interdependent relationship between agricultural and pastoral activities essentially aimed at achieving food and economic self-sufficiency. Several dramatic changes, such as collectivization drives under the communist regime, forced shepherds in several villages to increase their productivity in order to meet not only their needs but also the state's requirements. The fall of communism brought other major changes, with many young shepherds and farmers adapting to new conditions and work strategies by migrating to EU countries, using new types of grassland and cultivation techniques, and substantially decreasing the size of their flocks after 1990. As a result, the old vagrant pastoralist system has transformed into a more sedentary one. Nonetheless, local communities continue to consider the maintenance of transhumant practices as a way of conserving typical landscapes with their embedded memories. Moreover, the chapter stresses the value of a multidisciplinary approach to extensive pastoralism, a point also asserted in several other chapters (Chapters 1, 2, 3, 7, and 10).

Amidst such discontinuities and transformations, Špela Ledinek Lozej examines the progressive decline of mountain pasturage connected to the increase in lowland agriculture in another border area, the Italian region of Friuli-Venezia Giulia (bordering Slovenia). She finds that the introduction of new animal breeds less adapted to mountain pastures, intense depopulation and urbanization, and the abandonment of agropastoral activities has caused alpine pasturing and dairy grazing to become largely economically and socially unsustainable. Today, different models for the use and management of environmental and agropastoral resources in the region need to be integrated with broader economic and social frameworks such as productive intensification (forage and dairy production) while also addressing the changing governance of local areas, as also outlined in other chapters (Chapters 2 and 7). At the same time, this ethnography highlights multifunctional extensification (a shift to catering, accommodations), heritage revitalization, and new forms of cooperation and solidarity-based agriculture linked by a "new passion for work and life in the Alps" and the mountains more generally, a narrative that is currently one of the most successful (Bindi 2021; Chapters 1, 6, 7, and 13).

Seasonal reindeer transhumance in Finnish Lapland, a very traditional knowledge-practice system rooted in the Sámi homeland, is the focus of the chapter by Nuccio Mazzullo and Hannah Strauss-Mazzullo. The paper begins by describing the contemporary routines of Sámi reindeer herding juxtaposed with a brief overview of the historical and administrative limitations that have constrained old practices. At the same time, the authors outline some technological innovations that have enabled the Sámi people to adapt in different ways to modern environmental and socioeconomic transformations, such as new form of pastoral economy and reorienting their practices towards tourism. The chapter also identifies conflicts and forms of ambivalence in relation to various efforts to protect and valorize local biocultural heritage such as, for example, the risks and uncertainties caused by the increase of predators in grazing areas that seriously jeopardizes the economic sustainability of reindeer husbandry, as also outlined in other chapters of the book (Chapters 2 and 4).

Discontinuities are likewise at the center of Sebastiano Mannia's deep rethinking of transhumance as a cultural and touristic heritage item as well as a mode of production and a specific way to use rural landscape and spaces. Based on the dynamics of contemporary markets and policies, the ethnography focuses on two insular regions—Sardinia and Sicilia. In the first, pastoralism is still very widespread and thriving while in the second transhumance and extensive farming are presently affected by various critical issues as well as bureaucratic regulations that complicate the entire supply chain in many areas. Through his ethnographic accounts, the author outlines how transhumance continues to represent a traditional form of husbandry and milk, meat and wool production in both regions. He shows, moreover, that the increasing heritagization of the practice has recently led to its refunctionalization and tourist *"evénementalisation"* and brought it under the management of several stakeholders such as Local Action Groups, Social Promotion Associations, or other forms of local culture and identity activism. Similar trends can be seen in the ethnographic examples of the Beskids Mountains (Chapter 8) or, more recently and to a more partial extent, in Bosnia and Herzegovina (Chapter 13).

The last chapter, by Manca Filak and Žiga Gorišek, emphasizes the value of transhumance and traditional pastoralism as a special tourist resource and driver in the locality of Lukomir in the Federation of Bosnia and Herzegovina, located on the southern slopes of the Bjelašnica mountain massif. Using visual ethnography, the chapter shows how tourism represents an innovative strategy of survival and preservation for transhumant and traditional pastoral practices as well as a driver for modernization in the region and community. Through tourism, Lukomir is represented as a particularly authentic, traditional, isolated, and remote "ethno" village, thereby building a narrative about this almost mythic locality as a "bay of peace" (the literal translation of the name Lukomir). Everyday life and traditional gestures and practices thus become objects of video-documentation, a process that also engenders new agency in terms of constructing the meaning of places and identity self-definition. The varied reactions to tourism at the local level range from rejection to acceptance, but on Bjelašnica this has not led to the abandonment or rejection of transhumance. In some cases, the coexistence of traditional pastoralism and tourism represents an opportunity to continue practicing transhumance and to use this practice as a basis for promoting tourism in the area despite emigration and depopulation, the privatization of public spaces, legal constraints and, more recently, the kind of implications stemming from the pandemic also outlined in Marcol and Kurcz's and Mannia's final remarks (Chapters 8 and 12).

Between a Commodifying Gaze and "Responsustainable" Tourism

Modernization processes have progressively represented pastoralism and transhumant farming practices in particular as obsolete and backward life worlds, thereby pushing heritage-keepers towards sedentary and increasingly intensive livestock-rearing systems. In fact, the new market logics cast traditional pastoralism as a potentially unproductive technique and a hindrance to the economic growth of the regions in which it has historically been practiced. The very idea of extensive pastoralism is framed as somehow in contradiction with economic logics based on intensive, sedentary agro-food production, "extractivism," and the exploitation of local natural resources beyond considerations of environmental and social regeneration and the sustainability of production practices. In a neoliberal market, the local knowledge deeply rooted in both the experience and practices of traditional and transhumant shepherds are usually considered insufficient to guarantee necessary earnings given the decrease in production costs. Indeed, this view wholly overlooks the reproducibility of the landscape and environment and its permanence over time thanks to grazing farming, and grants even less consideration to the potentially higher quality of the food or other products produced in pastures. In addition, until recently little or no attention has been paid to the quality of life of the animals being farmed. Given these persistent blind spots, a focus on extensive pastoralism and transhumance currently represents an important cautionary tale and opportunity to reflect on both the limits of development and the need to rethink the entire agro-food production chain.

At the opposite end of this continuum lies the denigrating logic characterizing transhumance as an unsustainable hindrance in relation to contemporary life. Today in particular, traditional pastoralism and transhumance are increasingly represented according to the rose-colored, romantic cliché that casts them as a form of life existing in harmonious balance with nature and through exoticizing, folkloric rhetoric expressed through tropes of "the good savage," lost authenticity, and "structural nostalgia" (Herzfeld 2004). In this framing, transhumant shepherds are represented as the guardians of an atavistic, picturesque rural past that is bent and molded to the rhetoric of pro-tourist narratives. Increasingly, it is also embraced by projects for the sustainable development of peripheral mountainous and inner areas driven by the more or less formal organizations and groups involved in development processes. This narrative has gained even more prominence with the COVID-19 pandemic because of the focus on small villages and remote areas as a respite from the excessive crowding and unsustainability of metropolitan life (Bindi 2021).

The pastoral "other"—imagined as existing cozily in its supposed ahistorical immobility, the opposite pole to the civilization of urban and post-capitalist consumption—has become a good imaginary for the travel industry to narrate and evoke, in keeping with recent tourism trends focused on greater contact with nature and close relationships with local populations. Transhumant paths have thus been heritagized and commoditized as primarily an asset of the landscape and local areas after having been constrained and governed by national laws (environmental and cultural protection rules set by ministries). They have been subsequently transformed into tourist routes involving fewer shepherds and, above all, fewer animals; in this way they become traces of a lost landscape, signs of a narrated map of the past. The visual arts, media, and web designers have also adopted this approach, adapting the memory of pastoralism for the tourism industry and articulating it in the form of a narrative reenactment, an event, and an attractive destination.

Given these shifts, contemporary extensive pastoralism spaces can be considered "friction zones" (Tsing Lowenhaupt 2000). On one hand, in fact, modern intensive agro-farming and "extractivism" has led to the increasing abandonment and disuse of transhumant landscapes and has upset the kind of production conditions—and especial human-animal interactions—typical of extensive pastoralism. At the same time, in many cases these repackaged pastoral routes and sites are reshaped as recovered lands, deeply heritagized at the intersection of multiple, conflicting moral regimes and economic systems.

Some more recent multidisciplinary and interdisciplinary analyses approach responsible tourism (Mihalic 2016) as the cornerstone of a new framework in which the traditional agricultural sector might become multifunctional and potentially sustainable, characterized by mobile and minimal structures, sustainable production methods and raw material processing, and forms of distribution proximity involving short supply chains distinguished by agroecological and socially responsible consumption as well as growing attention on the welfare of farmed animals. Heritage communities and environmentalist groups and associations are at the forefront of supporting sustainable development as they seek to develop strategies to recover and use pastoral spaces and the transhumant past in new ways. Through political recovery and rhetorical valorization, this biocultural heritage can be redefined as a common good in the global value chain in keeping with its inclusion in the UNESCO ICH List. Such a redefinition generates contemporary discourses on transhumance preservation and valorization as part of the debate on the limits of neoliberal economics and new moral imperatives to safeguard biocultural heritage while also preserving the environment and landscape. Transhumance is thus re-signified to be consumed as a "spectacle," with its "biopolitical function ... within the neoliberal state ..., decommissioned, remediated, and repurposed as ground(s) for popular recreation" (Krupar 2016: 116). Viewed in this way, extensive pastoralism becomes a field at the intersection of multidisciplinary and multiscalar research with highly local questions; a field positioned between past and future predicaments unfolding through fragmentation, disembedding and reassembling processes in a "zone of awkward engagement" (Tsing Lowenhaupt 2004); a field comprised of ecological activism, cultural heritage enhancement, and tourism promotion in a contemporary conflicting scenario.

Acknowledgments

The concept and part of the researches and cooperations of this book were developed in the framework of the activities of the EARTH Erasmus + 'EARTH Project' (598839-EPP-1-2018-1-IT-EPPKA2-CBHE-JP) funded by EACEA Erasmus Program.

Letizia Bindi has been Professor of Cultural and Social Anthropology at several Italian universities and a visiting scholar at several European universities. She is a member of the major national and international societies of cultural and social anthropology. She is presently a professor at the University of Molise, Italy, where she directs BIOCULT, the research center on biocultural heritage and local development. The main ongoing projects she coordinates include: the Erasmus + Capacity Building Project—EARTH (Education, Agriculture, Resources for Territories and Heritage) and the Italo-Argentinian Project—TraPP (Trashumancia y pastoralismo como elementos del patrimonio Inmaterial).

References

Adger, N. W. 2000. "Social and Ecological Resilience: Are They Related?" *Progress in Human Geography* 24(3): 347–64.
Aime, Maurizio, Stefano Allovio, and Pier Paolo Viazzo, eds. 2001. *Sapersi Muovere. Pastori Transumanti Di Roaschia*. Rome: Meltemi.
Aisher, Alex, and Vinita Damodaran. 2016. "Introduction: Human-Nature Interactions through a Multispecies Lens." *Conservation & Society* 14(4): 293–304.
Angioni, Giulio. 1989. *I Pascoli Erranti. Antropologia Del Pastore in Sardegna*. Napoli: Liguori.
Arhem, Kaj. 1984. "Two Sides of Development: Maasai Pastoralism and Wildlife Conservation in a Changing World." *Ethnos* 49: 186–210.
Aronson, Dan R. 1980. "Must Nomads Settle? Some Notes Toward Policy on the Future of Pastoralism." In *When Nomads Settle: Processes of Sedentarisation as Adaptation and Response*, ed. Philip Salzman, 173–84. New York: Praeger Publishers.
Asad, Talal. 1970. *The Kababish Arabs. Power, Authority and Consent in a Nomadic Tribe*. London: Berg.
Barbera, Filippo, Joselle Dagnes, Angelo Salento, and Ferdinando Spina. 2016. *Il Capitale Quotidiano. Un Manifesto Per L'economia Fondamentale*. Roma: Donzelli.

Barth, Frederik. 1961. *Nomads of South Persia: The Basseri Tribe of the Khamseh Confederacy*. London: Allen & Unwin.
Battaglini, Luca, Giovanna Fassio, Valentina Porcellana, and Marzia Verona. 2017. "Jeunes Bergers 'De Souche' et Par Choix: Le Cas Piémontais." In *Cammini Di Uomini, Cammini Di Animali*, ed. Katia Ballacchino and Letizia Bindi, 57–72. Campobasso: Il Bene Comune.
Bender, Barbara. 2001. "Landscapes On-The-Move." *Journal of Social Archaeology* 1(1): 75–89.
Bendix, Regina, Anita Eggert, and Angel Peselman, eds. 2012. *Heritage Regimes and the State*. Göttingen Studies in Cultural Property. Vol. 6. Göttingen: Universitätsverlag.
Bindi, Letizia. 2013. "Alla Fiera Delle Identità. Patrimoni, Turismo, Mercati." *Voci* X: 13–27.
———. 2020. "Take a Walk on the Shepherd's Side. Transhumance and Intangible Cultural Heritage." In *A Literary Anthropology of Migration and Belonging*, ed. Michelle Tisdel and Cecilie Fagerlid, 22–53. New York: Palgrave Macmillan.
———. 2021. "Oltre Il 'Piccoloborghismo'. Comunità Patrimoniali E Rigenerazione Delle Aree Fragili." *Dialoghi Mediterranei* 48. Retrieved 1 March 2021 from http://www.istitutoeuroarabo.it/DM/oltre-il-piccoloborghismo-comunita-patrimoniali-e-rigenerazione-delle-aree-fragili/.
Brisebarre, Antoine M., Patrick Fabre, and George Lebaudy, eds. 2009. *Sciences Sociales: Regards Sur Le Pastoralisme Contemporain En France*. Maison De La Transhumance. Avignon: Cardère Editeur.
Calandra, Lina. 2019. "Pascoli E Criminalità in Abruzzo. Quando La Ricerca Si Fa Denuncia (L'Aquila, 30 Giugno 2019)." *Semestrale Di Studi E Ricerche Di Geografia*. XXXI(2): 183–87.
Campbell, John. 1964. *Honour, Family and Patronage. A Study of Institutions and Moral Values in a Greek Mountain Community*. Oxford: Clarendon Press.
Carnegie, Edward, and Stephen McCabe. 2008. "Reenactment Events and Tourism: Meaning, Authenticity and Identity." *Current Issues in Tourism* 11: 349–68.
Chatty, Dawn. 1986. *From Camel to Truck: The Bedouin in The Modern World*. Vantage Press: New York.
Cole, John W., and Eric R. Wolf. 1974. The *Hidden Frontier: Ecology and Ethnicity in an Alpine Valley*. Berkeley: University of California.
Costello, Eugene, and Eva Svensson. 2018. *Historical Archaeologies of Transhumance Across Europe*. New York: Routledge.
Davis, Dona, Anita Maurstad, and Sarah Cowles. 2013. "Co-Being and Intra-Action in Horse-Human Relationships: A Multispecies Ethnography of Be(Com)Ing Human and Be(Com)Ing Horse." *Social Anthropology* 21(3): 322–35.
Debarbieux, Bernard, Mari Oiry Varacca, Gilles Rudaz, Daniel Maselli, Thomas Kohler, and Matthias Jurek. 2014. *Tourism in Mountain Regions. Hopes, Fears and Realities*. Genève: FAO Editions.
Delavigne, Anne-Élène, and Frédérique Roy. 2005. "Anthropological Analysis of a Group of Films on Pastoralism." *Natures Sciences Sociétés* 13(3): 279–83.
De Rossi, Antonio. 2019. *Riabitare l'Italia. Le aree interne tra abbandoni e riconquiste*. Roma: Donzelli Editore.

Digard, Jean-Pierre. 1981. *Techniques Des Nomades Baxtyari d'Iran*. Cambridge, UK: Cambridge University Press.
Evans-Pritchard, E. E. 1940. *The Nuer: A Description of the Modes of Livelihood and Political Institutions of a Nilotic People*. New York: Clarendon Press.
Fabre, Patrick. 2017. "Comment défendre la transhumance? L'exemple de la Maison de la transhumance et de ses projets." In *Cammini di uomini, cammini di animali. Transumanze, pastoralismi e patrimoni bio-culturali*, ed. Katia Ballacchino, 133–50. Campobasso: Il Bene Comune Editore.
Folke, Carl. 2006. "Resilience: The Emergence of a Perspective for Social-Ecological Systems Analyses." *Global Environmental Change* 16(3): 253–67.
Galaty, John G., and Johnson L. Douglas, eds. 1990. *The World of Pastoralism: Herding Systems in Comparative Networks*. New York: Gilford Press.
Haraway, Donna. 2008. *When Species Meet*. Minneapolis: University of Minnesota Press.
Herskovits, Melville. 1926. "The Cattle Complex in East Africa." *American Anthropologist* 28: 230–72; 361–80; 494–528; 630–64.
Herzfeld, Michael. 1985. *The Poetics of Manhood: Contest and Identity in a Cretan Mountain Village*. Princeton: Princeton University Press.
———. 1990. "Pride and Perjury: Time and Oath in the Mountain Villages of Crete." *MAN* (N.S.)25: 305–22.
———. 2004. *The Body Impolitic. Artisan and Artifice in the Global Hierarchy of Value*. Chicago: University of Chicago Press.
Ingold, Tim. 1980. *Hunters, Pastoralists and Ranchers: Reindeer Economies and Their Transformations*. Cambridge: Cambridge Studies in Social and Cultural Anthropology.
Ingold, Timothy, and John L. Vergunst, eds. 2008. *Ways of Walking: Ethnography and Practice on Foot*. London: Routledge.
Kilburn, Nicole. 2018. "Culinary Tourism, the Newest Crop in Southern Italy's Farms and Pastures." *Anthropology of Food* 13. Retrieved 11 February 2022 from https://doi.org/10.4000/aof.8384.
Korf, Benedikt, Hagman, Tobias, and Emmenegger, Rony. 2015. "Re-Spacing African Drylands: Territorialization, Sedentarization and Indigenous Commodification in the Ethiopian Pastoral Frontier." *Journal of Peasant Studies* 42(5): 881–901.
Krupar, Shiloh. 2016. "The Biopolitics of Spectacle: Salvation and Oversight at the Post-Military Nature Refuge." In *Global Spectacles*, ed. Bruce Magnusson and Zahi Zalloua, 116–53. Seattle: University of Washington Press.
Lewis, Ian M. 1961. "A Pastoral Democracy: A Study of Pastoralism and Politics among the Northern Somali Ngorongoro, Tanzania." *Ethnos* 49(3–4): 186–210.
Maffi, Luisa. 2007. "Biocultural Diversity and Sustainability." In *The Sage Handbook of Environment and Society*, ed. Luisa Maffi, 267–78. London: Sage.
Magnaghi, Alberto. 2010. *Il Progetto Locale. Verso La Coscienza Di Luogo*. Torino: Bollati Boringhieri.
May, Sara, Debra J. Romberger, and Jill A. Poole. 2012. "Respiratory Health Effects of Large Animal Farming Environments." *Journal of Toxicology and Environmental Health*. 15(8): 524–41. Retrieved 12 June 2020 from https://www.ncbi.nlm.nih.gov/pmc/articles/pmc4001716/.

Mee, Kathleen J., and Sarah Wright. 2009. "Geographies of Belonging." *Environment and Planning A: Economy and Space* 1: 772–79.

Melotti, Marxiano. 2013. "Turismo Culturale E Festival Di Rievocazione Storica. Il Re-Enactment Come Strategia Identitaria E Di Marketing Urbano." In *Mobilità Turistica Tra Crisi E Mutamento*, ed. Romina Deriu, 144–54. Milano: Franco Angeli.

Mencini, Giannandrea. 2021. *Pascoli Di Carta. Le Mani Sulla Montagna*. Vittorio Veneto: Kellerman Editore.

Mihalic, Tanja. 2016. "Sustainable-Responsible Tourism Discourse—Towards 'Responsustable' Tourism." *Journal of Cleaner Production* 111(Part B): 461–70.

Monlor, Neus, and Emma Soy. 2015. *El Pasturisme: Un Producte Turístic Nou Per a Les Comarques Gironines? Anàlisi De Recursos I Propostes De Futur*. Retrieved 25 July 2020 from https://premisg.costabrava.org/wpcontent/uploads/2017/11/02_ybettebarbaza_pasturisme_issu.pdf.

Müller, Oliver. 2021. "Making Landscapes of (Be)Longing: Territorialization in the Context of the EU Development Program Leader in North Rhine-Westphalia." *European Countryside* 13(1): 1–21.

Müller, Oliver, Olga Sutter, and Sina Wohlgemuth. 2019. "Translating the Bottom-Up Frame. Everyday Negotiations of the European Union's Rural Development Programme LEADER in Germany." *Anthropological Journal of European Cultures* 28(2): 45–65.

Nori, Michele, and Ian Scoones. 2019. "Pastoralism, Uncertainty and Resilience: Global Lessons from the Margins." *Pastoralism* 9/10. Retrieved 14 July 2020 from https://doi.org/10.1186/s13570–019–0146–8.

Norris, Fran H., Susan P. Stevens, Betty Pfefferbaum, Karen F. Wyche, and Rose L. Pfefferbaum. 2008. "Community Resilience as a Metaphor, Theory, Set of Capacities, and Strategy for Disaster Readiness." *American Journal of Community Psychology* 41: 127–50.

Palladino, Paolo. 2017. "Transhumance Revisited: on Mobility and Process between Ethnography and History." *Journal of Historical Sociology* 31(2): 119–33.

Rapport, D. J. 2007. "Healthy Ecosystems: an Evolving Paradigm." In *The Sage Handbook of Society and Environment*, ed. John Pretty et al., 431–441. London: Sage.

Rock, Melanie, Bonnie Jay Buntain, Jennifer M. Hatfield, and Benedikt Hallgrimsson. 2009. "Animal–Human Connections, 'One Health' and the Syndemic Approach to Prevention." *Social Scence & Medicine* 68(6): 991–95.

Salzman, Carl P., and John G. Galaty, eds. 1990. *Nomads in a Changing World*. Naples: Istituto Universitario Orientale.

Sassen, Saskia. 2014. *Brutality and Complexity in the Global Economy*. Cambridge, MA: Harvard University Press.

Schlee, Günther. 1989. *Identities on the Move*. Manchester, UK: Manchester University Press.

Scoones, Ian, ed. 1995. *Living with Uncertainty—New Directions in Pastoral Development*. London: IT.

Steiner, Artur, and Marianna Markantoni. 2013. "Unpacking Community Resilience Through Capacity for Change." *Community Development Journal* 49(3): 407–25.

Tsing Lowenhaupt, Anna. 2000. "The Global Situation." *Cultural Anthropologist* 15(3): 327–60.
———. 2004. *Friction: An Ethnography of Global Connection*. Princeton: Princeton University Press.
Viazzo, Pier Paolo. 1989. *Upland Communities: Environment, Population and Social Structure in the Alps since the Sixteenth Century*. Cambridge, UK: Cambridge University Press.
Wilson, Geoff. 2012. "Community Resilience, Globalization, and Transitional Pathways of Decision-Making." *Geoforum* 43(6): 1218–31.
Wolf, Meike. 2015. "Is There Really Such a Thing as 'One Health'? Thinking about a More Than Human World from the Perspective of Cultural Anthropology." *Social Science & Medicine* March 129: 5–11.
Wright, Stephen. 2014. "More-Than-Human, Emergent Belongings: A Weak Theory Approach." *Progress in Human Geography* 39(4): 391–41.
Yuval-Davis, Nira. 2006. "The Politics of Belonging." *Patterns of Prejudice* 40: 197–214.

PART I

Pastoralism as a Biocultural Heritage?

CHAPTER 1

Transhumance in Greece
Multifunctionality as an Asset for Sustainable Development

Athanasios Ragkos

Introduction

Sheep and goat transhumance (SGT) describes the seasonal movement of flocks between specific summer and winter domiciles, with the key objective to take advantage of natural vegetation in highland rangelands during summer and of mild weather conditions in the lowlands during winter (Nyssen et al. 2009). Therefore, transhumance is a specific form of "pastoralism," as it is based on grazing natural vegetation (Farinella, Nori, and Ragkos 2017), but differs from "nomadism," because flocks perform circular predefined movements (Vallerand 2014). It is also a multifunctional production system as it produces a wide range of goods and services jointly with food (milk, meat, and dairy products, etc.).

The multifunctional character of SGT provides important contributions towards sustainable and inclusive development and provides significant ecosystem services (D'Ottavio et al. 2018). The socioeconomic role of SGT is of utmost importance for mountainous/marginal/inland areas where the economic activity is not sufficiently diversified. SGT has been the main/only economic activity for centuries, and today it is still an important source of income even in communities in which other sectors have emerged (Farinella et al. 2017). In addition, SGT protects rural livelihoods, reducing depopulation of marginal and remote areas and is in fact critical for maintaining life and productivity in the more marginal and fragile territories of the Mediterranean region. Tailored to local conditions, SGT plays a vital environmental role by enhancing biodiversity and providing ecosystem services (Varela and Robles-Cruz 2016; D'Ottavio et al. 2018). In a variety of Mediterranean settings, transhumant systems make efficient use of natural resources and have shaped unique natural landscapes (Caballero et al. 2009), while also protecting genetic diversity by rearing autochthonous breeds. In addition, pastoral systems effectively mitigate climate change effects (e.g., soil carbon stock in rangelands) towards ecosystem resilience. The cultural heritage of SGT and the heterogeneity of its sociocultural contributions have characterized territories up to now: tacit knowledge concerning the functioning of local ecosystems, farm and land management; habits and customs; traditions, norms and tacit rules; processing skills. SGT in Greece, Italy, and Austria was recently included on the UNESCO list of Intangible Cultural Heritage of Humanity (UNESCO 2021).

The European Union has addressed the issues related to the multifunctionality of agricultural and livestock production systems through a broad array of policy measures. Since the accession of Greece in the European Union (EU) in 1981 the Common Agricultural Policy (CAP) has been applied, replacing all previous national agricultural policies. The CAP generally uses two types of policy tools available to all farmers in the EU, as described in Regulations and Directives, which are generally referred to as first-pillar and second-pillar policies. The former type includes all the measures that provide income support to EU farmers, either in the form of coupled subsidies to production and/or to cultivated land and animals reared or of direct payments, such as the Single Farm Payment and decoupled payments. The latter policies support rural development and provide financial incentives to farmers who wish to ameliorate their performance and competitiveness; to undertake commitments regarding their production practices; or to supply society with particular environmental and social services. These payments include agri-environmental and mountainous areas payments, financial support for investments of various types and other types of structural measures. However, it has been documented that, in Europe, pastoral systems share common challenges which are poorly addressed by the current policy framework (Ragkos and Nori 2016).

Transhumance in Greece has been examined by several authors from a rather broad range of disciplines. The available literature can be categorized into two types. The first approaches transhumance from an ethnological point of view (Nitsiakos 1995) and examines its historic evolution (Gkoltsiou 2011; Ntassiou, Doukas, and Karatassiou 2015). These studies present various aspects of the life and social structure of transhumant societies in the past, the evolution of the system under volatile historic and economic conditions, and the process of integration of nomadic and transhumant communities within the modern Greek society. A part of these studies considers the sheep and goat farming systems currently prevailing in the country (extensive or semi-intensive sedentary systems) as the natural evolution of animal production systems (for instance Hadjigeorgiou 2011). The second type of studies examines transhumance in terms of its ecological and environmental implications (Sidiropoulou et al. 2015; Sklavou et al. 2017) or evaluates its economic performance (Galanopoulos et al. 2011; Ragkos, Siasiou, et al. 2014) and management practices (Gidarakou and Apostolopoulos1995; Siasiou et al. 2018). Since the sector is undergoing a modernization process and is struggling to integrate to market conditions without compromising its traditional elements, the studies of the second type are actually gaining increasing attention.

The purpose of this chapter is to provide an outline of the Greek sheep and goat transhumance sector. In particular, detailed structural data of the system are presented, which demonstrate its dynamics across the country. Then, the actual operation of the system is described in detail, focusing on the particularities of SGT regarding the management of the three basic production factors (land, labor, capital). The chapter closes with a critical presentation of the main problems of the system that hinder its development and of measures to increase its dynamics and potential.

The Structure of Sheep and Goat Transhumance in Greece

In Greece there are currently two different types of transhumances with notable differences. The first involves cattle and the second, sheep and goats. Cattle transhumance is becoming increasingly popular, as it is supported heavily through the implementation of CAP Pillar I measures (Koutsou, Ragkos, and Karatassiou 2019) and has considerably lower requirements in human labor and capital. These characteristics render extensive bovine production an alluring alternative to sheep and goat transhumance, and it is thus being expanded in recent years, standing, however for no more than 6.5 percent of total cattle raised in Greece (Ragkos et al. 2013). Nevertheless, this chapter focuses on sheep and goat transhumance, which has played a significant role in socioeconomics for decades.

Compared to data from previous decades, sheep and goat transhumance has shown a decline in Greece. However, during the last thirty years the population of transhumant small ruminants has remained stable while the number of farms is decreasing, which indicates the formation of larger, viable farms. This structural evolution has been interpreted as a shrinkage of the system, and sheep and goat transhumance has been considered as a system of trivial importance. In the general context of support for the intensification of the Greek farming sector—which was further strengthened by EU CAP policies—SGT was ignored as an anachronistic system of little market potential and unable to support the modernization of the Greek farming sector. Nevertheless, available data show that this is not the case.

Collecting data about transhumance in Greece is not an easy task for two reasons: one, the lack of cohesive statistical databases, and, two, the multiple facets of the system in Greece. Based on processed data originally retrieved from the Greek Payment Authority of CAP Aid Schemes (OPEKEPE) for the year 2011 (THALES 2015), SGT is practiced in most parts of Greece by 3,051 farms, and transhumant sheep and goats account for almost 7.5 percent of the national flock. The center of SGT is Thessaly where 805 farms (26.4 percent) rear almost 338 thousand animals (33 percent), closely followed by Central Greece with a slightly lower number of farms (787, 25.8 percent) and a significantly lower transhumant sheep and goat population (214 thousand, 20.9 percent). Peloponnese, in the southern mainland part of the country, is ranked third (19.7 percent of farms, 16.8 percent of animals), followed by Epirus in the western part, Macedonia and Thrace in the north, and the islands, including the island of Crete. Compared to statistical data quoted by Syrakis (1925), the percentage of farms spending winter in Thessaly is almost stable (28 percent in 1924) but the percentage of farms in Central Greece and Peloponnese was much lower in 1924 (10.1 percent and 13.6 percent respectively). Since the 1960s, the contribution of transhumant sheep and goats to the national flock has been reduced from 30 percent to 7.5 percent (Hatziminaoglou 2004). In the summer, almost all mountainous areas of the country are grazed, to some extent, by transhumant small ruminants. Southern parts of the country (Peloponnese and Central Greece) are home to almost 42 percent of transhumant flocks during summer accounting for 34 percent of SGT. Nonetheless, more than 26 percent of the transhumant sheep and goat population move to Macedonia during summer, especially in the western mountainous areas of Pindos.

The average size of transhumant farms as well as the importance of sheep and goats vary across regions. The average farm size is 335.5 sheep and goats, with smaller flocks prevailing in southern mainland regions (with average sizes of 272.3 and 286.5 animals in Central Greece and Peloponnese respectively), and relatively larger flocks in the northern parts: 419.8 animals per farm in Thessaly, 387.0 animals per farm in Macedonia, 333.9 animals per farm in Thrace. Sheep farms prevail in the central regions of the country yet the importance of goat farming is escalating. In Peloponnese, goat farms correspond to about 28 percent of the total—as they are well adapted to the rocky pastures of its southern part—while in Crete the percentage of mixed farms is very important (41 percent). In Macedonia the majority of farms are mixed or rear goats exclusively. The size of the average goat farm is significantly higher than the average sheep farm, both at the country level (403.3 animals versus 280.2) and at the regional level, with the exception of Crete.

Considering the distances of movements SGT flocks, "trasterminance" is the prevailing type (56.6 percent of all sheep and goat mobilities), which is also the case in Italy, as pointed out by Pardini and Nori (2011). Movements of up to 100 km are more common for the central/west part of the country and Peloponnese. Remote movements of over 100 km flocks—which usually exceed the conventional boundaries of Regional Units or Region—are the most typical in the eastern part of Thessaly and also very common in Etoloakarnania in the western part of Central Greece. Very remote movements exceeding 200 km and up to 350 km are also common to the central part of the country, of which the most interesting are those of flocks whose winter domiciles are in the capital district of Attica, around the urban web of Athens.

In the last decades, movements with trucks are the established means, especially in cases of movement over 100 km, but even for shorter distances; only short local movements are still performed on foot throughout the country. According to a survey of a sample of 551 transhumant farms, 27 percent of them moved on foot, 65 percent used trucks and 8 percent used both means (Lagka et al. 2015). Changing lifestyles and integration to the market economy with intensification trends are some of the reasons behind this shift. With truck movements, farmers achieve considerable time savings and higher milk production, but renting trucks is sometimes costly, especially under reduced liquidity and low revenues when product prices are low (Ragkos, Karatasiou et al. 2016). For this reason, in the last decade a "return" to movements on foot has been seen, reviving old transhumance routes (for road distances over 100 km), especially when returning to winter domiciles, where the downhill road and the fact that the animals are not milked make the movement easier, quicker, and more comfortable.

SGT in Greece in general is predominantly for milk production (dairy ewes). According to the results of a technical and economic analysis on a sample of sheep and goat transhumant farms in the region of Thessaly (Ragkos, Siasiou et al. 2014), milk was by far the most important product of transhumant flocks (56.4 percent), with an average milk yield of 96 kg/ewe(dam)/year. Siasiou, Galanopoulos, and Laga (2020) reported 97.45 kg/ewe(dam)/year for the whole country based on a sample of 551 farms, which is substantially lower than for other production systems. For instance, milk production in intensive farms rearing Chios-breed sheep was 226 kg/ewe/year (Theodoridis et al. 2012) and 218.6 kg/ewe/year in semi-extensive sheep farms in northeastern Greece (Ragkos, Koutsou, and Manousidis 2016). For goat farms, Gelasakis et al. (2017) reported an annual yield of 207±115.3 kg/goat/year in mainland Greece. Meat was also an important source of income (28.3 percent) while on-farm transformation of milk (cheese production) was of low importance for the overall income of the average farm (4.9 percent) because this activity is restricted to only a few farms. The farm income was supplemented through subsidies (10.4 percent).

Aspects of Management of Transhumant Flocks in Greece: Modernization and Tradition

SGT has not remained unaffected by the general trend for intensification in the Greek livestock sector (Ragkos, Koutsou et al. 2016; Karanikolas and Martinos 2012). The sector has developed a dualistic pattern, being (semi-)intensive in winter and (semi-)extensive in summer, with more use of concentrates for animal nutrition and modern buildings and machinery. This intensification has not altered, however, its predominantly traditional character. This can be seen in more detail based on previous research (Ragkos, Siasiou et al. 2014; Lagka et al. 2014; Siasiou et al. 2014; Galanopoulos et al. 2011; Loukovitis et al. 2016). The management practices of transhumant farms have been evolving through time and continue to adapt to technological advances. Indeed, SGT typically incorporates traditional elements appropriately adapted to today, along with practices that can be characterized as innovative.

Depending on soil and climatic conditions as well as local habits, management practices of transhumant farms vary across the country. In the northern and central areas (Macedonia, Thrace, Thessaly, Epirus, and Central Greece), the movement to the mountains starts in May and the flocks remain there for four to five months, returning to the lowlands usually by the end of October. In southern areas—parts of Central Greece and Northern Peloponnese as well as on Aegean islands—the stay in the mountains can be longer (from late April until mid-November), while in southern Peloponnese and Crete the weather conditions in the mountains are milder and flocks may graze there up to eight to nine months every year. An important characteristic of SGT in Greece is that in most cases the family moves along with the flock in the highlands when the distance between the lowlands and the highlands is not small.

The lives of transhumant farmers are intertwined with the needs of their flocks. The production period starts with the birth season of sheep in November. By that time, flocks in northern Greece are already in the lowlands, while in southern Greece it is not uncommon that births take place in the highlands. The weaning period lasts about forty-five days so that lambs are sold to markets before Christmas, in order to profit from higher meat prices. The birth season of goats and of ewes at first lambing takes place by the end of February or early March, to meet the high demand for meat during the Easter period.

On Christmas—or soon after—the milking season begins. Typically, animals are milked twice and—more rarely—three times a day. The milking period usually lasts for six to seven months—most commonly until late July—which means that animals are milked regularly during their stay in the mountains for about two months. Regarding nutrition, animals graze and/or are fed indoors, depending on weather conditions. In northern and central parts of the country, animals are kept indoors almost exclusively from November until March, and only by early spring do animals start to graze in natural or cropped pastures. In southern parts animals may graze throughout the year, depending however on weather conditions and the availability of rangelands. In the highlands, animals are fed exclusively in natural rangelands freely throughout the day or under the supervision of shepherds, if predators are present. The period from August to the beginning of October is the most relaxed for transhumant farmers, as they do not milk or feed the animals, until they start their return to lowland communities.

Sheep and goat transhumance in Greece exhibit important particularities which discern it from other—either extensive or intensive—production systems. These stem from the specific use of available resources (land, labor, and capital) and are outlined in what follows.

The Role of Land in SGT

Sheep and goat transhumance is highly dependent on land management compared to other livestock production systems, especially compared to intensive ones. This is due to the use of rangelands in both winter and summer domiciles, but also to the cultivation of land for the production of fodder and concentrates.

Rangeland management and the allocation of land constitute issues of high importance to the survival of transhumance. Until the early decades of the twentieth century, transhumant farmers were organized in specific collective actions named *tseligata* (singular, *tseligato*) within which transhumant families cooperated in rearing their animals and managing common resources (Karavidas 1931; Koutsou et al. 2019). Within *tseligata* farmers fully understood the value of natural resources as common goods and developed a practical land management system based on empirical observation and knowledge of vegetation and weather conditions. Based on these they assessed the grazing capacity of the areas they used, and each farmer grazed his animals in a specific place each year.

The individualization of transhumant farmers in the second half of the twentieth century combined with socioeconomic and political developments and a general trend of modernization and intensification brought about competition in rangeland use. Nowadays, livestock farmers are faced with a bureaucratic system of rangeland allocation, which increases uncertainty and—in some cases—is the cause of social conflicts in rural areas (Koutsou et al. 2019). In particular, rangelands are actually allocated to livestock farmers based on a rough estimation of grazing capacity by paying a small premium per animal. In other cases, some areas are allocated to farmers based on an auction process. In addition, the development of the road network, the expansion of cropland and the encroachment in mountain paths due to under-grazing have reduced accessibility to mountain terrains. Although there are local variations of this system across the country, this system has reduced the importance of traditional ecological knowledge, which resulted in environmental degradation in some areas (Ragkos, Koutsou et al. 2020) and brought uncertainties and sometimes conflicts among farmers and other land users (Koutsou et al. 2019).

Transhumant flocks are also dependent on the production of feedstuff (forage and concentrates) for winter. Although they have the option to buy feedstuff from markets, many prefer to cultivate land for animal feed, in order to reduce costs. It is not unusual to own land in the lowland but also to rent it either for the cultivation of feed or for grazing crops. In general, the cultivation of land by transhumant farmers follows the pattern of other sheep systems in Greece—that is, it is an activity of secondary importance, with less intensive use of purchased inputs compared to crop farmers and seldom is part of the production sold in markets.

The Role of Labor in SGT

Labor is a factor of crucial importance for livestock farms, while for pastoral production systems it is necessary for the expansion of farms, and its lack could lead to abandonment not only of the sector but also of whole rural areas (Nori 2017b). Nowadays, the debate over labor in the primary sector is part of the broader debate on the role of skilled or unskilled labor in the development process around the world. Transhumance differs from conventional intensive or semi-intensive livestock systems because it is in direct contact with nature. The production process is not automated/industrialized, as it is, for example, in intensive dairy farming, and therefore requires workers with specific practical skills, experience, and appropriate training, willing to adopt a way of life that is intertwined with nature and the productive cycle of animals. Labor in a transhumant farm represents multiple challenges: manual milking, sometimes under harsh conditions; surveillance of livestock in pasture to prevent predator attacks; protection against extreme weather conditions; treatment of diseases; repairing farm machinery and/or makeshift buildings and equipment, etc. These are combined with increased requirements for managerial work for the effective management of capital and decision-making for production and product sales as well as bureaucratic procedures (Ragkos and Nori 2016; Ragkos, Koutsou et al. 2018). A modern-day transhumant farmer in Greece is expected to combine the special "wealth" of traditional ecological knowledge inherited from his predecessors with knowledge of technological innovations, livestock science, and farm economic management.

Transhumant farms have traditionally been run by family members (Loukopoulos 1930) and this is also the case nowadays. Most Greek transhumant farms meet all six conditions set by Gasson and Errington (1993) defining the "farm family business," which constitutes the backbone of the European Model of Agriculture (Vermersch 2001). The family works on the farm either exclusively or occasionally, much like in the past (written testimonies by Loukopoulos [1930] and Syrakis [1925]) but also with notable differences which are described in the remainder of this section.

The allocation of tasks among family members generally maintains the basic principles (by gender and age) of previous decades. A male family member (usually the husband) is the head of the farm and is also involved in feeding, milking, animal health, and crop production as well as in economic management, decision-making, communication, and product sales. Women work in a supporting role—e.g., in milking or during birth season and weaning—and they are usually the ones producing cheese. Other family members have auxiliary roles; these include children who are away for studies but return home for summer and holidays, as well as grandparents. Another form of labor division is pluri-activity, through the horizontal or vertical development of parallel activities within the family farm. In areas where such activities are available—often taking advantage of financial opportunities of the CAP—farms expand to crop production for market (e.g., olive and citrus fruits in Crete and the Southern Peloponnese), tourism and household artisanal production (cheese/dairy products). In this form, the family members who are put in charge stop working on the livestock enterprise but still remain within the family business.

The problem of farm succession has made the pattern of labor organization in transhumant farms more complex. Throughout the Mediterranean, young family members tend to leave the family farm, for studies or seeking employment in other sectors, as working in the sector is generally not appealing due to its harsh requirements (Ligda et al. 2012; Nori 2017b).

This way, farms are deprived of an input of crucial importance, but, more importantly, they are also left with no successor (Farinella et al. 2017). The lack of labor has hindered the development of the Greek primary sector in general (Karanikolas and Martinos 2012; Kasimis and Papadopoulos 2005). In previous decades, the lack of family labor—for example, because of family member migration—led to a decline in the primary sector in general, as there were no alternative sources of hired labor, because the social perception of working in another farm was demeaning (Loukopoulos 1930).

The shrinking pattern of livestock production until the 1980s was reverted when—starting from the 1990s—cheap hired labor started to be extensively offered by migrants, mainly Albanians, who came to the country in large numbers—a phenomenon also recorded in Italy (Nori 2017a; Farinella et al. 2017).

Actually, Albanian workers are employed throughout the country; Bulgarians and Romanians work mainly in the north; Pakistanis and Indians are mostly found in central and southern Greece. These individuals, in many cases, had previous experience with livestock farming in their countries of origin and their employment costs were low (Nori and Ragkos 2017; Ragkos and Nori 2018), so they quickly became employed on livestock farms. Transhumant farms, however, resort to hired labor only when necessary, in fact the analysis of data from Thessaly (Ragkos, Siasiou et al. 2014) showed that about 27 percent of labor requirements were covered by hired workers. These people work all year round or are only recruited during peak periods and are engaged in simpler but harsher and more time-consuming tasks such as grazing, cleaning, and feeding and also in milking. The remuneration of hired workers includes their salary and benefits in kind, such as housing, food, and clothing.

The Role of Capital in SGT

Sheep and goat transhumance, as practiced today, has incorporated a higher level of capital use compared to the past. To some extent, capital has now substituted the other two factors of production (land and labor), especially in winter domiciles. On the other hand, this trend is not excessive, and the system maintains a traditional extensive character with less dependence on capital compared to semi-intensive or intensive systems (see for instance Theodoridis et al. 2012, 2013). CAP funding has played an important role in this process, providing the opportunity to modernize their equipment at relatively low financial costs.

Fixed capital is used in the form of buildings, machinery, and livestock. Buildings in the winter domiciles are usually stable and not always makeshift, made of a variety of materials (wood, stone, bricks, or concrete). Their values vary significantly depending on materials and size as well as the region, as in northern and central Greece more stable and functional facilities are required to deal with low temperatures, humidity, and snow. In summer domiciles, buildings are usually makeshift and traditional, with materials that are part of the landscape and natural environment, while in some areas there are municipal/community buildings for flocks (sheds or fences). The machinery of transhumant farms is relatively limited. It definitely includes feeders and watering cans (metal or wooden), a milk cooler for preserving fresh milk until it is collected by the dairy industry, and a farmer truck for the transport of inputs and products. In many cases the farm has a hammer mill (for preparation of mixtures of concentrates), a tractor, and more rarely, a truck. Portable milking machines that can be transported between winter and summer domiciles are becoming increasingly popular, however farmers who are experienced in milking by hand are sometimes reluctant to switch to machine milking (Lagka et al. 2014). If the farm cultivates feedstuff, it has, in most cases, the necessary tractor accessories (ripper, cultivator, disc harrow, fertilizer distributor, tank) and, less frequently, sowing and harvesting machines as well as mowers.

A significant number of transhumant farms rear local breeds of sheep and goats—either purebred or crossbreeds. These breeds are characterized by low yields compared to imported ones, but their milk usually has high quality characteristics (Ragkos, Koutouzidou et al. 2017). In addition, they are well adapted to grazing under the adverse conditions of mountainous areas, while they are resistant to illnesses and have low nutritional requirements (Ragkos, Koutouzidou et al. 2019). Particular examples are the Kalaritiki breed in Epirus, the Anogia breed in Crete, the Karystos breed in Evia, and the Vlachiko breed in Western Macedonia.

The use of variable capital mainly relates to liquidity for the purchase of consumables and affects managerial choices of farmers (e.g., use of purchased or on-farm production of feedstuff). The most significant variable capital expense concerns the purchase or cultivation of feedstuff which is very important especially in winter. With the exception of some islands, where animals graze freely all year long and are kept for meat, this expenditure stands for at least 35–40 percent of the total expenses of transhumant sheep and goat farms, especially in the Northern part where winter is harsher and longer (Ragkos, Siasiou et al. 2014). Thus, the five-month stay in the mountain pastures leads to significant cost savings. For flocks that spend winters in the lowlands of Thessaly, grazing in mountain rangelands substantially reduces feeding costs by 47 percent to 58 percent compared to intensive production (Ragkos, Siasiou et al. 2014). Such savings are even greater in Peloponnese and the islands, where feedstuff cultivation is limited and therefore the prices of basic feedstuff (corn, clover, cereals) are significantly higher. As mentioned above, on-farm feed production is more common in the north and contributes to further savings.

Challenges for Greek Sheep and Goat Transhumance

The specificities of SGT in Greece render the system vulnerable to numerous socioeconomic and policy-related problems. The lack of clear recognition of its multiple roles combined with the lack of a specific policy framework adapted to its particularities bring about exogenous challenges, while the balance of the system between modernization and tradition brings endogenous issues that need to be tackled. Some of these problems are common to extensive livestock systems across Europe while others are more severe in the Greek setting.

Problems in Land Uses

Problems in land use, as discussed earlier in the chapter, is perhaps the most important problem of the Greek transhumance sector, as farmers do not have easy and unobstructed access to rangelands both in the lowlands and in the highlands. In order to mitigate the negative effects of the current system of rangeland allocation, Integrated Rangeland Management Plans (IGMPs) are expected to be delivered in the following years. IGMPs will be based on a precise estimation of grazing capacities of rangeland "parcels" which will be allocated to livestock farmers for a considerable time period—and they will then be entitled to undertake all necessary activities to prevent degradation (Ragkos and Koutsou 2021).

Apart from the issue of rangeland uses, there are conflicts relating to the unclear ownership of some areas. This is the result of the lack of an integrated land registry—which is still under preparation. "Forest maps" were recently delivered with an aim to discern forest areas from other land uses, but still they have not been finalized. In addition, competition from alternative land uses (intensive agriculture and livestock in the lowlands, such as Peloponnese and Central Macedonia; tourism on islands and coastal areas; sustainable energy systems; expansion of habitats of predators) limits the possibilities of the operation of the system in certain parts of the country.

Problems Related to Infrastructure

The access to the specific mountain rangelands where animals graze from the highland communities where farmers live with their families during the summer is sometimes difficult, as they are not accessible with farmers' cars due to the poor quality of forest roads. The provincial road network connecting highland communities with main roads are also in need of repairs and upgrades to enhance accessibility. Public infrastructure for livestock is not always appropriate (e.g., makeshift barns and watering stations, huts for shepherds) or lacks maintenance. Furthermore, many mountain communities lack basic health and education infrastructure as well as entertainment and culture opportunities, despite the incentives provided by rural development policies.

Problems Due to Economic Performance

Transhumant sheep and goat farmers are faced with particularly high input prices, which have been increasing for decades (Karanikolas and Martinos 2012; Ragkos, Koutsou et al. 2016). These changes occur at a time when product prices (milk and meat) show significant decreases, especially during the past few years—a trend which is no different for pastoral farmers across Southern Europe. Although this is a problem of the Greek livestock sector as a whole, there is more pressure for sheep and goat transhumance because they often do not profit from the added value of their quality products, but are treated with the same pricing policy as intensive farms (Ragkos, Koutsou et al. 2020). Alternative marketing options are limited, thus leaving significant margins for marketing improvement through certification or the design of new products (Ragkos, Theodoridis, and Arsenos 2019). In addition, low liquidity is a major operational problem, as the system has developed a stronger dependence on capital compared to previous decades, especially in the lowlands, where animals are kept indoors during winter.

Problems Related to Human Capital

There are still significant issues relating to the aging transhumant populations that hinder the development of transhumance in Greece, following the general trend in the Greek and European farming sectors (see for instance Koutsou, Ragkos, and Botsiou 2015). In addition, due to a relatively low level of education among the farmers, there is an inability to cope with regular financial and bureaucratic management procedures. This is a problem of the Greek livestock sector in general, as the lack of modern educational programs corresponding to the actual problems and aspirations of farmers decreases opportunities for innovation and market integration. This issue is even more important for extensive livestock farmers—including transhumant—whose specific practices and exposure to natural constraints increases their need for training and information regarding particular practical issues.

A specific issue relating to the human factor is the lack of collective actions. This lack can be traced not only among transhumant farmers but also across the majority of Greek farms. This indicates the limited social capital, a common characteristic of the Greek countryside during the last decades (Koutsou, Partalidou, and Ragkos 2014). Actually, there is one Association of Transhumant Farmers in Epirus (Western Greece), which does not undertake commercial functions. Otherwise, transhumant farmers participate mainly in Agricultural Cooperatives and Associations of Livestock Farmers involving mainly sedentary or semi-intensive farmers. This lack constitutes one of the reasons why transhumant farmers find difficulties in solving problems which are particular to the sector or even contrast with the pursuits of sedentary and/or intensive farmers.

Problems Related to Bureaucracy and Lack of Recognition and Information

There are several problems reported by transhumant farmers related to the level of complexity of control and administration requirements at the central level. These requirements are often generic and transhumant farmers are expected to follow the same requirements as sedentary and intensive farmers. These issues sometimes impact the timely payment of CAP financial support. In this regulatory context and without a clear positioning in markets, transhumant farmers struggle to consolidate their identities. Actually, the system lacks recognition not only concerning its multiple contributions to society but also regarding its mere existence (Ragkos, Theodoridis, and Arsenos 2019)

Conclusions and Policy Implications

Drawing on the above presentation of SGT in Greece, it is evident that the system is an important source of employment and income, and it also has important potential for the future. In order to support its survival and assist its transition towards a more sustainable pattern, three types of recommendations can be made.

The first type includes general measures and actions to favor transhumance alongside with other pastoral systems in Greece and in the EU. The effect of these policies has been beneficial to Greek farms in general, as they boosted their viability (Karanikolas and Martinos 2012) but failed to address particular problems of transhumance and/or pastoral systems. There are already examples of measures which have benefited transhumant farms—as outlined in the introduction to this chapter. The new framework of CAP 2021–27 provides, however, opportunities for novel approaches for extensive (based on grazing) livestock production, including a more targeted implementation of Pillar II measures. A deeper and more coordinated support towards pluri-activity in pastoral households would increase and stabilize their sources of income, like the example of Crete and other islands. Given the important environmental role of pastoral systems a highly relevant example involves the introduction of certification schemes (eco-schemes) in the form of a "grass-fed" certification (Lampkin et al. 2020). In addition, the example of "Payments for Ecosystem Services" which has already been applied in other countries of the European Union could induce a more straightforward connection of CAP payments with the provision of benefits that society appreciates. It should also be stressed that considerable funds could be allocated to the development of targeted training and information programs for pastoralists. In other Mediterranean countries there are examples of relevant structures providing training and support to pastoralists (for instance in Spain), while in Greece there is still much room for improvement in this domain.

The second type of intervention is more specific to transhumance and involves market-based measures and actions to be undertaken. Measures of this sort could involve the development of marketing channels for transhumance-specific products, which will take advantage of the high quality of its products. Actually, most dairies and industries do not collaborate exclusively with transhumant farmers and do not produce transhumance-specific dairy products. This means that this premium quality milk which is produced during the summer period when flocks graze the rich vegetation in the highlands (Zdragas et al. 2015; Ioannidou et al. 2019) is paid the same prices as conventional. In addition, dairies and industries pay farmers mainly according to their volume of production and not according to the quality characteristics of their milk (Roustemis 2012). As described by Ragkos et al. (2019), transhumance-specific products could increase returns for farmers under specific conditions, while the effects of premium pricing of such products were found to be important for the economic performance of the system (Ragkos, Koutsou et al. 2020).

Finally, a third type of intervention is specific to SGT and involves the development of a policy toolkit which will accommodate its multifunctional roles and will transform these characteristics to opportunities to increase competitiveness and approach new markets. SGT in Greece requires an enabling environment to permit the system to flourish and live up to its true potential. In this domain, the diversity of new schemes proposed in the new CAP could provide additional certification opportunities for transhumance-specific products, as existing schemes have not been proven particularly successful for the case of transhumance, nor are they able to integrate the whole range of its multifunctional roles (Ragkos et al. 2019). Furthermore, an asset of importance for SGT is the development of old transhumance routes, not only for use by flocks but also as an attraction for alternative tourism (Ragkos, Karatasiou et al. 2016; Ntassiou and Doukas 2019).

Athanasios Ragkos is an agricultural economist currently employed as a researcher at the Agricultural Economics Research Institute of the Greek Agricultural Organization "ELGO-DIMITRA." He obtained his PhD in 2008 from the Aristotle University of Thessaloniki. Since then, he has taught several courses at Greek universities and technological institutions. He has participated in more than thirty research projects (four as Coordinator) and has published more than eighty-five papers in peer-reviewed journals and conference proceedings. His research fields include the economics of agricultural and animal production, environmental economics, the multifunctional character of agriculture and extensive farming systems, project appraisal, and sustainable rural development. He is experienced in designing interdisciplinary approaches by maintaining close collaboration with scientists of relevant fields. He works closely with international and domestic companies in the secondary sector (e.g., dairies and propagation material) and also with cooperatives.

References

Caballero, Rafael, Federico Fernandez-Gonzalez, Rosa Perez Badia, Giovanni Molle, Pier Paolo Roggero, Simonetta Bagella, Vasileios Papanastasis, Georgios Fotiadis, Anna Sidiropoulou, and Ioannis Ispikoudis. 2009. "Grazing Systems and Biodiversity in Mediterranean Areas: Spain, Italy and Greece." *Pastos* XXXIX(1): 9–154.

D'Ottavio, Paride, Matteo Francioni, Laura Trozzo, Elmir Sedić, Katarina Budimir, Pietro Avanzolini, Maria Federics Trombetta, Claudio Porqueddu, Rodolfo Santilocchi, and Marco Toderi. 2018. "Trends and Approaches in the Analysis of Ecosystem Services Provided by Grazing Systems: A Review." *Grass and Forage Science* 73(1): 15–25.

Farinella, Domenica, Michele Nori, and Athanasios Ragkos. 2017. "Changes in Euro-Mediterranean Pastoralism: Which Opportunities for Rural Development and Generational Renewal?" Conference: Grassland Science in Europe—Grassland Resources for Extensive Farming Systems in Marginal Lands: Major Drivers and Future Scenarios 22: 23–36.

Galanopoulos, Konstantinos, Zafeiris Abas, Vasiliki Laga, Ioannis Hatziminaoglou, and Jean Boyazoglu. 2011. "The Technical Efficiency of Transhumance Sheep and Goat Farms and the Effect of EU Subsidies: Do Small Farms Benefit More Than Large Farms?" *Small Ruminant Research* 100(1): 1–7.

Gasson, Ruth, and Andrew Errington. 1993. *The Farm Family Business.* Wallingford: CAB International.

Gelasakis, Athanasios, Ian Rose, Rebecca Giannakou, Georgios Valergakis, Alexandros Theodoridis, Paschalis Fortomaris, and Georgios Arsenos. 2017. "Typology and Characteristics of Dairy Goat Production Systems in Greece." *Livestock Science* 197: 22–29.

Gidarakou, Isavella, and Konstantinos Apostolopoulos. 1995. "The Productive System of Itinerant Stockfarming in Greece." *Medit* 6(3): 56–63.

Gkoltsiou, Aikaterini. 2011. Research Theme: Routes of Transhumance. Research Report for Greece. *Culture and Nature: The European Heritage of Sheep Farming and Pastoral Life.* Retrieved 7 January 2022 from https://ich.unesco.org/en/RL/transhumance-the-seasonal-droving-of-livestock-along-migratory-routes-in-the-mediterranean-and-in-the-alps-01470.

Hadjigeorgiou, Ioannis. 2011. "Past, Present and Future of Pastoralism in Greece." *Pastoralism: Research, Policy and Practice* 1(1): 24.

———. 2004. *Sheep and Goat in Greece and in the World* (in Greek). Thessaloniki: Giahoudis and Giapoulis Publishing.

Ioannidou, Maria, Maria Karatassiou, Athanasios Ragkos, Zoi Parissi, Ioannis Mitsopoulos, Paraskevi Sklavou, Vasiliki Lagka, and Georgios Samouris. 2019. "Effects on Fatty Acids Profile of Milk from Transhumant Small Ruminants Related to the Floristic Composition of Mountainous Rangelands." *Options Méditerranéennes A, "Innovations for Sustainability in Sheep and Goats"* 123: 183–86.

Karanikolas, Pavlos, and Nikolaos Martinos. 2012. "Greek Agriculture Facing Crisis: Problems and Prospects." Retrieved May 2015 from http://ardinrixi.gr/archives/3811 (in Greek).

Karavidas, Konstantinos. 1931. *Agrotika* (in Greek). Athens: Papazisi.

Kasimis, Charalambos, and Apostolos Papadopoulos. 2005. "The Multifunctional Role of Migrants in the Greek Countryside: Implications for Rural Economy and Society." *Journal of Ethnic and Migration Studies* 31(1): 99–127.

Koutsou, Stavriani, Maria Partalidou, and Athanasios Ragkos. 2014. "Young Farmers' Social Capital in Greece. Trust Levels and Collective Actions." *Journal of Rural Studies* 34: 204–11.

Koutsou, Stavriani, Athanasios Ragkos, and Maria Botsiou. 2015. "Newcomers in Agriculture: New Ideas, New Dynamics?" Presented at the International Conference Meanings of the Rural—Between Social Representations, Consumptions and Rural Development Strategies, University of Aveiro, Portugal, 28–29 September 2015.

Koutsou, Stavriani, Athanasios Ragkos, and Maria Karatassiou. 2019. "Accès à la Terre et Transhumance en Grèce: Bien Commun et Conflits Sociaux." *Développement Durable et Territoires. Économie, Géographie, Politique, Droit, Sociologie* 10(3). Retrieved 7 January 2022 from https://journals.openedition.org/developpementdurable/14969.

Lagka, Vasiliki, Anna Siasiou, Athanasios Ragkos, Ioannis Mitsopoulos, Vasileios Bampidis, Stavroula Kiritsi, Vasileios Michas, and Vasileios Skapetas. 2014. "Milking and Reproduction Management Practices of Transhumant Sheep and Goat Farms." *Options Méditerranéennes* 109: 695–99.

Lagka, Vasiliki, Anna Siasiou, Athanasios Ragkos, Ioannis Mitsopoulos, Aristotelis Lymperopoulos, Stavroula Kiritsi, Vasileios Bampidis, and Vasileios Skapetas. 2015. "Geographical Differentiation of Gestational Practices of Transhumant Sheep and Goat Farms in Greece." Presented at the 66th Annual Meeting of the European Association for Animal Production, Warsaw, Poland, 31 August–4 September 2015.

Lampkin, Nic, Matthias Stolze, Stephen Meredith, Miguel De Porras, Lisa Haller, and Dora Mészáros. 2020. *Using Eco-Schemes in the New CAP: A Guide for Managing Authorities*. Brussels: IFOAM EU, FIBL and IEEP.

Ligda, Christina, Edmond Tchakerian, Evangelos Zotos, and Andreas Georgoudis. 2012. "Tradition and Innovation in the Mediterranean Pastoralism: Recognition of Its Multiple Roles for the Sustainable Development of Rural Areas." In *New Trends for Innovation in the Mediterranean Animal Production*, ed. R. Bouche, A. Derkimba, and F. Casabianca, 264–69. Wageningen: Wageningen Academic Publishers.

Loukopoulos, Dimitrios. 1930. *Pastoralism in Roumeli* (in Greek). Athens: Historic and Folklore Library of the Society of Dissemination of Beneficial Books.

Loukovitis, Dimitrios, Anna Siasiou, Ioannis Mitsopoulos, Aristotelis Lymberopoulos, Vasiliki Laga, and Dimitrios Chatziplis. 2016. "Genetic Diversity of Greek Sheep Breeds and Transhumant Populations Utilizing Microsatellite Markers." *Small Ruminant Research* 136: 238–42.

Nitsiakos, Vasileios. 1995. *The Mountain Communities of North Pindos: In the Aftermath of Long Endurance* (in Greek). Athens: Plethron Press.

Nori, Michele. 2017a. "Migrant Shepherds: Opportunities and Challenges for Mediterranean Pastoralism." *Journal of Alpine Research* 105: 4.

———. 2017b. *Immigrant Shepherds in Southern Europe*. Berlin: Heinrich Böll Foundation.

Nori, Michele, and Athanasios Ragkos. 2017. "The Role of Migrant Workers in the Sustainability of Extensive Livestock Farming in Greece." Presented at the 6th Panhellenic Conference of Animal Production Technology, 3 February 2017. Thessaloniki, Greece.

Nori, Michele, Athanasios Ragkos, and Domenica Farinella. 2016. "Pastourism: Exploring Future Venues for Agro-Pastoral Systems in Mediterranean Islands." Presented at MIC: Mediterranean Islands Conference, 21–24 September 2016. Island of Vis, Croatia.

Ntassiou, Konstantina, and Ioannis Doukas. 2019. "Recording and Mapping Traditional Transhumance Routes in South-Western Macedonia, Greece." *Geojournal* 84(1): 161–81.

Ntassiou, Konstantina, Ioannis Doukas, and Maria Karatassiou. 2015. "Exploring Traditional Routes of Seasonal Transhumance Movements with the Help of GIS. The Case Study of a Mountainous Village in Southwest Macedonia, Greece." Presented at the 7th International Conference on Information and

Communication Technologies in Agriculture, Food and Environment, 17–20 September 2015, Kavala, Greece.

Nyssen, Jan, Katrien Descheemaeker, Amanuel Zenebe, Jean Poesen, Josef Deckers, and Mitlku Haile. 2009. "Transhumance in the Tigray Highlands (Ethiopia)." *Mountain Research and Development* 29(3): 255–64.

Pardini, Andrea, and Michele Nori. 2011. "Agro-Silvo-Pastoral System in Italy: Integration and Divestification." *Pastoralism: Research, Policy and Practice* 1: 10.

Ragkos, Athanasios, Maria Karatasiou, Zisis Georgousis, Zoi Parissi, and Vasiliki Lagka. 2016. "A Traditional Route of Transhumant Flocks in Northern Greece: Cultural Aspects and Economic Implications." *Options Méditerranéennes* 114: 345–48.

Ragkos, Athanasios, Georgia Koutouzidou, Christos Christodoulou, and Athanasios Batzios. 2017. "Which Orientation for Strategies and Policies for Local Animal Breeds? The Role of ICT and Novel Technologies." Presented at 8th International Conference on Information and Communication Technologies in Agriculture, Food and Environment (HAICTA 2017), 21–24 September 2017, Chania (Crete).

Ragkos, Athanasios, Georgia Koutouzidou, Stavriani Koutsou, and Dimitrios Roustemis. 2019. "A New Development Paradigm for Local Animal Breeds and the Role of Information and Communication Technologies." In *Innovative Approaches and Applications for Sustainable Rural Development*, ed. A. Theodoridis, A. Ragkos, and M. Salampasis, 144–68. Springer.

Ragkos, Athanasios, and Stavriani Koutsou. 2021. "Collective and Individual Approaches to Pastoral Land Governance in Greek Silvopastoral Systems: The Case of Sheep and Goat Transhumance." In *Governance for Mediterranean Silvopastoral Systems*, ed. Teresa Pinto Correia, Maria Helena Guimaraes, Gerardo Moreno, and Rufino Acosta-Naranjo, 260–78. London/New York: Routledge.

Ragkos, Athanasios, Stavriani Koutsou, Maria Karatassiou, and Zoi Parissi. 2020. "Scenarios of Optimal Organization of Sheep and Goat Transhumance." *Regional Environmental Change* 20(1): 13.

Ragkos, Athanasios, Stavriani Koutsou, and Theodoros Manousidis. 2016. "In Search of Strategies to Face the Economic Crisis: Evidence from Greek Farms." *South European Society and Politics* 21(3): 319–37.

Ragkos, Athanasios, Stavriani Koutsou, Alexandros Theodoridis, Theodoros Manousidis, and Vasiliki Lagka. 2018. "Labor Management Strategies in Facing the Economic Crisis. Evidence Form Greek Livestock Farms." *New Medit* 2018: 1.

Ragkos, Athanasios, Ioannis Mitsopoulos, Anna Siasiou, Vasileios Skapetas, Stavroula Kiritsi, Vasileios Bambidis, Vasiliki Lagka, and Zafeiris Abas. 2013. "Current Trends in the Transhumant Cattle Sector in Greece". *Scientific Papers Animal Science and Biotechnologies* 46(1): 422–26.

Ragkos, Athanasios, and Michele Nori. 2016. "The Multifunctional Pastoral Systems in the Mediterranean EU and Impact on the Workforce." *Options Méditerranéennes* 114: 325–28.

———. 2018. "Foreign Workers in Grazing Small Ruminants: Assessment of Their Practical Knowledge and Skills." Presented at the 9th Conference on Rangeland Science, October 2018, Larissa, Greece.

Ragkos, Athanasios, Anna Siasiou, Konstantinos Galanopoulos, and Vasiliki Lagka. 2014. "Mountainous Grasslands Sustaining Traditional Livestock Systems: The Economic Performance of Sheep and Goat Transhumance in Greece." *Options Méditerranéennes* 109: 575–79.

Ragkos, Athanasios, Alexandros Theodoridis, and Georgios Arsenos. 2019. "Alternative Approaches of Summer Milk Sales from Transhumant Sheep and Goat Farms: A Case Study from Northern Greece." *Sustainability* 11(20): 5642.

Roustemis, Dimitrios. 2012. "Design of the Breeding Goal for Chios Sheep." PhD diss., Democritus University of Thrace, Greece (in Greek).

Siasiou, Anna, Konstantinos Galanopoulos, and Vasiliki Laga. 2020. "The Future of Transhumant Farming: An Economic Analysis of Management Characteristics of Transhumant Greek Farms." *International Journal of Veterinary Science and Agriculture Research* 2(1): 9–15.

Siasiou, Anna, Konstantinos Galanopoulos, Ioannis Mitsopoulos, Athanasios Ragkos, and Vasiliki Laga. 2018. "Transhumant Sheep and Goat Farming Sector in Greece." *Iranian Journal of Applied Animal Science* 8(4): 615–22.

Siasiou, Anna, Vasileios Michas, Ioannis Mitsopoulos, Vasileios Bampidis, Stavroula Kiritsi, Vasileios Skapetas, and Vasiliki Lagka. 2014. "Nutritional Management of Transhumant Sheep and Goats at the Region of Thessaly—Greece." Presented at the 65th Annual Meeting of the European Association for Animal Production, 25–29 August 2014, Copenhagen, Denmark.

Sidiropoulou, Anna, Maria Karatassiou, Georgia Galidaki, and Paraskevi Sklavou. 2015. "Landscape Pattern Changes in Response to Transhumance Abandonment on Mountain Vermio (North Greece)." *Sustainability* 7(11): 15652–673.

Sklavou, Paraskevi, Maria Karatassiou, Zoi Parissi, Georgia Galidaki, Athanasios Ragkos, and Anna Sidiropoulou. 2017. "The Role of Transhumance on Land Use/Cover Changes in Mountain Vermio, Northern Greece: a GIS Based Approach." *Notulae Botanicae Horti Agrobotanici Cluj-Napoca* 45(2): 589–96.

Syrakis, Dimosthenis. 1925. "Nomadic Livestock Production in Greece" (in Greek). *Agricultural Bulletin of the Greek Agricultural Society* XII: 651–777.

THALES. 2015. *The Dynamics of Sheep and Goat Transhumance in Greece. Influences on Biodiversity 2015*. Final Report. Thessaloniki.

Theodoridis, Alexandros, Athanasios Ragkos, Dimitrios Roustemis, Georgios Arsenos, Zafeiris Abas, and Efthimios Sinapis. 2013. "Technical Indicators of Economic Performance in Dairy Sheep Farming." *Animal* 8(1): 133–40.

Theodoridis, Alexandros, Athanasios Ragkos, Dimitrios Roustemis, Konstantinos Galanopoulos, Zafeiris Abas, and Efthimios Sinapis. 2012. "Assessing Technical Efficiency of Chios Sheep Farms with Data Envelopment Analysis." *Small Ruminant Research* 107: 85–91.

UNESCO. 2021. "Transhumance, the Seasonal Droving of Livestock along Migratory Routes in the Mediterranean and in the Alps." Retrieved 3 August 2020 from https://ich.unesco.org/en/RL/transhumance-the-seasonal-droving-of-livestock-along-migratory-routes-in-the-mediterranean-and-in-the-alps-01470.

Vallerand, Francois. 2014. *Seasonal Movements and Transhumant Farms in European Mediterranean*. Retrieved 15 June 2019 from http://www.metakinoumena.gr/El/downloads/category.

Varela, Elsa, and Ana Belén Robles-Cruz. 2016. "Ecosystem Services and Socio-Economic Benefits of Mediterranean Grasslands." *Options Méditerranéennes* 114: 13–27.

Vermersch, Dominique. 2001. *Multifunctionality: Applying the OECD Framework. A Review of Literature in France*. Paris: Directorate for Food, Agriculture and Fisheries, OECD.

Zdragas, Antonios, Theofilos Papadopoulos, Ioannis Mitsopoulos, Georgios Samouris, Georgios Vafeas, Evridiki Boukouvala, Loukia Ekateriniadou, Kyriaki Mazaraki, Athanasios Alexopoulos, and Vasiliki Lagka. 2015. "Prevalence, Genetic Diversity, and Antimicrobial Susceptibility Profiles of Staphylococcus Aureus Isolated from Bulk Tank Milk from Greek Traditional Ovine Farms." *Small Ruminant Research*.

CHAPTER 2

The Conflict of Itinerant Pastoralism in the Piedmont Po Plain (Collina Po Biosphere Reserve, Italy)

Dino Genovese, Ippolito Ostellino, and Luca Maria Battaglini

Introduction

Piedmont is a region in northwestern Italy and its name in Italian (Piemonte) translates as "land at the foot of a mountain range." It is right here that the Po Valley—the largest plain in Italy, formed by the sediments of the Po River—begins.

The interconnections between the plain and the western Alps are pronounced, as mountains largely surround the regional borders. Historically, the geography of this landscape has allowed for the extensive breeding of both cattle and sheep (Lazzaroni and Biagini 2008; Battaglini et al. 2001).

The traditional practice of seasonal vertical transhumance—which consists in transferring flocks to mountain pastures during the summer period and bringing them to lower valleys in winter—still exists to this day.

In the high valleys, these cycles are still part of the identity of local communities and they represent a moment of territorial celebration (Battaglini 2007). After the summer, the herds return to the farms (nowadays only by truck), which are mainly located on the plains. Here, in most cases, the animals are housed in the barn and fed hay. The plains are also where most of today's intensive livestock farms are located: in terms of volume, said livestock represents a significant branch of production; however, it is not visible, as it is permanently kept indoors. Therefore, only the transhumant sheep farmers maintain the predominant grazing and shape the livestock landscape.

Itinerant pastoralism is a form of extensive farming that is based on the continuous movement of flocks in order to make the most of the plant production of different and complementary ecosystems (Nori and De Marchi 2015). Like any pastoral mobility activity, itinerant pastoralism is dependent on the breeders' personal ability to identify the availability and quality of grazing areas and on their knowledge of the physiology and health of farmed animals. These breeders do not have full availability of the land they use, since they are just owners of the herd. The social capital of the shepherds is based on a series of shared values, rules, and codes; these allow for different types of organization and bargaining that govern disparate interests, the use of available resources, and the management of any potential conflicts that may arise (Nori 2010).

Nevertheless, nomadic pastoralism suffers more and more from the logistical difficulties of extensive stockbreeding and the increasing physical and legal constraints of the territory involved. The itinerant shepherd must therefore be able to perform tasks in abandoned territories such as the mountains but also in highly industrialized and densely populated agricultural plains. "The success of its rambling practice depends on the ability of shepherds to find traditional pastures not only geographically (e.g., between wheat fields, cities, road infrastructure, private fields, protected areas), but socially and politically" (Aime, Allovio, and Viazzo 2001).

In Piedmont, sheep production is modest in percentage. In 2019, the flocks amounted to 120,744 animals (National Livestock Registry 2020), which represents 1.80 percent of the total Italian livestock. However, the presence of these sheep has some peculiarities, as some autochthonous breeds are found only in this region (e.g., the Biellese, Frabosana, Sambucana, Langhe breeds). There are indeed regional projects to secure and improve these local livestock and their products. Among them, there are 72,778 transhumant sheep (60.27 percent of the regional total), mainly used for meat production (Biellese and its mixed breeds, and Sambucana breeds). Most of the cattle subject to vertical transhumance stay on the plain during the rest of the year and do not move to a barn; some remain itinerant and practice a form of grazing called *pascolo vagante*.

The term *pascolo vagante* refers to a seasonal livestock system—mainly sheep grazing—characterized by continuous migration along very long trails. This system of nomadic pastoralism has been described as perpetual tramping and the herders as permanent wanderers (Buratti 1999: 52). Permanent buildings are usually not adopted, with the exception maybe of a caravan or the likes for shepherds and lambs. In northern Italy, this practice is carried out between autumn and spring, along shallow rivers or in regions with low hills. This practice is regulated by Italian law, and shepherds must be equipped with a logbook in order to record times and permits.

In the remaining period of time, the livestock is led to mountainous landscapes near the plain. Only the initial movement onto the plain (one to two hundred kilometres) is carried out by truck. After this period of time, flocks move autonomously every day along trajectories and stop according to grass availability. Routes are drawn within marginal areas, along rivers or resting fields. Nomadic pastoralism is one of the simplest livestock methods, and rustic and resilient sheep breeds are perfect for this.

Itinerant shepherds and their animals represent a recurring seasonal presence for the residential communities they traverse. Each shepherd and his flock take the same route every year, making changes based on the weather and vegetation conditions. The shepherds have permissions for a personal regional zoning. This is not determined by planning but rather by repeated annual steps. Spaces available for itinerant pastoralism, however, are increasingly reduced due to the development of urban areas (Bernardino and Zullo 2016).

The main logistical requirement is animal feeding. This type of cattle almost exclusively grazes on spontaneous green forage. The sheep feed on grass and crop residues allowed by farmers. Integrations need to be provided, especially in the case of snowfall, which makes pastures inaccessible. Routes along the rivers make it easy for the animals to drink along their journey. Any veterinary control and animal care operations such as shearing are carried out in the place where the flocks roam. These operators support nomadic breeders, working with the flocks in their resting areas (Fortina et al. 2001).

In Piedmont, itinerant pastoralism has declined during the last few decades. For instance, just in the high plain of the *baraggia* located in the Biella region, Buratti (Buratti 1999: 53) estimates a 90 percent reduction of transhumant flocks over the last twenty-five years of the last century. There are currently sixty-five itinerant shepherds active in Piedmont and they have flocks of four hundred to three thousand heads (Mattalia et al. 2018: 749).

From a historical point of view, these shepherds are the keepers of a centuries-old tradition of the sheep-breeding landscape between the Alps and the Po plain; similar forms are found in the northeast of Italy under the name *pensionatico* (Fioravanzo 2015), or in Provence, in the south of France, where the wandering shepherd is called an *herbassier* (Dupré, Lasseur, and Sicard 2018: 223–24). This is not always a poor practice that opportunistically fits into other agricultural economies. Sometimes, it is a planned activity which optimizes the promiscuous use of territorial resources by means of different governances, such as religious orders' and abbeys' (Rao 2014; Archetti 2011; Andenna 2005: 137–38).

With population growth in Europe since the seventeenth century, the thickening and intensification of the agricultural system has progressively limited and confined seasonal and transhumant grazing practices in the productive plains, and also threatened their very existence (Sereni 1961). Today, in the Piedmont Po Plain, itinerant pastoralism has the characteristics of an interstitial practice (Mattalia et al. 2018), which clashes mainly with the farmers of the trampled lands and the territorial authorities that deal with nature conservation policies along the rivers (Verona 2006).

It is undeniable that the conduct of the itinerant shepherd is not always appropriate and numerous cases against them can be found in historical documents. A historical document by Bajo (Bajo 1858: 41) about itinerant pastures in the Venetian Plain states that sheep are left to graze throughout the winter unsupervised and untrained; landowners complain, but no one really knows whether the animals have the right to graze. Shepherds take advantage of this, and others suffer the damage without knowing its source; the municipal authority indolently allows the abuse of *ad libitum* pasture practices, without the usual tribute.

The apathetic stance cited by Bajo, that is, imposing constraints and regulatory restrictions or sanctions instead of promoting a strategic vision, is often a consequence of institutional misconduct and is the result of institutions sadly refusing to play a role in the management of resources (Messina 2016: 116).

This nomadic activity is distrusted as the grass resources are not a shepherd's possession (Aime 1997; Aime et al. 2001). However, flocks seasonally graze on these fields after local harvest, or within abandoned areas.

Figure 2.1. Itinerant flock in a corn field after cultivation. Collina Po Biosphere Reserve, 2019. © Dino Genovese

In the Italian agricultural landscape, itinerant pastoralism is therefore presented as a seasonal practice, both superimposed and complementary to the standard agricultural use of the territory. Nevertheless, this promiscuous use of land is often perceived as a form of intrusive use and it is in conflict with the local governance of the agrarian landscape.

Itinerant shepherds aim to ensure their flocks are fed and maintained. Indeed, animals fertilizing does improve a grassland's quality, thus having a positive impact on environmental, landscape, and ecological functions. The acceptance of itinerant flocks is historically documented in the lowland farmhouses of Piedmont in exchange for sheep manure for the fields and vineyards (Aime et al. 2001) or expensive forms of compensation such as the payment of rent, either through money or produce such as milk, cheese, animals, meat, wool, and leather (Archetti 2011). Even today, itinerant pastoralism offers many ecosystem services both in economic and anthropologic terms, such as the utilization of marginal and often fragile areas (DG Agricoltura 2013: 14) and in sociocultural terms, by representing an opportunity for the nonrural population of learning about the pastoral practice and culture (Oteros-Rozas et al. 2014: 1280). Therefore, itinerant pastoralism in winter can be viewed as a good practice of environmental sustainability, as a cultural proposal to the citizenry (e.g., folklore, tourism, history, recreation), and as a possible local production. Nevertheless, the shepherds do not seem to take advantage of these social and economic opportunities.

In some cases, the public administration—oblivious of the value that others recognize in nomadic grazing and herding in general—tends to favor simplistic solutions: instead of intensifying the control of herders and flocks and acting as a facilitator of this territorial dynamic between farmers and shepherds, it prefers to adopt a generic stance and to ban grazing, which often makes it difficult even to cross the territory (Verona 2016: 80–81).

Disputes between shepherds and managers of parks or Natura 2000 areas are also frequent, especially along the great lowland rivers (Verona 2016: 85–88): their mutual distrust—surely compounded by historical stereotypes—could be overcome today thanks to finer entrepreneurial skills and the adoption of a more modern approach by the shepherd, who is seen as a "technician" in the grassland management. The EU legislation treats herding as a homogeneous activity, without considering the difficulties of smallholder breeders, or the diversification of traditional local objectives, strategies, and production. Today, working within a protected area is not always seen as an advantage by breeders, but there could be significant exceptions, when shepherds recognize the Park Authority as a subject capable of coordinating local interests and actors to preserve grasslands (Messina 2016: 116).

The Research Area: The Natural Reserves along the Po River

There are some natural reserves along the Po River; in Piedmont, these were established in 1991. They protect habitats and wildlife in a territory that has been highly industrialized and urbanized. Agriculture has also intensified and the river spaces have remained the main ecological corridors of the Po Plain. Even itinerant shepherds have been limited by the pressure of anthropogenic activities: their animals—whose grazing is intrinsically linked to the natural dynamics of the grasslands—have had to move along the rivers to where they find water available. As a consequence, this has resulted in a conflict between nomadic herding and nature conservation policies along the Po River.

For this reason, we looked with great interest at the participatory planning processes laid out by the technical service of the protected areas of the Po River in Piedmont: they wanted to create a spatial and temporal definition of the itinerant grazing routes. More specifically, these processes initiated by the Park Authority of the Po River were analyzed. The conflict related to the practice of itinerant pastoralism was particularly interesting. Winter nomadic peregrinations (of flocks and shepherds) interact with a great number of actors and are dependent on a large number of factors. Shepherds must request an authorization to have access to private fields, must try to establish good relations with local farmers and guarantee the animals' optimal sanitary conditions, and must conduct their activity within nature conservation policies. If itinerant pastoralism is managed correctly, it can represent a way of operating the landscape and determine a positive interaction between man and nature.

In terms of research, the metropolitan area in Turin is also especially important because in 2016 it was recognized as a Collina Po Biosphere Reserve, and natural reserves are its core zones. Shepherds' routes also intersect this material and conceptual context. These areas are recognized as territorial laboratories within the UNESCO MAB (Man and Biosphere) program, aimed at experimenting with good practices between human development and nature conservation policies. Some flocks of up to two thousand heads cross this region during their wandering. Because of habitat and protected species' conservation, the animals' need for water at the river and need to use sensitive areas has escalated the conflict between these parties, as well as between farmers and local inhabitants. In 2016, the Park Authority of the Po River seized on the opportunity of entering a dialogue with shepherds and coordinating with them pasture routes that would have the least impact upon nature conservation policies. This action of facilitating and coordinating the other territorial institutions involved has been a good practice and has innovatively favored the integration of itinerant pastoralism into activities entailing the use of the land (Genovese et al. 2018). A similar approach can be found in the guidelines on the wandering sheep pasture of the neighboring Lombardy region (DG Agricoltura 2013: 15–17).

The resulting debate highlighted the importance of starting a dialogue between the conflicting parties. It was an innovative approach to governance of river space through the contribution of all stakeholders, trying to choose the best solutions both for nature and the community, thus minimizing negative impacts. Many topics were discussed and not all problems were resolved, but the resulting dialogue was fundamental in the creation of a new herd landscape. The ecosystem services of this activity have also been considered and enhanced since then, especially in this historical period when transhumance was suggested to be on the UNESCO ICH list. The research purposes were to explore this dialectic through the role-playing tool.

Materials and Method

Natural science didactics have developed a role-playing methodology required to neutrally address some of the major topics of the environmental debate (Colucci and Camino 2000; Camino and Dogliotti 2004) that affect our everyday life and (often) cause emotional distress. This method gives space to a plurality of opinions arising from the debate, without initially privileging the experts' advice. The topic, if not widely discussed, risks not being understood and accepted in the final common decision (Camino et al. 2008). Given the critical issues of territorial management connected to itinerant pastoralism and the several actors involved therein, the decision was made to project and test a specific role-play.

For this research, a role-playing game was designed and implemented. It dealt with the conflicts associated with the practice of itinerant pastoralism. The case was inspired by existing grazing bans, infringed upon by shepherds, and mentioned in newspapers. It was set in an unspecified lowland municipality near the city of Turin, characterized by the presence of a river and habitats of naturalistic interest; this territory is traversed by a selection of itinerant shepherds and their flocks.

The flock under study is made up of two thousand sheep, who cross the municipal territory two-to-three times a year between late autumn and early spring.

The case being discussed in this role-play is inspired by a petition from a group of citizens who had previously asked the municipal administration to ban the practice of itinerant pastoralism in the local territory for a host of reasons. Such a petition was indeed received and implemented by certain Italian municipalities affected by the practice of nomadic pastoralism.

In this role-play, the city council gathers to discuss whether to sustain or uphold the ban.

The object of the discussion is therefore not the legitimization of the itinerant pasture over a large area, but the specific opportunity to have the presence of transhumant flocks on its municipal territory for a few days a year. University students in environmental and agricultural courses were chosen as participants in the role-play since the research aimed to evaluate this method in terms of professional training also (Davodeau and Toublanc 2019).

The role-play was carried out within three different contexts:

A. May 2017 with twelve Master's students in Forestry and Environmental Science (Torino University, Italy)
B. December 2018 with nine Landscape Engineering students of the Ecole de la Nature et du Paysage of Blois, Val de Loire (Institut National Sciences Appliquées, France)
C. March 2019 with fourteen Master's students in Animal Science (Torino University, Italy)

The Role-Play in the Analysis of Itinerant Pastoralism

This role-play is specifically designed to analyze existing opinions and stereotypes about itinerant pastoralism. This tool is an opportunity for reflection among participants and represents a synthesis possibility for the researcher, who can compare fictitious opinions that highlight students' personal knowledge and skills.

Here is a sample of some of the game cards that were randomly distributed among the participants.

> Card 1: Mr. Alberto Conti, forty-eight years old, mayor.
> Who is he? A manager in a small company that manufactures wire nets. From a young age active in volunteering, first, scout and then a volunteer in emergency services.
> What does he think? "I have received a request to ban itinerant pastoralism on the territory, but I think breeding and agriculture are a resource for the municipality. Although itinerant pasture is a form that puzzles me."
>
> Card 2: Ms. Roberta Costa, fifty-five years old, town council member.
> Who is she? An architect, she works in a design studio in Milan with her husband. She loves to spend her free time taking long horseback rides.
> What does she think? "Greenways are an important element in urban planning. The action of the shepherds is useful to keep these corridors clean and accessible."

Card 3: Mr. Fabrizio Leone, fifty-two years old, town council member.
Who is he? He has a degree in accounting, deals with accounting in a small wholesale paper distribution company. He is allergic to lactose.
What does he think? "I am quite indifferent to the pastoralism issue, but it does not seem right to know that they graze so freely while the local companies pay all the taxes."

Card 4: Ms. Annalisa De Giorgi, forty-six years old, town council member.
Who is she? A primary school teacher. Afer high school she enrolled in the World Wildlife Fund (WWF) and in her youth participated as a volunteer at WWF summer camps in Sicily for turtles.
What does she think? "I have heard that the itinerant flock is a problem for wildlife, but it too is an activity to defend. A compromise must be made."

The three discussions help us to understand how itinerant pasture is perceived. "Is it a relevant phenomenon? What is the fallout from this activity? Who is actually disturbed by the passage of animals?" Considering that itinerant pastoralism is a relevant topic for the city council, the implicit questions that participants asked themselves during the initial stages were: "What economic consequences are there for the country?" and "How much does this activity influence tourism activity?"

A territory's overlapping uses is also expressed in leisure and sports activities, and in interaction with nature conservation policies too. In one group, an interesting discussion concerned who is more entitled to use those lands (farmers, itinerant shepherds, or hunters), evoking and claiming the greatest right of one to practice over another, even recalling the evolutionary steps of human history. Of course, the itinerant pastoralism which takes place on the territory bears witness to "activities which have been practiced for several generations." Finally, "the evolution of the population on the territory" has to be harmonized with a complex "functional cohabitation between the parties."

Some participants pointed to respect for private property and the perception that the itinerant shepherd freely takes advantage of a resource which belongs to residents who take care of it throughout the year. Indeed, agreements for land use do not always exist and are not always fully formalized. Nevertheless, some players—acting as residents—refused the shepherds' attitude by stating that it is an "exploitation of the territory without paying taxes or anything," or putting forward the notion that they "steal the grass" or doubting the actual legitimacy of their actions: "but do the shepherds have an authorization?"

Although it often takes place in the marginal areas of the municipal territories, transhumance represents an economic activity which—due to its characteristics and manner of conduct—interferes with the social dynamics of the villages crossed. In this sense, the intervention of the municipal authority is required: "private individuals have not the right to take everything they want; it is the municipality that must control this through the police." There is a risk that stereotypes about the nomadism and the itinerant shepherd will prevail in political decision-making. In very small communities, the opinion of individuals can influence the decision of the municipality, excluding possible technical alternatives: "We are here to decide as a municipality, the shepherds have already made their decision: to graze everywhere!"

Shepherds require in-depth knowledge to find grass every day for the flock among the different vegetative cycles of crops and uncultivated lands. One participant in the role-play pointed out that "the shepherd is a figure intrinsically linked to the environment and ecological function." On the other hand, from the objections of others, it seems that at present "the shepherds are not able" to assume this role and we need a "shepherd who adapts to the reality of our times." "But who would choose shepherd as a profession?" What is required to support itinerant pastoralism is "an action for men, not for sheep!"

The mayors in the role-play are required to end the city council session and decide upon the ban on wandering grazing. The decision is left to the participant's improvisation. It requires compliance only with a closing time within which the city council must be terminated.

All three groups respected the deadline and the mayors ended the discussion of the item on the agenda with a decision. Mayor C opted for a vote by show of hands (nine voted against, three in favor, and the mayor abstained); in the other two cases, the mayors took into account the expression of the apparent majority which emerged during the stages of the debate. Here are the final speeches of the City councils delivered by the three mayors:

> Mayor A: "A solution must be found and the option to abolish grazing in our municipality must be ruled out. We must summon the petitioners, explain and negotiate a solution in order to allow grazing in the municipality while ensuring that these animals do not eat in people's gardens, so that the pastoral activity which has always existed in our municipality can continue."
>
> Mayor B: "The petition is being reviewed by our office. The framing of the profession of the itinerant shepherd on our territory requires better definition. We need a technical service to support this profession and a study to explore the potential of this practice, which should no longer be experienced as a constraint but as a strategic element for our municipality."
>
> Mayor C: "Pastoralism is part of our tradition but we live in a society that is very advanced and this type of breeding is at odds with the new farming methods. We must consider everyone, even those who came to live in our country from the city. We indeed lose typical products, but if there are no more itinerant shepherds it will be the settled breeding farms that will take charge of them. The strongest justification behind this decision I am making is linked to health and a biosecurity risk: this is a problem for all of us and

our children. That's why I'm presenting the petition by which the itinerant pastoralism will be banned."

After receiving the role card, the participants played the corresponding character during the city council meeting game. They argued their position in relation to itinerant pastoralism based on their knowledge, but also according to the information on their role card. This allowed them to detach themselves from the technical viewpoint which stemmed from the mindset of their professional training.

Thanks to the freedom granted by the role, the participants were able to highlight aspects of the practice of itinerant pastoralism, which contributed to the construction of a practical description of the activity. Many of the problems attributed to itinerant pastoralists emerged: for example, the difficulty of managing large flocks in areas unfit for breeding, damage caused to vegetation and the dirtiness that follows their passage, abandoned and exhausted animals, employees living in caravans, and little care or interest in the management of pastoral resources. The technical problems overlap with the stereotypical cultural associations made towards "vagrants": this triggers a contradictory relationship between agriculture and breeding, sedentary and itinerant, urban and rural. Nomadic shepherds remain on the fringes of communities, not only because of the problems their animal breeding approach entails, but also because of the social isolation inherent to their trade.

On the other hand, participants highlighted that today's urbanized generations no longer know how to relate to the rural context and its players. Despite belonging to families of peasant origin, most locals from the places affected by itinerant pastoralism are today poorly informed about agriculture. They don't recognize the fatigue and the work that breeding requires: "But how it is possible to enhance the products of the region if there is no knowledge of the countryside?"

At livestock shows and transhumance fairs, participation and recognition is high; however, the problems arise in the day-to-day life when people who have no experience of breeding are faced with—and negatively perceive—critical issues about nomadic pastoralism. This sentiment is generally felt in the plains and metropolitan areas, where it becomes difficult to define what animal "welfare" is. Many students were struck by a video contribution of a satirical newscast, which presented the extreme situation of a flock left in the open field in the winter season: "I never questioned the fact that the droppings could be a nuisance or that the animals could be cold: the role-playing game helped me to think about that."

Communication lies at the basis of each project: but in the case of the itinerant pastoralism, how can one communicate to the public something that the public opinion itself considers to be wrong? Many people, as role-play participants, felt that they had to intervene by directly supporting the shepherds and conducting public awareness-raising activities and providing voluntary support to assist the jobs and lives of these permanent walkers. In the Alps, new small farms are starting up thanks to new shepherds, some of whom also have a university degree. "Perhaps a new generation is possible" and with this new generation comes a tourist-cultural development which may also positively affect itinerant pastoralism. What are the chances of seeing farmed animals in the meadows of the peri-urban area of Turin? It does happen, and cars stop along the highways and roads to take a picture of the flock! Itinerant shepherds must strive to innovate and create a different, more respectful, image of their person and their job: "They must be the first to hone their skills and join in a lobby if they want to survive." The cultural promotion of transhumance is a great opportunity and shepherds must "concretize their cultural function on the landscape because they are among those who best know the land."

Therefore, if the desire is to reframe the perception of itinerant pastoralism in a cultural lens, how can we intervene in the ongoing political conflict, where the opinions of the settled inhabitants—as legal voters—dominate the point of view of the itinerant shepherds? Is it purely a matter of technical mediation? Will the shepherds, caught between constraints and ecological limitations, ever be able to become itinerant stakeholders of the greenway projects along the rivers? The candidacy of Transhumance to UNESCO ICH list could in this sense be a stimulus for the local populations and for the municipalities to recognize this practice that characterizes and defines their territory (Ballacchino and Bindi 2017).

Conclusion

Among the aims of this research is the attempt to summarize different discussions and identify common and recurring elements in order to better understand itinerant pastoralism as perceived in the social and territorial conflict. Through their role-play characters, many participants exemplified stereotypes, prejudices, and expectations about this breeding practice and also tried to suggest solutions. The different players—who adopted stances either in favor or against the adoption of the ban on itinerant grazing within the municipal territory—have sometimes come up with similar arguments or viewpoints; other times, they have put forward more original considerations. Thanks to the audio recordings, it was possible to analyze the various contributions (first, from the simulation of the city council session, then in the "out-of-role" discussion that followed), piecing together a final synthesis.

Itinerant pastoralism is an activity that generates many conflicts. Its very existence along extremely long routes leads to repeated, daily interations with a host of different players, which all belong to the same categories (Verona 2016). The shepherds must know how to lead the flock and how to relate to the many players and stakeholders they encounter: the farmers who own the land that the livestock moves across, the police officers who patrol the bridges and villages, the park rangers who protect habitats and wildlife, and the rural and urban citizens who will be encountered along the way. Only good relations and the proper sanitary management of their flocks allow shepherds to obtain permissions for grazing (DG Agricoltura 2013).

People are constantly talking about returning to a "slower pace": yet the shepherds—who practice a slow and ancient trade which follows the natural rhythms of life and seasons—struggle to harmonize this dimension with space-related activities (Verona 2016).

The sustainability sciences developed the methodology of role-playing games (Colucci-Gray 2007; Camino et al. 2008). This tool was devised to understand and analyze the tradition of the itinerant shepherd. It allows people to move away from the assumption of stereotypical roles and develop a mature opinion about this age-old trade. The game and its role cards highlight the limits and problems associated with itinerant pastoralism but, at the same time, they emphasize the cultural and anthropological values of these skills.

The role-playing game was directed at university students on different training courses; these students were near-graduates, that is to say, they will soon have a job as technical advisors (Davodeau and Toublanc 2019). Firstly, participation in the game allowed them to analyze and experience the logistical criticality of itinerant pastoralism and to challenge their basic knowledge and skills in a hypothetical scenario. However, most of the participants had never had direct experience with itinerant pastoralism; consequently, the participants' involvement in the game was informed by both an emotional response and a cultural perception of itinerant pastoralism.

Some students who are enrolled in the Animal Science degree come from families of sedentary breeders; during the role-playing discussion, these participants were more engaged in topic discussions than other students, who seemed more detached when sharing their opinion. Despite their animated engagement during the game dynamics, the students from breeder families showed a consistent and appropriate behavior during the role-plays and did not let their emotions condition them when debating issues.

What did condition all students, however, was the different degree course they attended; this had an impact on the solutions provided to address issues within itinerant pastoralism. In the game, the students of Forestry and Environmental Sciences opted for technical spatial planning; the students of the Ecole du Paysage of Blois preferred sociological solutions and recommended training for the shepherds and reaching out to the resident population; the students of Animal Science showed far more concern in addressing the logistical difficulties and health risks that this breeding practice can entail, particularly on very long itineraries.

In the discussion, the need for institutional subjects who can mediate the concerns of pastoral caregivers and people affected by itinerant pastoralism was emphasized by many. Through mediation processes, it is possible to reduce the territorial conflict associated with the itinerant pastoralism. The Park Authority of the Po River has understood the value of this approach and has been able to integrate grazing by itinerant flocks into conservation measures. In the eyes of the population, the legitimacy guaranteed by the Park Authority has ensured a new interpretation of the work of the itinerant shepherds, even if cultural stereotypes remain.

The need to understand this job and to interpret its needs and limits is essential for its survival. Despite growing difficulties, this practice is still functional to this day in certain forms of landscape governance and to preserve habitats that cannot be maintained in other ways. The zoning and concerted grazing solution adopted by the Park Authority of the Po River represents an innovative approach to managing the activity of the itinerant shepherd. If supported, it can also become a strategic practice in the management of the environment, and above all, it can act as an example of an integrated policy between man and nature as specified in the program MAB UNESCO, of which the territories of the Po River are Reserve of the Biosphere.

In an interview, Lora Moretto Albino, itinerant shepherd along the Po River, said to us: "If there is a conflict, for us everything is lost." The itinerant shepherds live among conflicts and, in certain cases, they themselves represent these conflicts. The role-playing game allows us to reflect on potential solutions, but even more importantly, it is a tool which enables us to become aware of an ancient figure whose periodic passages continue to tell the story of the changing seasons to the modern person.

Acknowledgments

Thank you to Enrichetta Valfrè for the language revision.

Dino Genovese has a PhD in Agriculture, Forest and Food Science with a research project about public-private agri-partnership model analysis in landscape governance practices. In particular, the goal of his research is the role of livestock farms in these systems. He graduated in Forest and Environmental Sciences and he holds a specializing degree in Architectural and Landscape Heritage. With twenty years of experience in natural protected areas of Po River and Collina Torinese, he specializes in management of forestry, environmental education, relationship with farmers, and hiking network planning.

Ippolito Ostellino was born in Turin on August 16, 1959. In 1987 he graduated in Natural Sciences and worked in the management and design of Alpine Scientific Gardens. In 1989 he participated in the foundation of Federparchi Italia. From 2007 to 2008 he was National President of AIDAP, the Italian Association of Directors of Italian Parks. In the Turin area he is the promoter of the Corona Verde project and teaches at the Polytechnic of Turin, where, in 2016 the recognition process of the reference territory brand in the UNESCO Man and Biophere program was coordinated. He curates the blog "Protected Areas and Biosphere" on the national magazine, *.Eco* (rivistaeco.it); collaborates with the platform "La Natura Returns to Art"; and is a member of the scientific committee on the book series on protected natural areas at the ETS publishing house in Pisa.

Luca Maria Battaglini is Professor of Animal Sciences and Technologies in the Department of Agricultural, Forestry and Food Sciences (DISAFA) at the University of Turin. His research, teaching, and public engagement activities concern livestock farming systems sustainability through an ethical and sociocultural framework. As a member of many national and international research groups, he is interested in environmental impact, ecosystem services, livestock biodiversity, and animal welfare, with reference to the Alpine region.

References

Aime, Marco. 1997. "La strada del pastore tra Alpi Marittime e Monferrato." *Lares* 63(4): 495–510.

Aime, Marco, Stefano Allovio, and Pier Paolo Viazzo. 2001. *Sapersi muovere. Pastori transumanti di Roaschia*. Milano: Meltemi.

Andenna, Giancarlo. 2005. "La rete monastica." History of Vercelli Conference, 18–20 October 2002. Vercelli: Società storica vercellese and Fondazione Cassa di Risparmio di Vercelli. Retrieved 20 July 2020 from http://rm.univr.it/biblioteca/volumi/vercelli/Andenna.pdf.

Archetti, Gabriele. 2011. "'Fecerunt malgas in casina.' Allevamento transumante e alpeggi nella Lombardia medievale." In *La pastorizia mediterranea*, ed. Antonello Mattone and Pinuccia Simbula, 486–509. Roma: Carocci editore. Retrieved 20 July 2020 from https://publicatt.unicatt.it/retrieve/handle/10807/8318/3462/Archetti percent20 percent28Roma percent202011 percent29.pdf.

Bajo, Pietro. 1858. *La servitù di pensionatico e l'ordinanza imperiale 25 giugno 1856. Cenni economico-giuridici*. Venezia: Tipografia del Commercio. Retrieved 20 July 2020 from https://play.google.com/store/books/details/La_servitu_di_pens ionatico_e_l_ordinanza_imperiale?id=t4tpAAAAcAAJ&gl=US.

Ballacchino, Katia, and Letizia Bindi, eds. 2017. *Cammini di uomini, cammini di animali. Transumanze, pastoralismi e patrimoni bio-culturali*. Campobasso: Edizioni Il Bene Comune.

Battaglini, Luca Maria. 2007. "Sistemi ovicaprini nelle alpi occidentali: realtà e prospettive." In *L'allevamento ovino e caprino nelle alpi: tra valenze eco-culturali e sostenibilità economica*, ed. Luca Maria Battaglini, Stefano Martini, and Michele Corti, 9–23. San Michele all'Adige: SoZooAlp. Retrieved 20 July 2020 from www.sozooalp.it/fileadmin/superuser/Quaderni/quaderno_4/1_Battaglini_SZA4.pdf.

Battaglini, Luca Maria, Riccardo Fortina, Sonia Tassone, Antonio Mimosi, and Marcello Bianchi. 2001. "Local Breeds Conservation and Typical Products in Piemonte (NW Italy)." In *Recognising European Pastoral Farming Systems and Understanding Their Ecology: A Necessity for Appropriate Conservation and Rural Development Policies*, 29–31. Bridgend: European Forum on Nature Conservation and Pastoralism.

Bernardino, Romano, and Francesco Zullo. 2016. "Half a Century of Urbanization in Southern European Lowlands: A Study on the Po Valley (Northern Italy)." *Urban Research & Practice* 9(2): 109–30. https://doi.org/10.1080/17535069.2015.1077885.

Buratti, Gustavo. 1999. "Les nomades de Piémont." *L'Alpe* 3: 52–55.

Camino, Elena, Carla Calcagno, Angela Dogliotti, and Laura Colucci-Gray. 2008. *Discordie in gioco. Capire e affrontare i conflitti ambientali*. Molfetta: Edizioni La Meridiana.

Camino, Elena, and Angela Dogliotti, eds. 2004. *Il conflitto: rischio e opportunità*. Torre dei Nolfi: Edizioni Qualevita.

Colucci, Laura, and Elena Camino. 2000. *Gamberetti in tavola, un problema globale. Un gioco di ruolo sugli allevamenti intensivi di gamberetti in India*. Torino: Edizioni Gruppo Abele.

Colucci-Gray, Laura. 2007. "An Inquiry into Role-Play as a Tool to Deal with Complex Socio-Environmental Issues and Conflict." PhD diss., Milton Keynes: The Open University. Retrieved 20 July 2020 from http://oro.open.ac.uk/59948/1/437809.pdf.

Davodeau, Hervé, and Monique Toublanc. 2019. "Les usages pédagogiques du jeu de rôle dans la formation des professionnels du paysage." In *Sur les bancs du paysage. Enjeux didactiques, démarches et outils*, ed. Anne Sgard and Sylvie Paradis, 129–48. Genève: Métis Presses.

DG Agricoltura. 2013. *La pastorizia ovina vagante in Lombardia*. Milano: Regione Lombardia.

Dupré, Lucie, Jacques Lasseur, and Julia Sicard. 2018. "'Berger, point barre.' Jalons pour une redéfinition pastorale de l'élevage bas-alpin." *Études rurales* 201(1): 218–39.

Fioravanzo, Daniele. 2015. "Il diritto di pascolo invernale nel Veneto sette-ottocentesco." *Studi Storici Luigi Simeoni* 65: 67–78.

Fortina, Riccardo, Luca Maria Battaglini, Sonia Tassone, Antonio Mimosi, and Alberto Ripamonti. 2001. "The Shepherd's Road: Pastoralism and Tourism in Piemonte (N-W Italy)." In *Recognising European Pastoral Farming Systems and Understanding Their Ecology: A Necessity For Appropriate Conservation and Rural Development Policies*, 26–28. Bridgend: European Forum on Nature Conservation and Pastoralism.

Genovese, Dino, Luca Battisti, Ippolito Ostellino, Federica Larcher, and Luca Maria Battaglini. 2018. "The Role of Urban Agriculture for the Governance of High Natural Values Areas. New Models for the City of Turin Collinapo." *Acta horticulturae* 1215: 345–50.

Lazzaroni, Carla, and Davide Biagini. 2008. "Perspective of Sustainable Piemontese Cattle Rearing in the North-West of Italy." In *Mediterranean Livestock Production: Uncertainties and Opportunities*, ed. Ana Maria Olaizola, Jean Pierre Boutonnet, Alberto Bernués, 121–26. Zaragoza: CIHEAM / CITA.

Mattalia, Giulia, Gabriele Volpato, Paolo Corvo, and Andrea Pieroni. 2018. "Interstitial but Resilient: Nomadic Shepherds in Piedmont (Northwest Italy) Amidst Spatial and Social Marginalization." *Human Ecology* 46: 747–57.

Messina, Simona. 2016. *Il paesaggio del Morso: integrazione dei pascoli residuali nel contesto periurbano contemporaneo*. Roma: Parco Regionale dell'Appia Antica.

National Livestock Registry. 2020. Anagrafe Nazionale Zootecnica, Statistiche. Retrieved 20 July 2020 from www.vetinfo.it/j6_statistiche/#/.

Nori, Michele. 2010. "Pastori e società pastorali: rimettere i margini al centro." *Agriregionieuropa* 6: 22.

Nori, Michele, and Valentina De Marchi. 2015. "Pastorizia, biodiversità e la sfida dell'immigrazione: il caso del Triveneto." *Culture della sostenibilità* 8(15): 78–101.

Oteros-Rozas, Elisa, Berta Martín-López, José Antonio González, Tobias Plieninger, César A. López, and Carlos Montes. 2014. "Socio-Cultural Valuation of Ecosystem Services in a Transhumance Social-Ecological Network." *Regional Environmental Change* 14(4): 1269–89.

Rao, Riccardo. 2014. "Le Alpi Marittime e l'invenzione bassomedievale della montagna." In *Uomini e ambienti dalla storia al future*, ed. Paolo Cesaretti and Renato Ferlinghetti, 33–46. Azzano S. Paolo: Bolis edizioni.

Sereni, Emilio. 1961. *Storia del paesaggio agrario italiano*. 22th ed. Bari: Laterza.

Verona, Marzia. 2006. *Dove vai pastore? Pascolo vagante e transumanza nelle Alpi Occidentali agli albori del XXI secolo*. Scarmagno: Priuli e Verlucca Editori.

———. 2016. *Storie di pascolo vagante*. Bari: Laterza.

CHAPTER 3

Between Two Different Worlds
Pastoralism and Protected Natural Areas in Provence-Alpes-Côte d'Azur

Jean-Claude Duclos and Patrick Fabre

Transhumant Pastoralism and Protected Natural Areas: The Meeting of Two Worlds

The implementation of protective status for natural areas, whether through natural reserves (1957), national parks (1960), regional natural parks (1967), the coastal conservation authority (1975), Natural Zones of Interest for Ecology, Fauna, and Flora or the ZNIEFF (1982), or Natura 2000 Zones (1992), to account for only the main French measures, has seen considerable advancement in the last sixty years. In 2020, the protected areas all together cover 20 percent of the national territory (Lefebvre and Moncorps 2013: 44).

At the international level, the UN recommends that 30 percent of land and sea areas be put under protection by 2030.[1] Therefore, with various forms and results, almost two hundred national governments, all signatories of the Convention on Biological Diversity (CBD), have committed to a worldwide policy of protection. The aim is to battle against the loss of biodiversity and to reach a sustainable use of natural resources. As a consequence of the convention, natural protected areas have a central place nowadays in the ecological strategies of the signatory states.

As much for the scientific knowledge provided by the monitoring of the ecosystems thereby protected as for the implementation of proper conservation, these classifications and their associated regulations form indeed the best way to preserve those natural environments that are deemed essential to the continuation of biological diversity. These classifications not only focus on research and conservation goals but also contribute to an enlighten territorial planning, since the protected areas are part of an economic and social totality that is guided by policies implemented on the regional, national and, in the case of the CBD, planetary levels.

The Development of the Protection of Nature and Pastoralism

In France, and in the world, a large number of natural protected areas are pastoral areas. Indeed, two national parks, eight—soon to be nine—regional natural parks, and several national reserves, including the Coussouls of Crau Reserve, have been created over extensively pastured land as part of a breeding qualified as "pastoral," all within the territory of the Region Provence-Alpes-Côte d'Azur alone. To be clear, we must recall that what is said to be "pastoral" breeding is that which favors the consumption of grass through grazing—or pasture—yet complementary intakes in the form of fodder and cereals can still be incorporated as long as they represent less than a quarter of the total feed of the animals.

As this form of breeding is present in most of the protected areas, we can wonder if pastoralism was not the main actor of their conservation before they were classified. Such an observation may be surprising and seem exaggerated. We will see that it is not as meaningless as it may first appear.

Over the past sixty years of experimentations, we have seen the multiplication and evolution of various models and strategies in regard to the protection of nature. If, in natural reserves and the central zones of national parks, the choice was made to give sanctuary status to the environment, excluding all human use, other forms of protection, which include human activity, have been experimented with. In those, interest from the local population was sought, and so was its involvement which has been sometimes obtained. This has been achieved in regional natural parks, which were in fact created with the purpose of protecting a territory in harmony with its inhabitants' activities. This approach is also experimented with in national parks, where, since 2006, local governments are involved in decision-making processes. It is even practiced in a few natural reserves, although the case of the Coussouls of Crau which we will develop, is probably unique. Sanctuarization is no longer recognized as a realistic option, except in a few, quite rare circumstances. Admittedly, those who advocate rewilding still support it fiercely. Yet, we would rather listen to people such as Luc Hoffman, cofounder of the World Wildlife Fund (WWF) among other things, who always acted in a humanist way, undertaking international initiatives and leading his entire life for the protection of nature, continuously asking the same question: "How can we ensure the Earth stays a viable environment for mankind?" (Hoffmann 2010: 211).

The enactment of the three types of classification we have seen—national park, regional natural park, natural reserve—has created the necessity for new skills. There needs to be implemented, between the decisional authority and, in the field, the most involved socioeconomic actors, a form of mediation. This function was given to engineers, project managers—sometimes called "project managers in pastoralism" in the area we are interested in—and technicians, in other words, men and women very often specialized in ecology. Mandated by their respective institutions, these agents suggest and, after receiving the approval of their governing bodies and very frequently of the scientific boards that surround these institutions, put into effect the measures of territorial management. It is through these agents, whose role as mediators has become central, that the protection structure, park or reserve, communicates with its inhabiting population. As we limit ourselves in this chapter to pastoral breeding, we will focus on the communication developing between the representatives of the protected natural areas and the pastoral world. We will come back on the definition of this distinctive world, made up of individuals united in their passion for breeding the animals that participate in their existence. Before that, a few more precisions on the specificities of the areas in and through which this pastoral world exists and persists seem necessary to consider.

In Mountain Pastures, Since Prehistoric Times

Let us start with the alpine mountain where archeologists have confirmed the existence of pastoralism between the second and third millennium BC (Walsh et al. 2006). Their research has even led them to observe that between the Late Neolithic and the Bronze Age this form of breeding so greatly modified the environment that mountain landscapes had already evolved to become similar to those we know today. Even though other activities, such as the search for various materials (silex, rock crystal, ore, etc.), were leading men towards the mountains, it was their pastoral activity there that transformed and shaped the environment permanently. Groups of semi-nomadic families took advantage of wide grasslands, traveling seasonally over several kilometers and soon after a lot more. Established around the pastoral use of a common space and likely a common flock, these "neighboring communities," as anthropologists call them, developed a know-how which, though evolving and adapting continually to circumstances, has been transmitted ever since. Over a very long period, knowledge on the ways to use the principal asset of the group, namely the permanent grass from which the herd is fed and which ensures its future, was established, refined, perfected, and passed on through action and example.

This use of grass through the regular and repeated pasture of domesticated herbivores, that are lead and kept, generated a singular flora and environment. This alpine pasture is called *alpage* in French, a term from which the Alps got their name. For these grasslands, that most will believe to be "natural," do owe their existence and their renewal to herbivores' teeth as well as to the shepherds who breed and lead them. Only prolonged grazing, thought to provide the vital needs of the human group through the welfare of bred herbivores, could enable the creation and upkeep of the altitude pastures. The expertise, which the pastures have become dependent upon, is still maintained today by pastoral breeders and shepherds and constitutes indeed a real "patrimony," what one generation leaves to the next in order to guarantee its existence and its transmission thereafter. This is the reason why alpine communities have long been stubborn about maintaining the collective ownership of their pastures. This ownership appeared vital to them as much for their own livestock as for the rent they earned from transhumance.[2] However, these mountain communities could not predict the dispositions put in place by the state in the nineteenth century, which were aimed at optimizing the use of forests and rebuilding the mountain soils through reforesting, but which would strip them of their pastures. Grazing and lumbering activities were forbidden on the very vast areas suddenly placed "under the forestry regime," areas that the communities had been using up to that point. Goats and sheep were specifically prohibited, judged by the administration of Water and Forestry to be the cause of the disappearance of forests and the deterioration of the mountain soils. These measures accelerated the desertification of the mountain areas and weakened the agropastoral activities so severely that the members of the communities had no other choice than to emigrate or invest in tourism. This is how quite a few of them gave up the ownership of their collective pasture in favor of the development of winter sports resorts and ski areas.

This brief review has no other goal here than to keep in mind the subdued state into which the mountain world and more generally the rural world are placed in when, in the name of public interest, a central power imposes their decisions with no negotiation. This is how the creation of the Ecrins and the Mercantour national parks were perceived locally, as an authoritarian decision of the state, infringing namely on their freedom to hunt. Some, as in the Valgaudemar, even felt that they had become unwanted. Therefore, individuals in the pastoral world, the majority of whom are from the mountains originally, have remained distrustful of externally dictated measures. All the more so as the nature on which protection is decreed is one they know well and which they have lived on for centuries. That is clearly the case in the Ecrins and Mercantour massifs where agropastoral activity has long been dominated, at least since the Middle Ages, by the summer stay of the transhumant herds of Provence.

These two alpine massifs are today managed as part of the national parks, dedicated, at least as a starting point, to the protection of the wild flora and fauna. Yet, 20 percent of the central area of the Ecrins national park and over 50 percent of the Mercantour National park are covered in grazing, which has long been used for pasture. When the parks were created, there was a tendency in policy to wait for the pastoral activity to disappear on its own, yet it had to change under the pressure of territorial collectivities that immediately demanded the continuity of "traditional pastoral activities" there. The governing bodies of these parks would have also recognized that supporting the perpetuation of pastoralism was a way to ease the tensions between them and the local population.[3] The least we can say is that the relationships between the pastoral sheep breeders and the teams of the parks were not easygoing at first and have become only more complex since the reappearance of the wolf in 1992. Emile Leynaud, who was an informed director of the Cévennes national park before becoming general environment inspector, declared about national parks: "their insertion in local communities greatly depends on the future of these institutions whose difficult mission is to succeed in turning the territory of others into the territory of all" (Leynaud 1985). We will consider the place of pastoralism within the national parks.

The Medium Mountain Areas and the Plain

The pastoral use of vegetal cover is not of course exclusive to the high mountain areas. All areas put to the use of pastoral breeding over a long period of time result in floristic and faunistic identities, caused by the regular pasture of domesticated ungulates under the lead and care of their experienced shepherds. Such is the case of the Verdon natural regional park spreading from the Durance to the Alps. From low altitude routes to alpine pastures, its territory has long been put to use by its inhabitants for the practice of an often-transhumant sheep pastoralism which, although it has known fluctuations, has seen quite an improvement since the park's creation in 1997. The governing bodies of the park recognized the interest of this form of breeding to help in the prevention of wildfire, to which the territory is particularly vulnerable, and the conservation of its biodiversity. Thus, they enrolled in the national and European network of "Green and Blue infrastructure" (TVB) whose goal is to tackle the loss of biodiversity by the preservation of "reservoirs of biodiversity," which are linked together through "ecological corridors," in line with the CBD's recommendations. The park encourages the presence of the breeders not only for their role in sustaining the "corridors," but also because of the economic activity they produce, and more broadly for the human presence they assure in an area that was, until recently, in the process of desertification. Today, the park is lending support to them in protecting themselves from wolf attacks which are common and put their activity in peril.

We will also consider, farther south, the cases of the Alpilles massif and the Crau plain, exposing the conditions in which this form of breeding has persisted. These two contiguous entities are today protected as a regional natural park (PNR) for the first, and a national reserve for the second, called the National Natural Reserve of the Coussouls of Crau (RNNCC).

The Alpilles PNR's web page describes its territory as follows: "Its landscapes owe as much to the deep forces of the earth as to the work of those who, over the centuries, have cleared the woods, brought up villages, planted vines and olive trees, dug mountains and plowed the land." Though in a lyrical way, the anthropization of this environment is acknowledged here, nevertheless, breeding was forgotten. It is true that the touristic purpose of this territory, incidentally the place of residence of wealthy individuals, might have caused the role of pastoralism to be overlooked when the PNR was created in 2007. Yet, it was only a short time before, in October of 1989, that about 1500 ha were destroyed by the flames in a few hours, mainly on the commune of Aureille; the catastrophic event brought back memories of the images of a time when herds roamed the hills. The prevailing conception of protection until then had mainly been that of the national forestry office, who had banned breeding activities from the area as they was considered detrimental to the development of the forest cover. Under the plan of Defense of the Forest against Wildfires (DFCI), trails had been created, yet, after the sudden spread of fire in 1989, this measure seemed no longer sufficient. Immediately afterwards, the intercommunal sylvo-pastoral Syndicate of the Alpilles was formed and it allowed about forty pasture areas to be attributed, through the mediation of an organization on which more will be said, the Center for Pastoral Studies and Implementations of the Alpes-Méditerranée (CERPAM). When it was created, the park relayed the syndicate's action and obtained a commission on pastoralism. Several of its agents go along with and, therefore, help pastoral breeders over fifty thousand ha of its territory, half of which are classified Natura 2000 Zone. Although some argue the park could do even more to promote pastoralism, there is progress.

South of the Alpilles, in the Crau plain where transhumant sheep breeding has maintained for a long time some vitality, the protection was put in place through a very different path. The naturalists who took the first botanical inventory of the Crau, in 1950, already noted that the flora of the *coussoul*,[4] "one of the richest in species of the Mediterranean region," was the produce of "a centuries-old pasture" (Molinier and Tallon 1949–50: 111). It could not exist, they scientifically proved, without the sheep grazing there. Therefore, it was evident for the ecologists, who would later expand on the knowledge of the coussoul's ecosystem, that the ecosystem is inseparable from the sheep that graze there from February/March to June, thanks to transhumant breeding. It was then admitted by all that the sustainability of the coussoul depends upon the continuity of this type of breeding. Naturalists also discovered that the seeds they find in the soil can stay viable for hundreds of years and that their identification can help reconstruct the past of this environment. Seeds that have been found near Roman sheep pens provide proof of the presence of the herds that used to be there 1,500 years ago! Naturalists and pastoralists go even further and assess the role played by the know-how of the shepherds in the interdependent relation linking the coussoul's vegetation to pasture. As a matter of fact, they observe how the shepherds' use of the *fin* and the *grossier* association in their herding sustains the coussoul's biodiversity. The *fin* is made up of a great diversity of short grasses—up to seventy varieties in a square meter—as well as the *grossier* of Mediterranean False Brome, the *baouco* in Provençal, and thyme. More simply: the first one feeds, while the second fills, which will constitute a perfectly balanced diet, on the condition that the herd is well led, and will guarantee, *in fine*, that the produced meat and wool are of great quality. The acknowledgement of the symbiotic relationship between the soil and the herd has played a major part in the organization and management of the protection of the dry Crau, which is incidentally recognized as one of the last steppe-like environments in Europe. The naturalists played an important role there; yet, the sheep breeders also managed to make their voices heard to the extent that they nowadays participate in the management of the national reserve via their representatives in the Chamber of Agriculture of the Bouches-du-Rhône. The case of the Natural Reserve of the Coussouls of Crau is unique to the best of our knowledge, and it is not comparable to the cases of natural reserves in general. About this, we will see that the distinction in the classification—natural reserve, national park, or regional natural park—is clear in its legislative perspective yet much less so in the field. Nevertheless, the case of the dry Crau, where a national reserve was created in 2001 over about 7400 ha, managed jointly by the Conservatory of Natural Areas of Provence (CEN PACA) and the chamber of agriculture, appeared necessary to us to investigate because of the important place it has in the practice of transhumant sheep breeding in the southeast of France.

From the Ecrins and the Mercantour to the Crau, through the Verdon and the Alpilles, which is to say from altitude pastures to coastal areas, all the levels promoted by pastoralism are thus represented in these sites which are today protected. We will now see, first from the point of view of the parks and reserves' agents, then from the point of view of the pastoral breeders and shepherds, how the perceptions are expressed and the exchanges are carried out. Beyond the knowledge we have of the different environments, we will base our arguments on the analysis of about twenty interviews, conducted by Vincent Dechavanne during his internship at the Transhumance House in 2019 (Dechavanne 2019: 54), with agents in charge of pastoralism in natural protected areas, as well as sheep breeders and shepherds using these areas.

The Agents of National Parks

In their work on the alpine national parks, Geographer Lionel Laslaz and his team have already analyzed well the relationship between these institutions and the inhabitants of the areas through pastoralism (Lazlaz et al. 2014: 416). As the academics suspect, the interest of the national parks for this form of sheep breeding would seem to rest nowadays much more in the possibility offered to ease tensions with the local population, than on its contribution to the biological diversity of the preserved environment. As a consequence, the quandary is permanent between the will to favor pastoralism with financial and material contributions, and the will to protect nature.

In the two national parks we are interested in, and maybe more so in the Mercantour park, the relationship with the sheep breeders can become authoritative. The Agroenvironmental and Climate Measures (MAEC), although accompanied by a financial compensation for the sheep breeders, may result in the park's agents in charge of pastoralism becoming the messengers of regulations, as their duty, for instance, is to ensure strict compliance to the pasturing calendar and the number of allowed animals. Sanctions are imposed in the event of an infringement. If sheep have been found grazing in any of the black grouse (*Lyrurus tetrix*) nesting areas, for instance, the national park agents may alert the competent authority (the Departmental Direction of the Territories and the Sea, DDTM) who can withhold the MAEC's support from the sheep breeder. The agents sometimes also count the ewes as soon as they arrive, as they climb down from the truck, and make regular visits to the herd to observe its evolution on the alpine pasture and make certain that the sheep breeders are in keeping with their commitments.

Figure 3.1. National park of Mercantour, 2019. © Patrick Fabre

As concerns, for example, the *queyrel*, a grass that animals will not eat when it hardens, the park's guideline is clear: "mandatory scrapping." A study showed that this vegetal species, formerly groomed by reaping, is an active part of an interesting floristic whole which is dominated, and then degraded by, *queyrel* unless it is grazed upon at the beginning of summer.

To keep to a few examples, the "unfavorable decision to the setup of impluviums in the center of National Park"[5] given recently by the scientific council of the Mercantour National Park has much more serious consequences. An "impluvium" is usually used by shepherds to trap water in the southern Alps where droughts are frequent in summer. They used to be made of stone but are now obtained by digging a trough in the ground and covering it with a waterproof tarp in order to collect, store, and redistribute rain and snow melt. They are used to water the herd. The scientific council opposes them due to the risks:

- for the Batrachia, by drowning [*sic*],
- linked to the plastic tarp's disposal,
- linked to the accumulation of organic matter facilitating the development of cyanobacterias,
- linked to the modification of the landscape,
- linked to their multiplication in case of a severe drought.

Such positions, associated on a larger scale to the protection of wolves, shows that the protection of nature and its opening to the public, as the scientific council's decision precisely mentions the "modification of the landscape," are still a priority for the park. But "protecting and opening" were the goals set by the national parks when they were first created. Should nothing have changed then?

We will not dwell here on the consequences of the wolves' takings on the herds, which is largely studied elsewhere,[6] we will, however, note that this issue is not treated everywhere in the same manner. This concern caused severe difficulties in the Mercantour where wolves reappeared for the first time and where the national park elected to take on their management, as if it were a victory, even though nothing was forcing them to do so. Since, the situation has evolved in a rather more favorable one, probably through a change in direction but also thanks to the local political will to work with the sheep breeders. The issue of the wolf appears to be approached with more calm in the Ecrins where the park has left the competent authorities in charge (the minister of the environment, through the National Office of Hunting and Wild Animals or the ONCFS). Furthermore, the consensual efforts to support pastoralism in the Ecrins, through the installation of shepherds' huts or the developments of access points to the alpine pasture among other things, have allowed for more peaceful relationships. Yet, wherever you are, the essential part of what is at play is happening on the level of interpersonal relationships between national park agents and actors of the pastoral world. Note that in national parks, guards seem to play a major role in those types of interactions.

The Regional Natural Parks' Method

The size of the role played by these agents is even more obvious in regional natural parks where, as we said before, protection can be conceived only with the involvement of the local population, or at least with their representatives. Jean Blanc, who used to be a transhumant shepherd, organized in 1966 the Days of Lurs for the DATAR (Delegation for Territory Planning and Regional Action), where the regional natural parks' doctrine was conceived. He explained it was aimed at giving an answer to the following question: "Are we capable, for some homogenous and sensible '*pays*' to move beyond real estate, industrial and touristic development, in order to 'preserve, prolong, develop,' in permanent thought, including all concerned, a 'frame of life' in harmony with quality of life?" Even with its utopian side—or maybe because of it?—such a challenge is still as relevant today. Is it to say that it is met with success everywhere? The testimonies of the agents of the regional natural parks of the Verdon and the Alpilles, who appear to have taken to heart the matter of pastoralism, seem to make it believable.

At the behest of the Verdon PNR's representatives, who were very eager to support pastoralism, the agents have put into action the directives of the "Green and Blue infrastructure," as part of the project "Campas," aimed at "regaining and bettering pastoral environments." Therefore, the defense of natural habitats of such rare species, animal or vegetal, and that of pastoralism appear to be part of the same purpose. One is beneficial to the other and vice versa. The work is time-consuming and complex as it involves, besides the mapping of the sites, obtaining the agreement of the owners, putting in place multi-year pasture conventions, and perhaps even setting up pastoral land consolidation associations (AFP), attributing them to one or several sheep breeders and following up on them. This backing has led the PNR to create the position of "support shepherd" who periodically comes to the aid of sheep breeders and shepherds in case of wolf attacks. With over ten wolf packs spotted in the park's perimeter, the situation has undoubtedly become difficult. Therefore, the creation of a second position of "support shepherd" is already being discussed.

At a lower altitude, in the Alpilles, where the wolves' presence is not yet a cause for concern, the elected representatives are rather worried about the danger of wildfire catching in the dry pine trees. They have all understood that sheep, goats, and bulls had to return to the hills and that their breeders should be received in good conditions when they come. Therefore, they have put in a lot of work on conciliations. Talks, without opposition, are still taking place with the hunters who do not want to see partridges or woodcocks leave the area or risk jeopardizing the crop they sow to attract wild game. Similar negotiations are under way with the agents of the ONF (National Forestry Office) who still worry about the grazing of sheep or goats, and with the private owners who must be convinced of their own interest in signing a pasture convention, etc. Confident that pastoralism is beneficial to the biodiversity of environments, these agents take for proof the scientific studies led on their territory about insects, birds, bats, and amphibian reptiles.

According to the park agent in charge of the Natura 2000 classified zones, who monitors with an utmost vigilance the evolution of "the sub-steppe course of annual grass," among others, the presence of the herd is a necessity. He would like this fact to be more largely acknowledged and hopes for a better communication on this point. He also wishes the park's charter, which is about to be renewed, could afford more space to pastoralism. The other agents—four in the team, each dealing more or less closely with pastoralism—share the same opinion. For them, all the op-

Figure 3.2. Natural regional park of Verdon, 2019. © Lionel Roux

erations funded by the European government (Life, Leader, FEDER, etc.) are opportunities to favor this form of sheep breeding. As they talk about the openings they have made by gyro-spinning before a sheep breeder and their herd finish the clearing, about the liaisons they plan between the pastoral "alveoli" maintained by pasture, about the settling of a young couple and their ewes, about the role the DFCI should play and about the negotiations they will start to ensure each of these actions succeeds, the benefits of pastoralism in the management of the Mediterranean forests is revealed. More broadly, it is the role it plays in the territory planning that is called into question. The regional natural park then becomes, as it was intended at creation, a "tool used for subtle land planning."

In any case, all the agents of the Verdon and the Alpilles PNR in charge of pastoralism rely on the expertise and support of the CERPAM. This organism, which we will return to, is in fact responsible for all matters of pastoral diagnosis as well as more technical evaluations.

Inside the National Natural Reserve of the Coussouls of Crau

In the plain of Crau, transhumant sheep breeding was always fragile because of the difficulties that breeders encounter on the markets for meat.

Yet, it seems to be enduring better than elsewhere in France and even beyond, in western Mediterranean regions, where almost everywhere the practice of transhumant breeding is scaling back.

Most of the irrigated Crau (around 13,000 ha) is composed of grassland whose hay is cut three times a year, in May, July, and September, to be sold under a "controlled designation of origin" (AOP) as *"Foin de Crau."* The fourth cut, in the fall, is left for the herds that have been brought back from the mountain and will remain in the meadows until the middle of February. They will then be led to the *coussouls* until June before going back up to the mountain for the next three or four months. Whether consuming grassland and hay, grazed on from October to February, or in the *coussouls* from February to June, the entire Crau, wet or dry, is put to use by sheep breeders over about 30,000 ha, of which 11,500 ha are *coussouls*. Part of the aim of the natural reserve, encompassing 7,400 ha spread over seven communes, is to "guarantee the future of transhumant sheep breeding and, at the same time, its jobs and economic activities." This is why it is managed, as was said, by naturalists of the Conservatory of Natural Areas of PACA (CEN PACA) as well as by representatives of sheep breeders, through the agriculture Chamber of the Bouche-de-Rhône. This double management is paralleled in the composition of the team working on the reserve which includes a technician from the Chamber in part-time employment on the reserve who was charged to see to the good relations between the team and the sheep breeders and shepherds.

Although she does not report major difficulties, she claims it is sometimes difficult to convince sheep breeders and shepherds that they should not install fences around areas of pasture. The use of stationary or mobile fences is actually a way to compensate for the lack of shepherds whose hiring has become more difficult nowadays, even more so since some people do not enjoy the shepherds' seasonal presence in Crau. The fencing refusal must, therefore, be a motivated decision. Though the naturalists observe for instance that sandgrouse preferably nest in open areas, their opinion is not firm on the issue of fencing. As a result, experimentations are under way. The technician also finds that, even if the intermediary position she is in, between the Reserve's naturalists and the sheep breeders, is not always comfortable, it is where she feels the most useful and she wishes to spend more of her time on mediation.

The only instructions given to the sheep breeders who pasture *coussouls* of the reserve, since it was created, is for them to keep doing things the way they have always done them. Yet, as one of the Reserve's agents observes, "the usual, the routine has never existed," for sheep breeders and shepherds are "in perpetual adaptation," always looking to overcome the constraints they must face, whether economic, social, or climatic. Consequently, she follows very closely the evolution of pastureland in order to assert the consequences of changes in conduct on fauna and flora, to testify of their influence, either positive or negative. In this last case, she says, contact must be made with the sheep breeders and shepherds to consider with them how their use of the *coussouls* can be modified to favor biodiversity. The ban on use for a sector or the change of pasture calendars, for example, are done through such negotiations. This is the reason why this agent wishes to be able to communicate more with the sheep breeders and shepherds, at the least through an annual meeting. She notes that shepherds take an interest in her research, when she has the opportunity to communicate with them, yet, she regrets that she often is unable to find them again the next year as they move so fast from one place to another. She is conscious of her contribution to a form of protection implemented in Crau different from that of national parks. She believes in the collaboration of protection and pastoralism as a favorable agroecological model to aim for.

An Original Professional Organization in Provence-Alpes-Côte d'Azur

Before we come to the pastoral profession itself, the role held by the Center for Pastoral Studies and Implementations Alpes-Méditerranée (CERPAM), to which we have already made several allusions, must be mentioned. In between two worlds, this association was created in 1977 under the impulsion of the agriculture chambers of the six departments making up the Provence-Alpes-Côte d'Azur Region. Following the political decision of the PACA Region, CERPAM ceaselessly defends pastoralism there, turning its actors into credible and constructive partners. Thus, the CERPAM has proven very useful each time pastoral pasture was put to use in protected areas. We can mention, among other such examples, the action carried out since the 1990s between the Luberon Regional Natural Park and the INRA (National Institute of Agricultural Research) on "modeling active relationships between the management of biodiversity and the activities of sheep breeding" (Lasseur et al. 2010: 90–96). Environmentalists, who for the most part are academics working for institutions for the protection of nature and the establishment of an equalitarian dialogue between the involved parties, required a way to translate in their language and with their own references the pastoral know-how in all the variety of its practice and all its effects on the environment.

As the illustrations are numerous, and often hard to summarize in a few words, we will limit ourselves here to the tool developed by the

CERPAM in coalition with the National Institute of Research for Agriculture, Food and the Environment (IRSTEA), pioneer in the French field of agroecological research, and with the alpine pasture Federation of Isère (FAI), a tool which aims at precisely evaluating the quantity of fodder removed by a herd on its pasture. The exploitation of data obtained through their tool, on the degree of removal, the circuits of pastures, and the management choices, notably, leads to an accurate evaluation of the use of the pasture. This enables the thorough monitoring of the MAEC and, on the long term, of the consequences of climate change. Its results allow for the most objectivity in judging the state of the pasture, from good to deteriorated, and by extension to judge the conditions of life of the fauna living there (birds, reptiles, insects, etc.). Both the managers of the protected areas and the sheep breeders and shepherds are interested in the results. The great complexity of pastoral know-how[7] is mastered by sheep breeders and shepherds through the force of habit, observation, and the constant search for the benefit of their herds. It is now available and accessible, becoming readable and useful to other people. As one of the engineers of the CERPAM noted, the sheep breeders' interests rarely align with those of the protection organizations. Therefore, their role must be put forward indirectly. On this subject, the engineer is sorry that breeders and shepherds failed to regroup in an association to defend their interests in the Alpilles. Happily, there are exceptions, but there are still few pastoral breeders who are ready to give some of their time to defend the trade.

The Pastoral World against the Managers of Protected Natural Areas

"Breeding is a very difficult activity, and when a sheep breeder meets too many difficulties and gives up, it's final!" warns a transhumant sheep breeder. The unease is real in an occupation where people feel "mistreated and unloved" or even "left behind as others reinvent the world." In the image he holds of the protection of nature, the world is reinvented into one in which pastoral breeders do not have a place to exist. But pastoral breeders are the inheritors of a way of life which used to have no one to answer to, or maybe only had to answer to their animals, following long tracks they have ceaselessly traveled, from plain to high mountain, covering a territory they believe they know better than anyone. They have owned their knowledge and often their livestock for generations and they have trouble tolerating new constraints.

By surrendering the transhumance on foot to the livestock vehicles, by submitting to sanitary rules, by taking the financial supports and benefits without which they could not exist today, by following the recommendations on protection from a wolf attack, the sheep breeders have changed nonetheless. They even have shown a surprising capacity for adaptation. And, yet, they must still obey the orders of managers who often appear convinced of knowing better than they do how to pasture an area, how many animals to put there, and what precautions to take to prevent the environment from being damaged from one year to the next.

Another transhumant sheep breeder, having spent many of his summers in one of the emblematic alpine pastures of the Mercantour National Park feels the same way: "There are two completely different worlds, where you can feel pastoralism is not a priority." He also deplores to have to park his ewes at night when it used to be so profitable for them to go find their own *couchade*, as he puts it, to spend the night, instead of forcing them to stand in the same over-pastured and manure-filled pen. He also believes the interdiction to set up impluviums is one more measure designed to push them to abandon the locations. He acknowledges the capability of the park's agents, and he respects them. Yet, he understands how "contempt has finally taken over both sides," and feels sorry for it. He also wonders why the park's project managers, who have done so much studying, are not trying harder for things to go well. He finds it unfortunate that the managers change so often, yet grudgingly concedes that "all the protected areas are located in pastoral sectors. The sheep breeders will have to deal with it."

"We would like to take our sheep to places that are not protected, to have a little more freedom," declares another sheep breeder, still wondering why the park reduced, on the pasture she rents, the allowed size of livestock from 1,900 heads to 1,700 without any explanation. Why should her shepherd not grill/cook in front of his hut anymore? Why are donkeys and goats prohibited? "The issue," she goes on, "is that they make us follow rules we don't understand and have no real effect. For them, everything must be done ideally, but in nature there is no ideal!" Obviously, the two perspectives are in conflict with each other, and there is no sign of the beginning of a mutual understanding of the other's interests or expectations.

Relationships are different according to whether the interlocutor of the manager of protected areas is a sheep breeder or a shepherd, which is to say, an owner or employee. We will leave out here the differences, which often causes discontent, pitting one against the other, and we will focus on what brings them together most of the time: their passion for breeding.[8] But we must also note that the young shepherds and shepherdesses, who are often the product of an urban environment, were formed in shepherds' school and maintain different a relationship with the protection organizations: first, because it is not rare to meet young people who have studied for years after high school; and secondly, because they have a preexisting interest in ecology and are already aware of the need to preserve biodiversity. Some even use this reason to justify their decision to become shepherd or shepherdess.

For instance, in the mountain, a shepherdess was outraged by the sight of Pyrenean mountain dogs, or *patous*, devouring lagopus' chicks, or, in Crau, passing time hunting ocellated lizards. She wonders if the mandatory ownership of a *patou*—without them, wolf attacks are no longer financially compensated—might not be more detrimental than helpful. She believes that the repeated attacks of the wolf will drive the poorest sheep breeders to abandon the mountain, for the benefit of the owners of the larger herds who have better means of defense and are used to the formalities. She perceives a real difference between the alpine pasture of Ristolas (Hautes-Alpes), classified as a Natura 2000 zone, where the herd she pastures there in summer is, according to her, watched from morning to evening via satellite pictures, where everything, from the dates of arrival and departure to the way of tending the sheep to the number of animals, is rigorously controlled, unlike the *coussouls* of the RNNCC "where they trust us blindly." There, she dislikes the incursion of tourists who "drive around us and our herds, five or six cars at a time. They hit the ewes and huts, take pictures of everything." However, hunters do not seem to cause her trouble. She would like to have more contacts with the RNNCC agents but "the Reserve doesn't ask us for anything," she admits, disappointed not to be put more to use or even to be considered more useful.

The regret of not having any return on the experimentations they take part in is expressed in several of the sheep breeders' and shepherds' testimonies, particularly as concerns the MAEC. Except for a few rare exceptions (in the circuit of "alpine pasture sentinel" and the tours organized by the CERPAM at the end of summer, noticeably) times for sharing are, indeed, nonexistent. Part of them at least would like to be associated more closely with the protection of the areas they pasture and several park or reserve agents wish to multiply the opportunities to communicate with them. What makes this meeting so problematic?

Improved Communication Needed to Benefit the Two Worlds

The previously mentioned testimonies were selected with the purpose of providing an overview of the main positions expressed. In regard to the protection structures, whether parks or reserves, we observe that their differences lie more in the circumstances surrounding their creation and the personality of the people who represent them on the field, than in the nature of their administrative statutes. Most of them wish to have more time available to communicate with sheep breeders and shepherds. On their side, the actors of the pastoral world dislike not being a bigger part in the decision process of the protection and dislike the lack of information about the effects of the measures they are asked to follow. Would it be idealistic to attempt to balance this relationship by considering that the expertise of the sheep breeders and shepherds is as valuable as the one of the naturalists, field agents and members of scientific counsels? The question must be asked, for the managers of protected natural areas always possess the power of decision. It would be reckless to take that power away. Yet, could we not find more understanding from each side, through frequent discussions, activity reports, and regular meetings at the beginning and the end of the season? All this expressed in a clear and comprehensible language, accessible to the actors of the pastoral world, as the CERPAM knows to do? A recent workshop started to prepare minds for the idea that dispositions may be taken in this respect (Duclos, Fabre, and Garde 2017: 165). A second conference, initially scheduled for May 2020, was cancelled due to the pandemic, and would have developed the detail of the plan. We can hope nonetheless that the reflection will continue, one way or another, and that a constructive dialogue based on trust will finally begin between the two sides.

However, we can hardly conclude without linking this conflict to the division that opposes our contemporary over the idea they have of their relationship to nature, through the modes of protection they defend. Who may pretend to know the truth, between the supporters of a protected and rehabilitated nature in what will be left of its wilderness, and the others to whom nature and culture are part of an acknowledged whole, and to whom local knowledge and practices that have proven their sustainability should be encouraged and supported? We would obviously not have conducted our analysis in this manner if we were not more inclined towards this last suggestion. We must furthermore observe, as we have witnessed in innumerable debates on the return of the wolf, that confrontation is a dead end, dialogue is a necessity. We will conclude, although the effort might seem worthless to some, by conveying a newly recorded proposition, which is to register transhumance to the intangible cultural heritage of humanity. The almost one-hundred pages long registration sheet may surprise by its length and the hundreds of referents and references it contains.[9] It is nevertheless enlightening regarding the idea of heritage which emerges rendering the cultural inseparable from the natural.

Acknowledgments

This chapter was translated from French by Alice Balique, PhD student in English Literary Translation at Aix Marseille Université.

Jean-Claude Duclos, Chief Curator of Honorary Patrimony, has directed the Dauphinois Museum (Grenoble) since 2011. Experienced museum and exhibit designer, he has also written numerous works and articles about the ethnology of the alpine world and museology.

Patrick Fabre, agricultural engineer, is Director of the Maison de la Transhumance, at the Center for Interpretation of Mediterranean Pastoral Cultures (Salon-de-Provence). He has written numerous works and articles, and designed several traveling exhibits and interpretation tools (paths of discovery, teaching kits) around the profession of shepherd and transhumance.

Notes

1. In 2011, the UN estimated that the surface of protected areas will represent 12.9 percent of the planet (2011–2020, Décennie des Nations Unies pour la biodiversité).
2. The case of the community of Abriès (Queyras, Hautes-Alpes) is a good example of this phenomena. See Rosenberg 2014: 191.
3. See, in particular, Laslaz 2008.
4. Coussoul: from the Late Latin *cursorium*, referring to the grazing course of ovine herds in this plain.
5. *Avis du Conseil scientifique du Parc national du Mercantour au sujet des propositions d'installation d'impluviums en cœur du parc*, 15 February 2019.
6. See, among others, Vincent 2011: 450.
7. See, among others, Meuret 2010.
8. See, in particular: Bonnet, Teppaz, and Vilmant 2020: 24.
9. *Fiche d'inventaire du patrimoine culturel immatériel—Les pratiques et savoir-faire de la transhumance en France*, 10 May 2020: 98. Retrieved 1 June 2020 from file:///Users/lizziemartinez1/Downloads/Les%20pratiques%20et%20savoir-faire%20de%20la%20transhumance%20en%20France.pdf.

References

Bonnet, Olivier, Clément Teppaz, Julien Vilmant, eds. 2020. *Bergers des Alpes: une vaste enquête sur le métier, les profils et les attentes des bergères, bergers et vachers salariés des Alpes*. Paris: Editions du CERPAM.

Dechavanne, Vincent. 2019. "Pastoralisme et espaces naturels protégés en Provence-Alpes-Côte d'Azur—Etat des lieux et prospectives." Thesis presented for professional license of farm management, Montpellier SupAgro.

Duclos, Jean-Claude, Patrick Fabre, and Laurent Garde, eds. 2017. *Élevage pastoral—Espaces protégés et paysages en Provence-Alpes-Côte d'Azur*. Cardère: Maison de la transhumance/Cerpam/Arpe Paca.

Hoffmann, Luc. 2010. *Luc Hoffmann, l'homme qui s'obstine à préserver la Terre. Entretiens avec Jil Silberstein*. Paris: Phébus.

Laslaz, Lionel. 2008. "Terre d'élevage ou 'nature préservée' en zone centrale des parcs nationaux français des Alpes du Sud?" *Méditerranée* 107/2006. Retrieved 25 July 2020 from http://mediterranee.revues.org/462.

Laslaz, Lionel, Christophe Gauchon, Mélanie Duval, and Stéphane Héritier, eds. 2014. *Les espaces protégés entre conflits et acceptation*. Paris: Ed. Belin/Littérature et revues.

Lasseur, Jacques, Jean-François Bataille, Bénédicte Beylier, Michel Etienne, Jean-Pierre Legeard et al. 2010. "Modélisation des relations entre dynamiques des territoires et des systèmes d'élevage dans le massif du Lubéron." *Cahiers Agricultures*. EDP Sciences. 19(2): 90–96. Retrieved 23 July 2021 from https://hal.inrae.fr/hal-02667410/document.

Lefebvre, Thierry, and Sébastien Moncorps, eds. 2013. *Les espaces naturels protégés en France*. Comité français de l'Union internationale pour la conservation de la nature (UICN). Retrieved 27 July 2020 from https://uicn.fr/wp-content/uploads/2016/08/Espaces_naturels_proteges-OK.pdf.

Leynaud, Emile. 1985. *Les parcs nationaux, territoire des autres. L'espace géographique*. CNRS 14(2): 127–38.

Meuret, Michel, ed. 2010. *Un savoir-faire de bergers*. Paris: Educagri Éditions/Editions Quae.

Molinier, René, and Gabriel Tallon. 1949–50. "La végétation de la Crau." *Revue générale de botanique*. T. 56–57. Paris: Librairie générale de l'enseignement.

Rosenberg, Harriet G. 2014. *Un monde négocié—trois siècles de transformations dans une communauté alpine du Queyras*. Translated from English by Jean-Pierre Brun et Jean-Claude Duclos. Coll. Le Monde alpin et rhodanien.Grenoble: Musée dauphinois.

Vincent, Marc. 2011. *Les alpages à l'épreuve du loup*. Paris: Ed. de la Maison des sciences de l'homme/Ed. Quae.

Walsh, Kevin, Florence Mocci, Stéfan Tzortzis, and Josep-Maria Palet-Martinez. 2006. *Dynamique du peuplement et activités agro-pastorales durant l'âge du Bronze dans les massifs du Haut Champsaur et de l'Argentierois (Hautes-Alpes)*. Documents d'archéologie méridionale. Retrieved 15 May 2020 from http://journals.openedition.org/dam/460.

CHAPTER 4

Reintroducing Bears and Restoring Shepherding Practices
The Production of a Wild Heritage Landscape in the Central Pyrenees

Lluís Ferrer and Ferran Pons-Raga

> Why is there a problem? This is the question. And I do not have the answer. But I am sure it is not . . . because we [public administration] pay late, underpay, and do not protect [the flocks] well. I am sure it is not about this, because this is not the point . . . there are more things.
> —Catalan's Director of Environment at the Catalan Parliament,
> 2 February 2019

Introduction

The French and Spanish governments signed an agreement in 1993 to launch a European Union (EU) LIFE project (Chandivert 2010) that aimed to *restore* the presence of brown bears in the Pyrenees once they were considered almost extinct (Camarra et al. 2011). Since then, the bear population has been increasing as a result of four waves of releases of translocated individuals from Slovenia. The reintroduction program also fostered a regrouping policy for sheep flocks to prevent bear attacks. This measure entailed the *return* of shepherds and livestock guardian dogs (LGDs), as well as restructuring previous local farmers' shepherding practices.

This chapter questions the notion of return through which both the reintroduced bears and the new regrouping policy—and more specifically the reappearance of shepherds—have been framed by proponents of the bear program. The program's proposal to restore a certain mountain landscape composed of bears, shepherds, and LGDs is flawed by the very essence of landscapes, which are studied from the past, but necessarily thought about from a presentism bias (Hirsch and Stewart 2005; Ringel 2016; Pèlachs et al. 2017). Consequently, translocation rewilding strategies (Nogués-Bravo et al. 2016) such as the bear reintroduction program stand on an unsolvable paradox. While they consist of an active environmental engineering endeavor, they also attempt to remake "formerly productive landscapes . . . both materially and semiotically through the practices of 'ecological restoration'" (Castree and Braun 1998: 2). In doing so, they fully engage with the construction and contents of the before-after succession: the temporal politics of the past (Ringel 2016).

Based on our ethnographic fieldworks in the Catalan (Spain) and Ariège (France) Pyrenees and inspired by Tania Li's standpoint of making "improvement strange" (2007: 3), we have tackled the notion of return with estrangement. This distancing view directed us towards, and has been reinforced by, two ethnographic studies on the history (see Hirsch and Stewart 2005) of the shepherding practices on both sides of the Pyrenees to criticize the very idea of return. Local and transhumant farmers have been continuously adapting to varying ecological, social, economic, political, and legal contexts. From this vantage point and inspired by Karl Jacoby's concept of moral ecology to criticize environmental conservation (2019),[1] we contend that the bear program must be read through notions of change and adaptation, rather than those of return and conservation. The accounts gathered through interviews with local farmers highlight a set of historical changes in the shepherding practices in the Central Pyrenees before and after the implementation of the bear reintroduction program. Following Jason Moore (2015: 28), we approach these historical changes "through the dialectical movements of humans making environment, and environment making humans."

In contrast to the above, the notion of return claimed by the bear program's proponents is based on Western historicism (Hirsch and Stewart 2005; Stewart 2016), in which the past is separated from the present, and on the Cartesian Nature/Society dualism (Moore 2015), in which the ontological status of entities is imposed over relationships. In this vein, the environment and the humans, epitomized by the bears and extensive farming, appear as two separate entities *from the past*, while the program's proponents and the farmers emerge as representatives of each one of them confronting amenity vis-à-vis production-based capitalist views of natural resources (Walker 2003). A critical insight into the program's chronology (Ingold 1993) allows us to challenge both Western historicism and Nature/Society dualism as well as to map out the ensuing political hierarchy of environmental conservation policies over extensive livestock farming.

Based on the two ethnographic studies on the history of shepherding practices presented here, we will show to what extent the bear reintroduction program was preceded by a set of changing shepherding models, each one of them strikingly different from the one this program claims to restore. As a result, we question the accuracy of framing as a return the reintroduction of brown bears and the reappearance of shepherds as well as LGDs. Rather, should we not consider them as a newly designed landscape, in which only certain features of a past time have been retrieved, whereas many others have been added for the sake of producing a certain wild landscape?

We divide this chapter[2] into three main sections: 1) an outline of the bear program's chronology and its imposing consequences for local farmers, highlighting the Nature/Society divide and the political hierarchy between these two separate realms; 2) two ethnographic historical accounts from the Bonabé and Biros valleys, in the Pallars Sobirà (Spain) and Ariège (France) districts respectively, to illustrate how the notion of return, as used to describe shepherding practices, crumbles in the face of historical changes in flock management on both sides of the range; and 3) a reflection on how the bear program unfurls in a twofold *naturalization* process via the idiom of heritage that leads to the production of a landscape through an imposition/salvation conundrum in which the renewed presence of bears is *naturalized* or taken-for-granted and the bear is presented as though it would make the Pyrenees a *more natural* or wilder place.

What Comes after the Bear? Imposing Wilderness and Shepherding Practices

While the *return* of bears as the quintessential environmental hallmark for biodiversity conservation is advocated as a way to *recover* lost Pyrenean natural values from the past, the reappearance of shepherds and LGDs—and the ensuing *restoration* of certain shepherding practices—is claimed to safeguard vanishing sociocultural ones. This narrative of *return* conceals, however, an ontological division and a political hierarchy between the natural and the social in which the bear (Nature) has been first detached and thereafter hierarchically conceived in relation to the shepherding practices (Society). The reintroduction of bears was conceived as a priority, whereas the implementation of a regrouping policy with a set of protective measures—shepherds, LGDs, and electrified enclosures—to mitigate sheep casualties was considered only when conflicts arose among local farmers. Both the ontological division, based on the classic Cartesian dualism, and the political hierarchy, are part of the very essence of the program's hegemonic ideology (Comaroff and Comaroff 2006). This ideology lays the foundations for, and is countered by, local farmers' feelings of imposition.

Taking the program's chronology as a prompt, we argue that Nature—bears—and Society—shepherding practices—have been approached as bounded and detached realms. This Western ontological premise paved the way for the ensuing sequential hierarchy between environmental conservation goals and the challenges that spring from the interplay between livestock and wildlife. Returning to the opening quote, the ontological separation and the political hierarchy between these two realms is key to tackling the question of "why is there a problem" around the bear program. In effect, part of the problem relates to how the idea of return is built upon the politics behind the bear. A brief context of the bear population and its shifting status in the Pyrenees will serve to introduce the state politics or practice of governments (Li 2007) that underpins the bear program.

The native bear population plummeted from a few hundred in the early twentieth century to barely five individuals in the 1990s (Casanova 1997; Marliave 2008). Even though bear hunting was formally forbidden both in Spain and France in the 1960s, poaching and isolated hunting accidents continued to occur. This situation and the spread of the second wave of worldwide environmentalism (Adams 2003; Anderson and Grove 1987; Guha 2000) led to the implementation of EU (e.g., Bern Convention 1979) and state initiatives that provided bears with full legal protection between the 1970s and 1980s (Casanova 1997), when the Pyrenean brown bear population was estimated at around twenty (Marliave 2008). In a short time span, bears shifted status from a hunted species to a protected one, ultimately leading to a cross-border large-scale reintroduction program to recover their plummeting population, which by that time was located only in the Western Pyrenees (Caussimont 2013; Parellada, Alonso, and Toldrà 1995). The political arena and the social tensions among the Institution Patrimoniale du Haut-Béarn (Mermet and Benhammou 2005)—a cluster of local councils, hunters, farmer organizations, and ecological and tourist associations in dialogue with the state in the western region of France—advised against releasing the first translocated individuals in this area, but rather setting the reintroduction in the Central Pyrenees, where bears had been already extinct. An institutional network, composed of a few ecological organizations founded specifically for this task, and four municipalities had been established in the French Haute-Garonne district (Benhammou 2007), where the first reintroduction took place. Since 1996, eight females and three males have been released in the mountain range in four waves (1996/97, 2006, 2016, and 2018), setting the current population over fifty bears (Réseau Ours Brun 2020), most of them currently dwelling in the Central Pyrenees. The success in raising the bear population from a handful of individuals settled in the Western Pyrenees to more than fifty, mostly dwelling in the Central Pyrenees, contrasts with the program's failure to gain acceptance from the farming sector. The local population's lack of participation in the process—local farmers in particular—due to the top-down approach ingrained in this program coupled with the relocation of the releases from the Western to the Central part of the range, brings us to underline the political character of this conservation program behind its alleged biodiversity conservation rationale. Both the ontological division and the hierarchy of Nature over Society are constitutive to state politics or the practice of governments deployed through the bear program.

In line with this perspective, we contend that the program has resulted in two sorts of impositions for local farmers: imposing wilderness (Neumann 1992) through the reintroduction of bears underpinned by what we call a twofold *naturalization* process; and imposing upon farmers' conducts through a sheep regrouping policy that has changed previous shepherding practices. Both the bear and the shepherding practices have been ideologically—and hence politically—claimed as natural and cultural heritage assets respectively, to be restored from the past and transposed to the present as an opportunity or even a salvation for the dire farming sector. On the one hand, we stress that the program spearheads, paradoxically resorting to the past via heritage narratives, a twofold *naturalization* in the Pyrenees, in which the bear would allow for the production of a wild heritage landscape (Baird 2017). Yet, the naturalized presence of bears and the naturalizing effect of such are both conceived of by farmers as impositions.

On the other hand, the notions of environmentality, as a public governance of natural resources that aims to conduct local farmers' conduct in myriad forms (Agrawal 2005; Fletcher 2010), and territoriality, as "the unfolding of a society into a territory" (Vaccaro, Dawson, and Zanotti 2014: 3), are key to understanding the feeling of imposition experienced by local farmers in the wake of the state-driven sheep regrouping policy. Farmers' forms of dwelling (Ingold 2000) have varied over time, but more recent ones, prior to the reintroduction of bears in the Pyrenees, have been questioned by the program's proponents for paradoxically resorting to, and aiming to recover, age-old shepherding practices. The renewed presence of bears has been followed by a state-driven regrouping policy for sheep flocks. This new shepherding system consists of gathering several sheep groups in a single flock with the abovementioned set of protective measures. This policy was fostered and funded by public administrations in order to protect flocks from bear attacks. The resulting collective flocks, and more specifically the renewed presence of professional shepherds and LGDs, are deemed as restoring the social and cultural heritage values of the Pyrenees by allegedly returning to an old shepherding model—the village flock—in which a local shepherd surveils private flocks owned by different farmers from the village. Schools of Shepherds, created in the 2000s, have become a way to preserve this heritage. New shepherds have gained social status through a professionalized career while their salaries have been paid by the state and their work conditions have improved in recent years. Inasmuch as the presence of shepherds is claimed as a way to restore waning shepherding knowledge and practices, institutional narratives have fostered the use of a specific breed as an LGD due to its contribution to the preservation of the Pyrenean heritage: the Great Pyrenees. This breed was historically prevalent in the Pyrenees as the flocks' protection dog until the early twentieth century when the decline of large carnivores along the range meant that these dogs were no longer used in the Central Pyrenees (Ferrer i Sirvent 2004).

Back to What? Tracking the Historical Changes among Shepherding Practices in the Bonabé and Biros Valleys: Parallel Paths for a Common Present?

> I had never seen a bear before. It was completely extinct. After this reintroduction, everything changed. Before, our sheep benefitted from the mountains. Now, they are not free and so they do not benefit from these pastures as they used to and do not become as round and fine as before. You know, we used to breed and graze the *broutard*, the young sheep, in the mountain pastures but we no longer can do this. It is too dangerous [due to bear attacks]. We had to change our method, we can no longer work as usual. (Farmer, Biros Valley)

Local farmers have been pressed to change their shepherding practices due to the renewed presence of bears. Indeed, bears have brought shepherds, LGDs, and electrified enclosures, but also a sense of constraint and discontent that has permeated into the core of the farming sector. The protective measures implemented in the Central Pyrenees after the bear program were seen by conservation advocates and political institutions as ways to restore a previous shepherding management system, presumably shared since time immemorial across the entire mountain range. However, the ethnographic data gathered on the multiple transformations among shepherding practices throughout the twentieth century in the Bonabé and Biros valleys, in the Pallars Sobirà (Spain) and Ariège (France) districts respectively, challenges these notions of restoration, continuity, and uniformity, by providing an insightful overview of the shepherding practices through the lens of historical changes.

In 2019, on the southern slope of the range, the "core bear area" of the Catalan Pyrenees, northeastern Spain—covering the Val d'Aran district and the northern regions of the Pallars Sobirà district (around 1,300 square kilometers in total), including the Bonabé valley—held six collective flocks comprising of around seven thousand sheep. Since 2012, and as a result of the consolidation of the bear population in the region, the village of Isil at the entrance of the Bonabé valley has hosted the only museum on the southern slope of the Pyrenees devoted to the brown bear: The House of the Bear. In 2019, on the northern slope, the mountain pastures of Ariège (southwestern France) held around fifty-nine thousand sheep, extending their territory over around 1,400 square kilometers. The *return* of bears in this area has also entailed a recent process of labeling, redefining this area as Pays de l'Ours (Bear Country) by several Pyrenean ecologist organizations to garner attention within the purview of local green tourism. The Bonabé and Biros[3] valleys and their respective mountain pastures become especially pertinent to assessing shepherding changes and adaptations due to their long-standing and current pastoral life alongside the large number of bears present in these two areas.

The Collective and Individualized Management of Flocks (Catalan Pyrenees)

"Along this road here [the main entrance to the Bonabé valley], which then was not a road but a track, 'el Tort' [the most popular and powerful farmer from that epoch settled in Alós d'Isil] brought six thousand sheep. And Pubill [another *strong house* from this village] had around three thousand. Look, the mountains were not leased then!" Pau,[4] born in 1933 in Alós d'Isil, vaguely remembers seeing from the balcony of his house the Tort's flock passing by from the plains to the mountains in transhumance. He also recalls his parents' time, when sheep were so abundant in their village that most of the collective pastures along the Bonabé valley, whose use rights were shared with the adjacent village of Isil, were not leased to foreign transhumant farmers. "There were enough livestock here!" Pau cries out, recalling the presence of three private flocks—two from Alós d'Isil and one from Isil—as well as the village flocks from the two villages, for a total of nearly fifteen thousand local sheep. The village flocks used to be tended by the main shepherd (*majoral*) and some assistants (*rabadans*) at a time when almost every household in the village kept livestock. The Bonabé's main pastures were not leased, but each flock, whether private or collective—the two village flocks—had a specific parcel (*partida*) assigned for grazing.

The dismantling of the Tort's flock with the outbreak of the Spanish Civil War in 1936 gave way to the *first historical change* of the twentieth century in the Bonabé valley. This transformation mainly relates to the origins and ownership of the flocks that were grazing over those pastures. By that time, Alós d'Isil and Isil held fewer than one thousand sheep each, which were grouped into the respective village flocks. Most sheep belonged to transhumant farmers, who came from the plains and leased the same pastures that three local farmers had almost monopolized throughout the previous decades. According to Roigé (1995), in the entire Pallars Sobirà district transhumant sheep flocks amounted to fifty thousand animals in the 1950s. Given the geological characteristics of the Bonabé valley, with its calcareous soil that allows for better grazing lands, in contrast with the predominance of granitic soils over the rest of the northern parts of the district, it is plausible to infer that a relatively high number of those fifty thousand sheep would come to this valley in transhumance. Two shepherds were responsible for tending and grazing each village flock, whereas each one of the numerous transhumant flocks of up to five thousand sheep were tended by five or six shepherds apiece. The money collected from each municipality, once the collective pastures were leased, partially served to pay the shepherds and *rabadans* of the village flock. The rest of their salaries derived from the *taxa*, a tax collected from every household that had stakes in the village flock in accordance with the number of sheep held by each.

"Peasantry was three months of hell and nine of winter." Family memories from the mid-twentieth century, such as these ones shared by Jesús, born in 1965 in Alós d'Isil, tend to recall a harsh period in which time was mostly devoted to working the land to make a living. This was a period during which machinery was not used, and every plot of land was valued for its potential contribution to the yearly harvest. Pau illustrated this mindset through the following sentence: "Where there wasn't even room for a car, you would sow some wheat." Although Alós d'Isil's sheep stock plummeted to fewer than one thousand animals with the dismantling of the Tort's flock, two shepherds were hired year-round to make sure that the village flock did not ruin the privately-owned cultivated fields surrounding the village. These daily concerns about the land materially produced a certain landscape. A mosaic of yellow, brown, and green composed of fields to harvest (*terres*) and meadows to mow (*prats*) kept the forest's advance at bay. The lack of trees and the geographical scope of the fields that stand out from this period conjured for Jordi, a local farmer from Isil born in 1992, a sense of dwelling and pride that strikingly contrasts with the notion of abandonment and resignation that emerges from today's landscape. By that time, regardless of the substantial differences in class and power among villagers, the village flock and the ensuing collective management of all flocks, provided households with a sense of community. It "was like opening the boundaries . . . a way of making community," according to Antoni, who did not live in those times, since he was born in 1984, but who has absorbed them through the stories told by his uncle, born in 1949. The village flocks were mainly tended by shepherds, but local farmers would organize themselves in turns to do *jadilla* with the sheep, that is to move them across private fields during the night in order to fertilize those fields in the fall before the first snowfalls. Shepherding practices were organized in a fragile equilibrium to enhance the productivity of the fields, pastures, and sheep. Within this balance, the bears stopped playing a significant role since their numbers were already much reduced and local farmers eradicated them from the scene. According to Casanova (1997), the last bear killed through legal hunting in the Pallars Sobirà district dates back to 1948, and Pau confirmed that the *last* bears settled in the Bonabé valley—although he also recognized that "there have always been bears here"—were poisoned by the local population when he was a child (early 1940s) after two horses were allegedly killed by them.

According to Jesús, the village flock remained until the beginning of the 1970s. By that time, although he was then a child, he remembers in detail the livestock count in his village: "Here, there were about eight hundred sheep . . . and one hundred and fifty goats . . . one [household] had sixty the other thirty, the other twelve." Some years before, a "migratory epidemic," as it was described by Pau, spread over the region. The 1960s were characterized by both an urban industrial boom in Catalonia that demanded a lot of labor power from the peripheral rural areas, and the mechanization of agriculture. The arrival of tractors displaced one of the main sources of income for local inhabitants. Traction horses became worthless overnight. Outmigration flows following new economic opportunities were recalled vividly by Pau, who was afraid that "nobody [was] going to remain in these lands . . . And almost no one was left here!" According to Jesús, the dismantling of the village flock was closely related to these demographic shifts: "Through the 1970s people had already gone and the [village] flock was completely dismantled . . . because, you know, many people moved to Barcelona [and other cities] . . . and then it went to hell! It was only our house [sheep farm] left [in Alós d'Isil]." The *second historical change* in the Bonabé valley then took place, since the dismantling of the village flocks brought about crucial consequences for both the management of the few remaining private flocks, and the use of collective pastures. Even though the total number of sheep in Alós d'Isil stayed progressively close to previous years, as long as the two remaining sheep farmers—Pau and Jesús—began to raise their numbers, livestock concentration in so few hands led to the disappearance of a social figure that had, up until that point, been of paramount importance: the shepherds of the village flock. "When the village flock was over [disbanded], we released them [sheep] to graze freely. There was no shepherd anymore," as pointed out by Jesús. The 1970s thus gave way to a new era in which a handful of local farmers began to let their flocks graze freely over the collective pastures. The previous collective management of the flocks, based on farming governance over pastures that were strictly bounded in both spatial and temporal terms, turned into a completely new shepherding model. Local farmers began to practice an individualized loose management of their private flocks under what Antoni called "shepherding without boundaries." This overarching transformation translated into the blurring of the previously strict territorial grid from the times of the village flocks, in which "the limits were the limits" and "everything was well set," in Jesús' words.

In the meantime, although the livestock numbers of the remaining local farmers rose steadily while the number of farms waned, the total of transhumant flocks also dwindled, and the overall number of sheep grazing over those pastures shrank strikingly. Jesús, for instance, began to run his own farm with one hundred sheep in 1985 and reached a maximum of four hundred in 2004. According to official data from the Catalan government, while in 1979 there were twenty-five households with agrarian activity among Isil and Alós d'Isil, in 2006 the number plummeted to thirteen, and in 2019 there were only six, three of them devoted to sheep breeding and amounting to one thousand animals (Manel Torres, personal communication, email sent on 1 April 2019). In the Pallars Sobirà district, transhumance drastically dropped from fifty thousand sheep in the 1950s to sixteen thousand in 1993 (Roigé 1995). Disparities in sanitary regulations among different counties in Spain made transhumant farmers refrain from coming to these mountain pastures in the early 2000s (Espinós 2014).

Sheep numbers during the summer season had been fluctuating in the Bonabé valley over the span of a century, but they had always been in thousands. The collapse of sheep transhumance in the Catalan Pyrenees at the beginning of the twenty-first century (Còts 2002; Estrada, Nadal, and Iglesias 2010) resulted in the *third historical change* to pastoralism, since it represented the first time in one hundred years that the sheep grazing in the Bonabé valley totaled no more than several hundred. Since then, expansion of vegetation has been followed by a rapid growth of forest-related fauna, either human-induced or natural, especially cervids—deer, fallow deer, and roe deer—but also bears.

In this context, the year 2011 marked a turning point in the management of flocks. Local farmers from Isil and Alós d'Isil and the public administration reached an agreement to group three private flocks into a single collective one of one thousand sheep tended by one shepherd and guarded by three LGDs following the triad of protective measures. After this pioneer program was tested, the reduction in sheep casualties caused by bear attacks motivated the Catalan government to extend the new regrouping policy over the rest of the "core bear area."

From *Escabot* to Shepherds as Sentinels (Ariège Pyrenees)

Anselme was born in 1925 in Sentein, the main village in the Biros valley (Ariège district, southwestern France). He evokes his grandparents' lifetime—the age of border-crossing transhumance—when flocks were so abundant in the village that farmers had most of their animals grazing over the Val d'Aran pastures (the district adjacent to Pallars Sobirà, in the Catalan Pyrenees) during the summer season.[5] Farmers constituted collective pastoral groups named *pariés* for each of the three mountain pastures that they leased on the Catalan side (Cabannes 1889). As Anselme recalls, "my grandfather used to go to the Varradòs mountain to graze his sheep flock from the 23rd of June to the 20th of September." Each chief of the pastoral group (*majouraou*) was a farmer from the Biros valley. These *majouraous* used to make agreements to lease these mountains and were responsible for conducting and guarding the animals belonging to the collective group. According to Anselme, while the largest sheep flocks crossed the border in transhumance to the Catalan Pyrenees, the smallest ones—between fifty and eighty sheep each—remained in the mountain pastures (*estives*) of the Biros valley. In turn, similar collective pastoral groups named *pariaus* (Chevalier 1951) were established among the remaining owners, who took turns surveilling the sheep flocks in the summer pastures. These flocks grazed in *escabot*—the usual shepherding system in the valley—in which farmers themselves acted as shepherds. The main role of these farmers-shepherds, however, was not to herd these flocks but to give them a direction (*largade*) and let them move in conditions of semi-freedom: instead of a single large flock each group grazed freely at different corners of the mountain. These turns were already agreed upon in the assembly held in spring and depended on the number of animals per owner. As Anselme explains, "twenty sheep were equivalent to a mountain turn. They called it *faire le tour* and it meant surveilling these flocks for one day." Two or three farmers would go up to the mountains until their turn was over, and then other local farmers from the village came up to relieve them. Customary rights inherited from the Ancien Régime ensured access to the mountain pastures associated with each village.

World War I (1914–18) brought an end to this dual shepherding system. Local flocks no longer ascended in transhumance to Val d'Aran, whereas the scenario in the Biros valley remained the same in terms of shepherding practices. The reduction of flocks due to the impact of the war and certain inner bureaucratic issues within the Catalan district (Chevalier 1956) led to the *first historical change* of the shepherding practices in the valley through the disappearance of transhumance and the two pillars of the pastoral system: the *pariés* and the figure of *majouraous*. Pastoral groups continued to organize the summer season collectively in the valley and flocks were surveilled in *escabot* by local farmers-shepherds following the same rule as *le tour de montagne*. However, Guillaume, born in 1933, recalls how his great uncle used to be hired as a shepherd in the mountain pasture of Bentaillou (Sentein). "He had a partner and the two of them surveilled sheep flocks in *escabot* until 1938. They got paid almost nothing." On the contrary, as Anselme explains, sheep used to graze freely with sporadic controls by local farmers or another member of the household in the mountain pasture of the Artigou, in the village of Bonac. Taking care of the animals was Anselme's job, since his father worked in the mines of Bentaillou extracting zinc and silver-bearing lead. His family had eight cows and around fifty sheep. During the summer, he would ascend to the Artigou every Sunday morning to check on sheep and treat their hooves. Although the figure of the farmer-shepherd as well as the collective management remained present until the 1950s, the managerial formats of the Artigou and Bentaillou mountain pastures show the heterogeneity of shepherding practices that existed in the valley during this period.

After World War II, the mining industry collapsed in the valley. This led to overarching transformations such as depopulation and the progressive abandonment of farming through the late 1950s and 1960s. Higher sheep flock concentration between fewer local farmers coupled with the end of bear predation led to a new managerial format. On the one hand, flocks began to graze freely over the *estives* with sporadic controls by farmers; on the other, the collective management was dismantled and replaced by an individualized model. Jean-François, a farmer born in 1939, remembers the last bear killed in the valley in 1951, hunted by his father and other farmers. He also recalls how the sheep used to ascend with no guidance to the *estives* nor permanent surveillance around 15 May and came back around 1 November. Hence, this period entailed the disappearance of the collective pastoral groups, the farmer-shepherd figure, and the *tour de montagne*. Given this management of the flocks, it became common to breed and have a flock while holding a position of public employment at the EDF—the French state-run hydroelectric company still operating in the valley today—or as a postman. However, the valley became progressively depopulated over the years and only a few local flocks grazed in the mountain pastures at the end of the 1960s. Pastoral lands became manifestly under-used to a critical point. Vegetation expansion and natural reforestation rapidly and radically modified the landscape of both mountain pastures and formerly cultivated private fields near the villages, resulting in what might be deemed as the *second historical change*.

Groupements Pastoraux (GPs) or Pastoral Groups came to be the response to that given scenario—pastoral land abandonment and vegetation expansion. These pastoral groups revitalized and restructured the Ariège mountains and pastoralism in the late 1970s and early 1980s, after the approval of the 1972 Pastoral Law, resulting in the *third historical change* of pastoralism. Through an institutional arrangement between local farmers and the state, GPs turned out to be a fundamental tool for shepherding and land management as long as they ensured and regulated the access to the *estives* as well as fostering the arrival of transhumant flocks from the foothills and the plains. Most of the GPs were created during the 1980s and early 1990s, driven by state subsidies that aimed to recover grazing pastures in public lands. Bear predation was extremely rare in the valley after the 1970s and local flocks still grazed in *escabot* with sporadic controls by local farmers. Thus, although GPs represented a new form of collective action in which the state started to play a major role, the shepherding model remained mainly individualized.

Since 1996, the presence of bears has progressively transformed shepherding practices in the valley once again. As a result of bear predation, flocks are no longer sporadically controlled by farmers nor do they graze in *escabot*. This old shepherding practice has disappeared in the valley, where sheep are now guarded by professional shepherds within single large flocks. The new system has also shortened the grazing season. When Thomas, a local farmer born in 1961, compares the current situation with the one in the late 1990s he displays a clear discontent coupled with the sense of having lost something precious. "We had the benefit of letting our sheep go free in the mountains from mid-May to November 1st. They used to graze the lower parts of the mountain in the spring and the fall. And all this is over. We have lost two months of the grazing season."

The transformations of the shepherding practices among the six GPs managing the current seven *estives* of the Biros valley has resulted in the surveillance of around 8,700 sheep by ten shepherds and twenty-two LGDs, and the use of electrified night camps in two *estives* during the summer season of 2019. Several local farmers from the Biros valley remember the attacks of *Hvala*, one of the Slovenian females released in the second wave. As Batiste pointed out, "I remember her well. I had several attacks in one week. And she has taught several generations of cubs how to hunt, so they will have the same attitude." Some transhumant farmers no longer ascend to these mountains, as a result of bear predation. In this regard, two out of the three *estives* of a GP were abandoned in the summer of 2019. The abandonment of these parts of the mountain would lead to reforestation, bush expansion, and probably an increase in ungulate populations. But more importantly, if this abandonment becomes a general tendency, a state-led conservation endeavor such as the bear program will have countered the main purpose of a previous state-subsidized plan that created GPs to revitalize mountain pastures.

Protection measures funded by the government have been gradually set out in the *estives* to prevent bear attacks. However, this implementation has been asymmetrical with some level of heterogeneity, since some GPs have been reluctant to use them. Farmers are hesitant about the effectiveness of the set of measures against bear attacks. Fabian, a young farmer recently installed in the valley, claims that bears adapt their behavior both to the presence of LGDs and the installment of electrified night camps, based on his own experience using these measures, and the attempts of his colleagues in the French Alps to prevent wolf attacks. "There is an evolution, bear attacks are occurring during the day now because we keep

Figure 4.1. A single large flock guarded by a professional shepherd grazing in a mountain pasture, Sentein, 2019. © Lluís Ferrer

the sheep inside fences at night in our *estive*. In the Alps, all the studies are showing this: half of the wolf attacks occur during the day because the animals are inside the fences at night." In a nutshell, these measures serve only, as Batiste contends, to train bears to become more skilled: "We are afraid of being pushed to an arms race where more and stronger LGDs will be required to adapt to new bear predatory strategies."

The ethnographic data reflects an ongoing set of parallel, though not simultaneous, historical changes in shepherding practices on both sides of the range. Each one of these changes brought about not only an important shift from the preceding period, but so great were these shifts that none of the aforementioned changes resemble the current scenario in the least. These transformations and the dynamism of these shifts challenge the idea of conceiving both a single past detached from the present and the new shepherding practices as a return to the old ones. Besides the dissonances between past(s) and present, the depiction of the recent wildlife expansion by local farmers as "the new squatters of the rural world" portrays the power struggles between farming and conservationist sectors today.

Conclusion: Looking Back, Thinking Forward

In this chapter, we have identified a set of local farmers' narratives that challenge the way in which a landscape composed of bears, shepherds, and LGDs, and the ensuing transformation of previous shepherding practices in the Central Pyrenees, have been framed by the bear program's politics. The program's claim to *restore* these human and nonhuman elements to the Pyrenean mountain landscapes based on notions of return and conservation is countered by the ethnographic historical accounts from the Bonabé and Biros valleys, which reveal two stories of pastoral change and adaptation. The so-called *return* of bears, shepherds, and LGDs are also presented as an opportunity for the farming sector. In this concluding section, we aim to expand our critique of the notion of return by linking it to the twofold *naturalization* processes engendered by the renewed presence of bears.

This critique has revolved around two premises: the separation of the past from the present following the precepts of historicism; and the detachment and ideological hierarchy between Nature and Society following the Cartesian dualism. How do these two Western ontological separations intersect, and how are they channeled through the bear program? We contend that heritage as a hegemonic idiom (Franquesa 2013) and the concept of time-tricking (Ringel 2016) help to provide a generative answer to these enquiries.

Heritage has been defined as a hegemonic idiom since it not only articulates "hegemonic and counterhegemonic projects," but also frames "conflicts in terms that, by concealing their connection with broader issues of political economy, are advantageous to dominant groups" (Franquesa 2013: 347). Keeping in mind this definition, while the use of heritage might recognize the value of old-age local farmers' knowledge and practices, it also appears as an idiom of the hegemony that is not recognized by some farmers. Recognition, approached from a radical justice perspective that questions "who is given respect (or not) and whose interests, values and views are recognized" (Svarstad and Benjaminsen 2020: 1), turns out to be a key word to tackle the conflicts between the bear program's proponents and the farming sector by critically engaging with heritage as both an idiom of the hegemony and a hegemonic idiom (Franquesa 2013).

Besides this preliminary note, deeming the bear reintroduction program as a heritage-making process allows us to reveal how it consists of isolating parts of the past and transposing them to the present following one of the main tenets of Western historicism, that is, the separation of the past from the present (Hirsch and Stewart 2005). Through the bear program, "the pastness of the past is crystallized in efforts to present [the bear and shepherding practices] as objects separated from the present" (Gordillo 2014: 8). Heritage thus highlights the "pastness" of the past vis-à-vis the "presentness" of the present through resignification and revaluation processes (Kirschenblatt-Gimblett 1998). These processes are not devoid of power, but rather fraught with politics. In concrete terms, the temporal politics of the past as a form of time-tricking through which the before-after succession on the interplay between wildlife and livestock is reshaped. Heritage also connects with the second ontological premise of the bear program, since the canonical classification between natural and cultural heritage values allows for the establishment of a hierarchical order between Nature and Society. Natural heritage values—epitomized by the renewed presence of bears—have taken a much more prevalent political stance over cultural ones—represented by the regrouping policy and the shepherding practices it claims to restore.

Despite the two aforementioned main problematic aspects of framing the bear program through the idiom of heritage, the program's proponents claim that the return of bears and the consequent restoration of former shepherding practices should be conceived as an opportunity or even a salvation for the farming sector. This discourse undermines farmers' complaints and feelings of imposition in the wake of this rewilding strategy and the ensuing transformations of shepherding practices by asserting that what the program asks farmers to do is nothing more than what their ancestors had done since time immemorial. Critically engaging with the politics of time devised by the bear program, we argue that, instead of representing a link to the past, the bear itself spearheads a twofold *naturalization* process in the Pyrenees. On the one hand, the program considers that its presence, as the ultimate symbol of wilderness, produces a more natural or wilder—and hence better—landscape. On the other it claims that the renewed presence of bears must be considered as a naturalized, unavoidable, and incontestable asset of current landscapes that will make local farmers thrive under an environmental conservation paradigm.

Both *naturalizations* de-politicize the bear program and may well explain the recurrent rejections of such by most local farmers, as they perceived two sorts of impositions: the top-down, outright imposition of wilderness; and the more subtle and insidious imposition of shepherding practices. The contradiction spurred by the bear program between imposition and opportunity is better understood in light of the idiom of heritage and the ensuing political hierarchy of natural over sociocultural heritage values. Indeed, the bear has been framed by the program's proponents as something of superior value within the public interest to be preserved, and, under the idiom of heritage, a pathway to salvation for the dire condition of the farming sector. This cross-border conservation project has forced sheep farmers to adapt to a novel scenario, in which a sequence of changes has been implemented by multiscalar political institutions. These changes aimed to safeguard natural, first, and cultural, thereafter, heritage values that had been vanishing throughout the twentieth century. Therefore, Nature and Society have been first devised as separate realms and then politically animated within a hierarchical order. In other words, bears were released and, only after testing and proving the negative impacts of the reintroduction of these large carnivores on sheep farmers, the reappearance of shepherds and LGDs within a sheep regrouping policy was fostered. Therefore, this political hierarchy constitutes the foundation of these two impositions—bears and shepherding practices—that local farmers resent, which ultimately results in a growing discontent and the increase in social conflicts.

The restoration of the bear population, shepherds, and LGDs in the Central Pyrenees should be framed through notions of change and design rather than those of return and conservation. This replacement emphasizes the transformative nature of shepherding practices through different historical stages in which farmers have been progressively losing control over their own flocks when they graze over the high mountain pastures in the summer season. The implementation of the sheep regrouping policy through the bear program was claimed to have reestablished the abandoned collective management of the flocks. However, this state-driven policy led, in fact, to a new high-modernist territoriality (Scott 1998;

Vaccaro et al. 2014) based on a public environmentality (Agrawal 2005; Fletcher 2010) over farmers, sheep, bears, and pastures. It led to the production of a wild heritage landscape in the Central Pyrenees.

Lluís Ferrer is a PhD candidate in Anthropology at McGill University (Canada). His research has focused on understanding the historical evolution of livestock farming, collective management of mountain pastures, and land tenure systems in the Catalan High Pyrenees (Spain). His current doctoral research studies how the reintroduction of brown bears transforms pastoralism, farmers' livelihoods, and mountain pastures in the Ariège Pyrenees (France). His dissertation has been funded by the Fonds de Recherche du Québec—Société et Culture (FRQSC).

Ferran Pons-Raga is a PhD candidate in Anthropology at McGill University (Canada). Under the framework of political ecology, his research has studied the transition over recent decades towards a leisure-based economy in the high mountain settings of Western Europe. Based on this overarching interest, his dissertation examines the production of green and common landscapes in the wake of the (in)compatible interplays of an alpine ski resort and the bear reintroduction program with extensive husbandry in the Catalan High Pyrenees (Spain).

Notes

1. Karl Jacoby stressed that he had coined the term "moral ecology" in order "to unsettle the earlier neologisms of conservationists—including, of course, the very term conservation, coined by Gifford Pinchot in 1910 in a brilliant turn of phrase that cast the movement as a conservative, common-sense measure to protect a self-evident nature rather than a new form of control over the environment that often brought radical ecological and social change in its wake" (2019: 291).
2. This chapter is the product of two years of fieldwork in the Catalan and Ariège Pyrenees. The results presented here have been elaborated on through: in-depth interviews and informal conversations with local farmers, shepherds, environmental conservation experts, and local politicians; participant observations on the shepherding practices, and the management and implementation of the regrouping policy; attendance to symposiums and conferences on the bear program in the Pyrenees; and the analysis of press news and the main politicians' interventions in the media.
3. In the Bonabé valley there are two villages (Isil and Alós d'Isil) that belong to the same municipality (Alt Àneu) that extend over 131 square kilometres, while the Biros valley has five municipalities (Antras, Balacet, Bonac-Irazein, Bordes-Uchentein, and Sentein) with a territory of 173 square kilometers total.

4. Names have been changed for the sake of anonymity.
5. This pastoral agreement was part of a larger historic treaty between these two territories (Roigé, Ros, and Còts 2002).

References

Adams, William Mark. 2003. *Future Nature: A Vision for Conservation*. Rev. ed. London: Earthscan.

Agrawal, Arun. 2005. *Environmentality: Technologies of Government and the Making of Subjects*. New Ecologies for the Twenty-First Century. Durham, NC: Duke University Press.

Anderson, David, and Richard Grove, eds. 1987. *Conservation in Africa: People, Policies and Practice*. Cambridge, UK: Cambridge University Press.

Baird, Melissa F. 2017. *Critical Theory and the Anthropology of Heritage Landscapes*. Gainesville: University Press of Florida.

Benhammou, Farid. 2007. "Crier au loup pour avoir la peau de l'ours. Une géopolitique locale de l'environnement à travers la Gestion et la conservation des grands prédateurs en France." PhD diss., Paris Institute of Technology for Life, Food, and Environmental Sciences.

Cabannes, Hyppolite. 1889. "Les chemins de transhumance dans le Couserans." *Revue de Comminges et des Pyrénées Centrales* 14: 104–23.

Camarra, Jean-Jacques, Pierre-Yves Quenette, Frédéric Decaluwe, Ramon Jato, Jokin Larumbe, Santiago Palazón, and Jordi Solà de la Torre. 2011. "What's Up in the Pyrenees? Disappearance of the Last Native Bear, and the Situation in 2011." *International Bear News* 20(4): 34–35.

Casanova, Eugeni. 1997. *L'ós del Pirineu: crònica d'un extermini*. Lleida: Pagès Editors.

Castree, Noel, and Bruce Braun. 1998. "The Construction of Nature and the Nature of Construction: Analytical and Political Tools for Building Survivable Futures: Nature at the Millennium." In *Remaking Reality: Nature at the Millennium*, ed. Bruce Braun and Noel Castree, 2–42. London: Routledge.

Caussimont, Gerard. 2013. "El oso pardo en el Pirineo: situación y perspectivas de conservación." *Lucas Mallada* 15: 49–65.

Chandivert, Arnaud. 2010. "Pastoralism and Heritage in the Central Pyrenees: Symbolic Values and Social Conflicts." In *Social and Ecological History of the Pyrenees: State, Market, and Landscape*, ed. Ismael Vaccaro and Oriol Beltran, 127–41. Walnut Creek: Left Coast Press.

Chevalier, Michel. 1951. "France rurale d'aujourd'hui." *Annales. Economies, Sociétés, Civilisations* 6(1): 15–22.

———. 1956. *La vie humaine dans les Pyrénées Ariégeoises*. Paris: Éditions M. Th. Génin.

Comaroff, Jean, and John L. Comaroff. 2006. "Introduction to Of Revelation and Revolution." In *Anthropology in Theory. Issues in Epistemology*, ed. Henrietta L. Moore and Todd Sanders, 382–95. Malden: Blackwell Publishing.

Còts, Pèir. 2002. "Evolucion deth bestiar e er amontanhatge ena Val d'Aran pendent era darrèra decada." *Annals del Centre d'Estudis Comarcals del Ripollès 2000–2001* 13: 247–78.
Espinós, Nicolas. 2014. *Projecte d'ordenació de pastures de la conca hidrogràfica de Moredo. Terme municipal de l'Alt Àneu (Pallars Sobirà)*. Lleida: Treball Final de Grau.
Estrada, Ferran, Eli Nadal, and Juan Ramón Iglesias. 2010. "Twenty-First Century Transhumants: Social and Economic Change in the Alta Ribagorça." In *Social and Ecological History of the Pyrenees: State, Market, and Landscape*, ed. Ismael Vaccaro and Oriol Beltran, 105–26. Walnut Creek: Left Coast Press.
Ferrer i Sirvent, Joan. 2004. "El gos de muntanya dels Pirineus. Garant de la biodiversitat Pirinenca." *Annals del Centre d'Estudis Comarcals del Ripollès 2002–2003* 15: 103–24.
Fletcher, Robert. 2010. "Neoliberal Environmentality: Towards a Poststructuralist Political Ecology of the Conservation Debate." *Conservation & Society* 8(3): 171–81.
Franquesa, Jaume. 2013. "On Keeping and Selling: The Political Economy of Heritage Making in Contemporary Spain." *Current Anthropology* 54(3): 346–69.
Gordillo, Gaston. 2014. *Rubble. The Afterlife of Destruction*. Durham, NC: Duke University Press.
Griffin, Carl J., Roy Jones, and Iain J. M. Robertson, eds. 2019. *Moral Ecologies: Histories of Conservation, Dispossession and Resistance*. Palgrave Studies in World Environmental History. Cham: Springer International Publishing.
Guha, Ramachandra. 2000. *Environmentalism: A Global History*. New York: Longman.
Hirsch, Eric, and Charles Stewart. 2005. "Introduction: Ethnographies of Historicity." *History and Anthropology* 16(3): 261–74.
Ingold, Tim. 1993. "The Temporality of the Landscape." *World Archaeology* 25(2): 152–74.
———. 2000. *The Perception of the Environment: Essays on Livelihood, Dwelling and Skill*. New York: Routledge.
Jacoby, Karl. 2019. "Afterword: On Moral Ecologies and Archival Absences." In *Moral Ecologies: Histories of Conservation, Dispossession and Resistance*, ed. Carl J. Griffin, Roy Jones, and Iain J. M. Robertson, 289–97. Palgrave Studies in World Environmental History. Cham: Springer International Publishing.
Kirshenblatt-Gimblett, Barbara. 1998. *Destination Culture. Tourism, Museums and Heritage*. Berkeley: University of California Press.
Li, Tania. 2007. *The Will to Improve: Governmentality, Development, and the Practice of Politics*. Durham, NC: Duke University Press.
Marliave, Olivier De. 2008. *Histoire de l'ours dans les Pyrénées: de la Préhistoire à la réintroduction*. Bordeaux: Ed. Sud-Ouest.
Mermet, Laurent, and Farid Benhammou. 2005. "Prolonger l'inaction environnementale dans un monde familier: la fabrication stratégique de l'incertitude sur les ours du béarn." *Ecologie Politique* 31(2): 121–36.
Moore, Jason W. 2015. *Capitalism in the Web of Life: Ecology and the Accumulation of Capital*. London: Verso.

Neumann, Roderick P. 1992. *Imposing Wilderness: Struggles Over Livelihood and Nature Preservation in Africa*. Berkeley: University of California Press.

Nogués-Bravo, David, Daniel Simberloff, Carsten Rahbek, and Nathan James Sanders. 2016. "Rewilding Is the New Pandora's Box in Conservation." *Current Biology* 26(3): R87–91.

Parellada, Xavier, Marc Alonso, and Lluís Toldrà. 1995. "Ós bru." In *Els grans mamífers de Catalunya i Andorra*, ed. Jordi Ruiz-Olmo and Àlex Aguilar, 124–29. Barcelona: Lynx.

Pèlachs, Albert, Ramon Pérez-Obiol, Joan Manuel Soriano, Raquel Cunill, Marie-Claude Bal, and Juan Carlos García-Codron. 2017. "The Role of Environmental Geohistory in High-Mountain Landscape Conservation." In *High Mountain Conservation in a Changing World*, ed. Jordi Catalan, Josep M. Ninot, and M. Mercè Aniz, 107–29. Cham: Springer.

Réseau Ours Brun. 2020. "Info ours. Rapport annuel du Réseau Ours Brun 2019." Office National de la Chasse et la Faune Sauvage.

Ringel, Felix. 2016. "Can Time Be Tricked? A Theoretical Introduction." *Cambridge Journal of Anthropology* 34(1): 22–31.

Roigé, Xavier (coord.). 1995. *Cuadernos de la trashumancia*. Vol. 13. Madrid: ICONA.

Roigé, Xavier, Ignasi Ros, and Pèir Còts. 2002. "De la communauté locale aux relations internationales: les traités de lies et passeries dans les Pyrénées Catalanes." *Annales Du Midi: Revue Archéologique, Historique et Philologique de la France Méridionale* 114(240): 441–99.

Scott, James C. 1998. *Seeing Like a State: How Certain Schemes to Improve the Human Condition Have Failed*. The Yale ISPS Series. New Haven: Yale University Press.

Stewart, Charles. 2016. "Historicity and Anthropology." *Annual Review of Anthropology* 45(1): 79–94.

Svarstad, Hanne, and Tor A. Benjaminsen. 2020. "Reading Radical Environmental Justice Through a Political Ecology Lens." *Geoforum* 108(January): 1–11.

Vaccaro, Ismael, Allan Charles Dawson, and Laura Zanotti. 2014. "Negotiating Territoriality: Spatial Dialogues Between State and Tradition." In *Negotiating Territoriality. Spatial Dialogues Between State and Tradition*, ed. Allan Charles Dawson, Laura Zanotti, and Ismael Vaccaro, 1–17. Routledge Studies in Anthropology. New York: Routledge.

Walker, Peter A. 2003. "Reconsidering 'Regional' Political Ecologies: Toward a Political Ecology of the Rural American West." *Progress in Human Geography* 23(1): 7–24.

CHAPTER 5

Transhumance in Kelmend, Northern Albania
Traditions, Contemporary Challenges, and Sustainable Development

Martine Wolff

Brief Historical and Geographical Focus

Certified since remote times, transhumance has never ceased to exist in Kelmend. What will become of it in the future depends on each one of us. This book reminds us that "pastoralism is one of the most widespread and ancient forms of human subsistence" as Letizia Bindi writes in the introduction, so it is also the case in Albania. In order to have a better understanding of the situation I would like to focus briefly on the geography and history of Albania. Albania consists of around 80 percent of mountains, with a short littoral plain along the Adriatic Sea, north of the town of Vlorë and south of the Ionian Sea. These two seas are recognized as being the oriental part of the Mediterranean Sea. Pastoralism in Albania belongs to this space of traditional Mediterranean pastoral culture. Archaeologists are assessing that transhumant pastoralism has been present in Northern Albania since extremely ancient times, dating its origins at the very beginning of the Neolithic period, and nomadic pastoralism since the Palaeolithic period. All over Albania, shepherds have determined most of the social and cultural values which characterize Albanian traditional culture (Tirta 2016; Palaj 2018). But I must underline that pastoralism in Northern and Southern Albania is very different and cannot be compared.

Kelmend, situated in the extreme north of Albania, south of the Montenegro border and northwest of Kosovo, has been historically the cradle of transhumant pastoralism, and the shepherds in Kelmend, up to this day, are the most respected persons in the local communities. Kelmend belongs to the municipality of Malësi e Madhe and the Shkodër region. But Kelmend is also part of the Albanian Alps. Up to the end of the Communist period, transhumant pastoralism existed all over the Albanian Alps. Since the 1990s, transhumant pastoralism has fallen dramatically. Nevertheless, it is still surviving in some zones from the Shkodër and Tropoje regions. Kelmend is one of them.

Kelmend contains four valleys. The alpine plains cover 40 percent of the territory, which is where spring and autumn transhumance occurs. Most of the pastures are over two thousand meters high (Progni 2000). Actual transhumance occurs from the littoral plain to the mountains, as well as from these alpine plains towards the high pastures where there exists very high biodiversity, with endemic flora and fauna and a wide variety of vegetation and medicinal plants. These landscapes have been shaped since very ancient times by transhumant pastoral farming. It is in these harsh natural conditions that a unique way of life within extensive transhumant pastoral activities has been constructed, in conjunction with self-sustaining family agriculture, incorporated within the demographic and historical aspects of traditional culture. Although no support has been given up to now by the government, every year, this area welcomes many transhumant shepherds, coming from the plain (in Lezhe, Milot, this year, I even met one shepherd coming from Durres). Most of these shepherds are of Kelmend origin, and for them, it is extremely important to come to their homeland pasturage. They are added to the herds of local breeders present in Kelmend throughout the year. In Kelmend, transhumant pastoralism is really a co-construction, a story of men and women in union with their animals and the surrounding nature. It belongs to the identity of the region and this fact is crucial to them. This book is about remembering how traditional pastoralism is reconsidered in terms of sustainability, human health/animal welfare, and environmental impact.

Regarding the Albanian Alps, and especially the zone of Kelmend, I like to name them the Illyrian Alps because especially in Kelmend there are ancient Indigenous Albanian populations who have refused all forms of assimilation from successive invaders, and who have tried to preserve and transmit all their autochthonic traditions and customs in these remote and isolated valleys (Rama 2020). The Albanian ethnologist Mark Tirta recognizes some of these traditions as having their origin in Illyrian time (Tirta 2016; Durham [1909] 2013). This culture cannot be understood outside the context of transhumant agropastoral dynamics. Let us look at some historical considerations. The territory of the Albanian state today is very reduced in comparison with historical Albanian territories. When Albanian independence was proclaimed after five hundred years of Ottoman occupation, Albania lost two thirds of its territories, due to the western geopolitical diplomatic games. Furthermore, historical Kelmend does not correspond to the present region of Kelmend in Albania. A part of the historical Kelmend is now in Montenegro and Kosova (Neziri 2012–13). Over time, the territory of Kelmend narrowed more and more, which is problematic for the pastoral community, as the habitants were cut off from their pastures, water, and herbal sources for the animals, from their transhumance paths and spiritual spaces. In ancient times, their transhumant moving existed over vast spaces up to Bosnia and Croatia in the north and to Bulgaria in the east (Durham 1928).

Throughout history, due to wars and political changes, these transhumance territories have moved and been reduced in size. Unfortunately, till now, shepherds from Kelmend are not allowed to cross the borders towards Montenegro and Kosovo with their herds.

In ancient times, every family in Kelmend was involved in transhumant pastoralism. In Communist times, despite collectivization by the state, transhumance was still very developed. The families secretly transmitted their traditional way of transhumant pastoralism.

Starting in the 1990s, the families tried to take back their traditional way for transhumant pastoralism. They were not encouraged at all by the government who up to that point recognized only intensive farming. The dramatic events of 1997 (collapse of pyramid schemes and widespread corruption, causing big anarchy and rebellion throughout the country) induced a break in this process, and each year the number of transhumant herders was diminished. Exact statistics are unknown; I have heard so many different numbers, depending on who is talking. I have now been in Kelmend for four years. Four years ago, there were still six to eight thousand sheep undertaking transhumance in summer. I cannot say any numbers for the goats. In 2019, only 2,400 goats and 4,700 sheep were registered. In 2018 I evaluated 145 families moving for transhumance to the high pastures. In 2019, they were less than 100 families. The main reason that many did not go this year was climatic difficulties, as explained to me by the shepherds.

From the Dictatorial Times up to the Modern Period

Let us mention the fifty years of totalitarian dictatorship when Albania was totally isolated, living in total autarchy. All those who resisted the dictator, refusing to accept his ideology, were severely oppressed. This was the case for most families in Kelmend. Today, it is terrible to see how in each Kelmend family, some members were killed or tortured in concentration camps and prisons. Many children were born and/or grew up in concentration camps. Collectivization of the lands, animals, and persons under state control must also be mentioned as a traumatizing experience for the population. The entire economy was centralized by the dictatorship. (Bardhoshi 2018). The traditional structure of the villages was destroyed, and the transhumance was reduced and organized by the cooperative and state farms. This resulted in loss of knowledge about the practice and paths of transhumance, and the breakdown of relationships between pastoral families in the valley asking for transhumance care and the shepherds living in Kelmend. There were tremendous long-term negative consequences to the multimillennial pastoral ecoculture within Kelmend. Before the Communist period, each family was involved in pastoral activities (Durham [1909] 2013, 1928). Still now the children are having a special moment of family education living in summer with all the family on the high pastures, helping regardless of their age possibilities on pastoral activities.

During Communist times, pastoral traditions specific to the region were undermined. However, although pastoral traditions were reduced, they have never ceased to exist (Bardhoshi 2018). Communism failed with Perestroika beginning in the 90s. Since then, traditional pastoralism has been renewed, with no support from the central or local government, the shepherds working only with the sweat of their brow. Taking back this lost and forbidden tradition became an essential priority for the inhabitants of Kelmend. The same occurs in Romania but nonetheless the maintenance of transhumant practice is considered a way of conserving typical landscape reflecting embedded memories (Chapter 9).

Pastoralism has been declining since 1997, due to the dramatic events that occurred in Albania. From 2013, due to massive emigration of the young, and the rise of mass tourism in some parts of the Albanian Alps, the decline of pastoralism has been strongly accelerated. Having no support each year it becomes more difficult for these transhumant shepherds to go on with the practice of transhumance.

But no one in Kelmend can imagine transhumance ceasing altogether. For the inhabitants of Kelmend, as well as some other parts of the Albanian Alps, transhumance must endure. Albania is now a candidate, together with France and Spain, in asking this nonmaterial cultural heritage to be protected by UNESCO. Therefore, we are working on heritagization of the transhumance process. This is giving a new hope and dynamic to the shepherds presently organizing themselves within networks of transhumant shepherds. They are taking a renewed awareness of their traditional pastoral culture, understanding that they are not alone, and that this culture is shared all over the world.

Transhumance Pastoralism, Still a Living Culture in the Twenty-First Century

In the Gheg language we say, *dale në bjeshkës*, going to the High pastures; *Merr rrugen per bjeshkë*, taking the road towards high pastures; and *jem tu bjeshkua*, I am in the process of grazing. These expressions say a lot about the importance of the transhumance phenomenon and the synergy involved. Transhumance in Kelmend concerns mainly sheep but also goats, led by shepherds from one "humus" to another, or from the plain to the mountain in tune with the rhythm of the seasons, in the organization of a pendulum movement of herds. They are accompanied by horses and mules, transporting human beings and goods, as well as dogs leading and protecting the herds. During the transhumance period, shepherds also collect wild berries and medicinal plants. This results in close complementarity and circulation between the plain and the mountains. Many cultural exchanges flow from this phenomenon. The mountain is a fabulous reservoir of grass, cool even in the heart of summer. By the end of May, beginning of June, the shepherds with their families and animals, leave the grasslands of the plains and the littoral, to find rich grass in the mountains and pure air (*ajër i pastër*). The shepherds tell me, "The health of the animal is asking us to take them to the high pastures, because the pastures clear up all the diseases."

Although nowadays in a very precarious balance, this practice still exists, distinguishing the big transhumance in summer and smaller flows of transhumance on the mountain plains (*vrri*), according to the different seasons (*bjeshkimit pranverore and vjeshtore*). The annual cycle of moving herds up to the mountain areas and then down to the lower plains has become a significant ritual, with celebrations which bring together the whole community. Its history and function shed light on the symbolic relationships that have for so long linked humans and animals to their natural environment, showing how transhumance is still persisting and resisting as an efficient form of farming deeply influencing landscapes (Brisebarre 2007). The Albanian shepherds believe that the maintenance of cultural landscapes and natural (ecological) balance is dependent on them. But also, a whole way of thinking, being, and acting (beliefs and representations) have determined social and cultural life in Kelmend throughout history, perfectly in line with what Tim Ingold outlines in the Foreword, about transdisciplinary cooperation and communication among humans, environmental and animal health, and recognizing this cultural practice as an important form of diversity thriving on the biodiversity of the environment in which it operates.

The practice of transhumance requires a great capacity for adaptation. It enables the development of an agriculture of self-sufficiency combined with breeding of the herds that are complementary to each other. It is a link between the coast and the mountains. It is based on the search for the most nourishing grass, according to the meteorological conditions and the altitude, and on the appropriation of the mountains by man. It is the challenge of a whole system of economic and social relations structuring a vast space, concerning everyone. And from there are delineated all the paths dedicated to transhumance which are called *stigjet* in Gheg vernacular language and *drailles* in French. *Lieux de passage* (trails) are places of meeting, communication, sharing, and exchanging since time immemorial; they are paths that warrant a sociocultural context. Transhumant pastoralism is deeply rooted in the identity of Kelmend. It is also combined with life organization since very ancient times around the multiple sources of water, rivers, and some resting places, which are considered sacred from generation to generation.

Pastoral Knowledge, a Living Intangible Nonmaterial Heritage

The essence of pastoralism is the knowledge of cooperation with heterogeneous ecosystems and a variable climate, in order to feed animals who have themselves learned the mountain regions. This results from a shared reciprocal, balanced listening and understanding (knowledge) between the shepherd and his animals, being one with them, as well as between themselves and their environment. It implies a deep affective relationship where human and animal well-being depends on each other. The shepherd has not only an intimate knowledge of the animals but also a deep complicity (closeness). The shepherds find themselves in a reciprocal listening, where human and animal well-being is closely linked. The animal is a fully-fledged partner of the shepherd in his way of life and production. It is in this relationship to the animal and also to the environment that the shepherd finds balance. He uses his empirical knowledge of the animal's ability to adapt to the environment, and thus they have the ability to move together whenever necessary. Pastoral knowledge includes not only herd composition and management but also specific practices related to animal care, mobility of animals and families, transhumance practices of herds, appropriate search for pastures and forage resources. The shepherd must know how to take care of a herd and lead them, to understand their behaviors, allow them to feed themselves, driving the herd where the best food is located. He must know the mountain and be able to adapt to it. Pastoral management relies on the knowledge of the shepherds who have a fine understanding of the behavior of their animals, the daily and seasonal rhythms as well as spatial organization of relief and vegetation. He has knowledge of both wild and cultivated plants. By his fine and precise observation, he knows the role these plants play in the nutritional status and health of the animals. He takes into account the diversity of plant species and the resilience of natural environments, or more specifically ecosystems, the resilience of natural habitats, and the irreversibility of certain processes of grassland degradation. Each aspect of his knowledge leads to ways of valuing the heterogeneity of environments, especially through herd mobility, and the nature of the production methods. The shepherds with their herds have maintained and restored these landscapes by ensuring a balanced use of meadows and pastures. The heterogeneity of the environment is valued by the herd's mobility which facilitates a mosaic of diversified vegetation (ecosystems and natural habitats) around which is articulated "a cultural corpus of typical landscapes." The shepherd is looking to identify and preserve them. From these pastoral issues come many skills such as assessing the quality of the meat and dairy products and traditional pharmacopeia, and making traditional local crafts which the shepherds defend fanatically. The shepherds strongly affirm this when they tell me, "all this intangible natural and cultural heritage has to be protected and transmitted." This is the core of their motivation to continue in spite of all the difficulties and adversity encountered day after day.

Living among the shepherds in Kelmend, I can describe them as a community with real social capital, making solidarity work in face of the dangers which happen on a wild mountain that becomes hostile in bad weather. The shepherds have a respectful relationship to their animals, the vegetal world, their whole environment, as well as a sensibility which comprises the mental and spiritual universe of the shepherd. Pastoralism is a co-built harmony between the shepherd and his family, his dog and his horse, his herd and the environment. It is this rich relationship between human beings, animals, and nature which is at the heart of the agropastoral system, with all its behaviors, passions, responsibilities, interdependencies, and freedoms. Another very important thing for shepherds is the genetic selection of animals, which they utilize to reinforce the hardiness of the animals in order to produce breeds which are productive but also suitable for their territory, a selection based on the shepherd's knowledge, his way of life and beliefs. All these aspects color this pastoral society as a unique world. As an anthropologist this fascinated me, I wanted to understand it in a better way. These inhabitants have the resilience to face the worst difficulties alone, without any support from anyone, tenacity to go on, and the heart to continue to transmit their heritage; this is what convinced me that this way of life has to be protected.

The Maintenance of a Balanced Natural Landscapes Depends on This Dynamic of Transhumance Life

Pastoralism determines the life of the landscape. By driving the animals from the plain to the mountain and from the mountain to the plain, cultural landscapes have been slowly shaped. The landscapes in Kelmend are the work of shepherds and their herds throughout history, resulting from the interaction between man and nature, testifying to the appropriation of nature, adaptation, and modification. Whether it is along streams, forests, or pastures, these landscapes are spectacular, containing a rich and rare biodiversity. There is truly environmental resilience that is entirely dependent on pastoral management and that seems to me a rare and significant model of harmony between man and nature. These cultural landscapes are composed of a succession of strata throughout history, the origin of which dates back in Kelmend to the Palaeolithic period, a strong ancient culture rooted in a nomadic way of life. The oral literature, especially the shepherd songs, testify the survival of the original Illyrian culture (ancestor of the Albanians). Deeply attached to their territory, this environment is of great importance for the inhabitants of Kelmend. They consider their territory as Mother Earth, venerating her and dedicating their deep homage to her at every moment and through each of their actions. It is in this context also that their relationship and deep respect to the animals, whether they be their herds or the wild animals living there, should be understood. This appropriation and the maintenance of these natural landscapes in balance are not limited to their material aspects, they include both cultural and spiritual aspects which must not be forgotten when working on heritagization of the transhumance process.

Pastoralism Is Cultural Heritage

These cultural landscapes have not only an economic significance but also an ecological, cultural, and spiritual significance. These material and intangible heritage elements are keys for reading a land occupation, contributing to a better understanding of the lifestyle and practices of the territory. These cultural landscapes participate in the foundations of the territory's own cultural identity. These cultural landscapes are the place where all traditional knowledge is developed and maintained, which includes, beyond the custody of the herds and the care of them, many other areas, such as excellent knowledge of the environment, plants, and wildlife, traditional medicine, and traditional use of many materials.

The ecological functions of these cultural landscapes play a fundamental role in maintaining the general stability of the human environment, and they also have a cultural function. The natural environment determines the self-expression of human beings, that is, culture. Their traces have to be evaluated and listed. They are "living cultural landscapes," which includes the artistic productions that express it and the ethical and spiritual values that are bearing it. These cultural landscapes of great diversity and richness are dynamic and connected to customary law code encompassing a resilient livelihood, as Ingold has highlighted in his Foreword. Kelmend inhabitants state that, "Our culture is to live together with the animals." Many of their practices belong to what I call "a whole pastoral cosmogony," which includes the bells and other decorations placed on the animals and the hairstyles on the animals' heads or tails. By these adornments, the animal is identified with its environment and its social group. These accessories are specific aesthetic markings through which the pastoral knowledge and beliefs of the shepherds are valorized. All these elements symbolize very specific cultural and religious practices which are in a strong syncretism with all the natural environment. There is also music and dance that illustrate great richness of expression and originality. It is the shepherds who are the creators, it is their voice that resonates from the mountains. Also of significance are the epic songs of the *Lahuta, kengë I kresnikëve*, which originated in the Homeric era, and the shepherd's flute *fyelli* and his lyrical songs; these songs describe the various heroes throughout history, and the *kengë krahu* is the song or call conveying information from one valley to the other, such announcements as births or deaths. These songs reflect the souls of these inhabitants and represent the memory of these mountaineers. They are a veritable encyclopedia that is transmitted from generation to generation since ancient times. They connect the social community to its lineage and its affiliations. They have a magical-spiritual component and accompany all the rituals and celebrations in a richly institutionalized, socializing enchantment. They facilitate through these moments of sharing, in moments of imbalance, the creation of a new balance, the restoration of communication and a circulation of life. They can become moments of regeneration and healing. And therefore, it is important to underline all these moments of feast where the social community meets together, experiencing restorative moments of joy and solidarity, encouraging them to go on living in these arid and wild environments. All that belongs to cultural heritage in links with transhumance.

These Cultural Landscapes Determine a Whole Social Organization

There is a clan organization in Kelmend. The population is exogamous, but the marriage alliances are determined by the pooling of pastures and by the work required by the pastoral culture, deeply linked to local and traditional social structures, kinship relations, and symbolic representations and settlements. These mountaineers have many children. Although their number are getting lower now, it is not rare to find a family with seven to ten children. They represent a work force that makes the pastoral culture possible. There is a knowledge that is transmitted within the family, from the elders to the youth, generation to generation since ancient times. For this pastoral community, they are links to the ancestors, therefore this transmission cannot be broken at any price. This link is sacred. The transmission of pastoral knowledge is carried out within the family, as within village communities and pastures. The art of conducting a transhumant herd is conveyed by gesture and experience. The tasks are clearly delineated between men, women, and children. The whole family takes part in transhumance, and we could say in some way, are therefore some kind of shepherds. The different events of the cycle of life, and the customs that animate them, profoundly influence the pastoral culture. Their social community includes, beyond the family and the village, the animals, and the natural and spiritual environment.

The landscapes and the life lived there have also determined, over time, a set of customary laws that are collected throughout northern Albania in different *Kanun*. In Kelmend, it is the "*Kanun* of the mountain" which predominates. These laws were determined by a council of elders (sages), according to the social reality encountered in the field. Many of them were shepherds and still now, they are the shepherds who assume a mediation role in conflicts, bringing back dialogue, peace, and harmony. This pastoralism of transhumance has determined a whole social and cultural life and many spiritual values that constitute the identity of the region. The animal is the center of everything. The way in which the shepherd apprehends the animals he is conducting in the environment, characterizes certain cultural components of agropastoralism and its strong impact on the landscapes as well as on the population. Because of his activity, the shepherd is confronted with the natural environment he exploits, and the animals he raises. The animal, in particular, is both the purpose of these activities, the operating tool, and the management auxiliary of an environment that is to ensure sustainability, valuing in the best way its potential foraging. The paths of transhumance are real corridors and first trails, passage for the herds and the men, but also for the animal and plant species associated with them. They are fundamental cultural and ecological exchange tools. They have been known for centuries and are the origin of cultural practices conveyed by transhumant shepherds. All this needs to be protected. The shepherds in Kelmend walk now where their ancestors have always been walking. Many generations have drunk at the same sacred water sources and have rested at the foot of the same trees which, too, are imbued with holiness. Pastoral culture generates many customs and traditions, including *mirëpritja*, the famous Albanian hospitality; *la parole donnée*, word of honor; and *Besa*, the pledge of honor. These societal values have been determined by transhumant shepherds and can only be understood through the eyes of this multimillennial pastoral culture (Palaj 2018).

Subsistence of Transhumant Breeding in the Twenty-First Century

Transhumant livestock has survived until the twenty-first century because the shepherds knew how to adapt to the demands of a constantly changing world (Fabre, Molénat, and Duclos 2002). But nowadays, in this postmodern period, what will become of them? The shepherds are totally left alone working and can defend themselves only by their own very hard work. Neither state nor local authorities are taking any interest in them. "They are just not considered by anyone in power," say the shepherds. "They do not merit any interest at all," are the words of some development NGOs spoken to me some years ago. Del'homme (2016) in his article confirms this attitude. Transhumance practices are totally denied by them and even condemned as primitive, stemming from prehistoric times. Shepherds work in such difficult conditions that few young people want to become shepherds, rejecting the harsh life of their parents. Not believing in any future within their own region, they attempt to emigrate at any price (Mema, Aliaj, and Matoshi 2019). The exodus to the plains and emigration has become massive and is a hemorrhage to the pastoral organization.

Albania is in economic transition and the country is struggling to recover from the destruction of a communist dictatorship. Reprivatization is accomplished with great difficulty and many errors are perpetuated, which detract from good agropastoral development. There is no land tenure plan. There is a total lack of infrastructure, including school and sanitation. Life becomes increasingly impossible. After the fall of the regime, the practice of transhumance began again but with several problems: loss of knowledge of the traditional practices of transhumance, privatization of land, despoilment of the pastures and trails in order to build roads, and illegal hydropower plants built against the will of the population, destroying the region's ecosystem, creating many difficulties for further pastoralism.

The Ministry of Agriculture has undertaken important economic reforms and promotes industrial farming with few considerations for transhumance practices. Management of traditional pastoralism is not taken in account. This seriously threatens the cultural landscapes that have been valiantly shaped for millennia. The shepherds are deeply worried about all these issues (Nori 2016). Modernity suddenly erupted when communism fell in Albania; the people were unable to prepare themselves, or adapt materially, psychologically, socially or culturally to the changes. Losing their identity, the population tries to survive as they can. The inhabitants of Kelmend state that, "We do not have the codes of operation of this modern world, we do not know how to deal with it." Albania, in this modern period, is in a very chaotic situation. We are witnessing a crumbling of the social body and a collapse of traditional village structures. The inhabitants, finding it difficult to cope with galloping globalization, are caught between two extreme ways of life in a context which is imposed on them, and which they cannot yet master, feeling torn apart between the forces of tradition and modernity. This results in ambivalence and fear, operating in crisis mode and feeling a drift between the two. These Albanian shepherds no longer possess the keys of interpretation from their ancestors and feel incapable of reproducing the social structures inherited from the past. Nor can they cope with this present society and do not recognize themselves in globalization. They feel locked up in an insecure modernity and seek in the mythical figure of the shepherd and his flock a source of identity, a re-anchoring to a regained identity that can be shared.

The Kelmend region is also presently facing a lot of ecological problems. One of them is its rivers. Cemi is the main one, and half of it is in Montenegro. Kelmend was to become a national park, but the lobbying for hydropower construction was stronger and this project has been silenced by the Albanian government. The historical part of Kelmend around the Cemi river is in Montenegro and is a UNESCO-protected national park, but in the Albanian part, non-scrupulous, illegal initiatives of hydropower construction are destroying the environment and the culture. The Kelmend inhabitants have saved this ecoculture with their blood through centuries. They are convinced of the urgent necessity that their region becomes, as promised, a national park and that they work for heritagization of the transhumance culture in their own region within UNESCO. Such a fact constitutes a deep wound for these inhabitants who find themselves very worried about the future of their region. Therefore, working on heritarization of transhumance is of the highest importance.

A Spiritual Foundation of the Local Population from which a Sustainable Development of Transhumant Pastoralism Can Arise

Nevertheless, transhumance is still alive in Kelmend. The inhabitants are firmly convinced that transhumance needs to endure. This is their force of life, which may enhance further sustainable development. They are conscious that without shepherds, Kelmend, with all the culture that animates it, will not survive. The government is holding a politic of massive tourism within the Albanian Alps, but the shepherd community is mainly against the development of such tourism. Tourism expansion is not their priority. In spite of the fact that these projects are rejected by the whole Albanian pastoral community, some of the shepherds' families began to propose to welcome some tourists within their families in their high pastures, giving them the opportunity to share for a moment their own life, enjoying a strong contact with the animals and surrounding nature.

Furthermore, the transhumance dynamic is very important to the pastoral community. The shepherds are mythical figures within the entire society and are seen to bring social stability and peace. They will always have the support of the inhabitants of the region. These traditional populations gather around a real ecoculture, their beliefs and customs are aligned to this. They are conditioned by this traditional pastoral life towards a deep listening to and respect for their environment. (Descola 2005, 2018). This harmonious relationship with nature and cosmogony gives rise to archetypes that are found within all spheres of society and radiates a whole culture of peace. Living with them for more than four years, I must admit that it was often in close contact with the shepherds and their flocks that the population found calm, stability, and regained peace which animated them through conflicts. They found among them the strength to continue to be resilient in spite of all the difficulties.

The Shepherd, Emblematic Figure, Guardian of Tradition, and a Culture of Peace

The inhabitants of Kelmend are characterized by their inner nobility, their values, their bravery, and their courage. These mountaineers have a particular relationship to the cosmogony. Their Christian faith is part of a great syncretism of animist *beliefs*, and the oral traditions they maintain bear traces of a rich mythology that dates back to ancient times but which is still alive. All this results in many ritual celebrations. These traditional populations carry a real "Ecoculture" and defend many values and customs dating back to ancient times. Their link to nature and to cosmogony is expressed in all spheres of life. It expresses itself also in oral traditions that accompany their children's development through the whole life cycle. This link regulates the inscription of everyone as a social and cultural being within the community to which they belong. The inhabitants still seek, despite a threatened ecosystem, to live in interaction and harmony, and this is where the shepherd carrying this multimillennial culture plays a major role as regulator and stabilizer within the society. This is even more valid for the uprooted sections of the population living on the plains, lacking benchmarks, being hostages of a brutal burst of modernity. The inhabitants do not welcome new modes of being, thinking and acting. They are in a negotiation of their here and now, while still struggling with tradition (Breda, Derridder, and Laurent 2013).

The shepherd plays a central role within Kelmend society, alongside the priest. Everyone listens to him, respects him, and honors him. He is often a mediator in conflicts, taking a central place at the assembly of elders. This pastoral people of the mountains of northern Albania maintains a close relationship to the cosmogony, in resonance with the visible and invisible world, in an intimate relationship to the environment that surrounds them, to others and to themselves. Their sense of the sacred in all things, the spiritual values they defend, their understanding of the therapeutic, of disease and healing, imbue them with many capacities of resilience, despite the worst difficulties (Brisebarre, Lebaudy, Gonzalez 2018). It is because they are aware of their identity, and of what animates and inhabits them, that they have been able to go on living, mobilizing all their creativity in face of adversity. It is in their attachment to tradition, to the long lineage of ancestors and to Mother Earth, in close relationship with the whole cosmogony, that these populations have found the strength to be able to resist and innovate in a better way.

The Shepherd, a Mythical Figure That Embodies Tradition and Stability

The inhabitants explain to me that the shepherd "is like the rock of the mountain." He has a very special importance within the society. As he has met many people, walked many paths, he has many things to tell. Living in nature that he knows perfectly and being keeper of the herds, he is the guarantor of balance and a peacemaker. He maintains a quasi-mystical relationship with his flock and the nature that surrounds him. This also includes the other beings living there. He thus has a very special function, being both a priest and a popular healer, allowing the creation of "new social landscapes," bringing together young and old, transmitting the traditional knowledge necessary to survive in the mountains. It is around the shepherd that the inhabitants, in this world full of social and cultural change, in deregulation and reconfiguration, try in community to recreate spaces of sociality. They try to reorganize, regain coherence, and rebuild a self-identity, a living together and, while facing adversity, to seek together how to navigate this passage. It is around the shepherd who has always been able to demonstrate adaptation and invention that the population seeks to make their way in the face of the chaotic situation caused by the abrupt eruption of modernity (Charbonnier and Romagny 2012). The shepherd and his herd are playing a central role in social and cultural terms. It is through their contact and oneness with nature, that everyone can regain a sensitive world, a vibratory link with nature and life, an intelligibility of the world. (These words have been said by local habitants.) Shepherds are living witnesses of a culture of the bond, and a participatory ecology. As some inhabitants explained to me: "They [shepherds] are the guardians of a balance to maintain and share. They remind us of the necessity of the beautiful, the orderly and the harmonious. It is their human qualities that will save us, in a relationship of intimacy, respect and peace with nature, the visible and invisible world." The shepherd is the artist around whom, in the light of contemporary issues, new contexts can be invented and recreated, which will generate tuning and cohesion and bring a renewal. This is coherent with the idea of transhumant pastoralism as a form of husbandry ecosystem services, contributing to the protection and regeneration of mountain environments and biodiversity.

What Can Be Done So That This Transhumant Pastoral Culture Survives?

Pastoralism modulates important aspects of cultural, historical, and traditional heritage, established and developed by the history of these populations, but which today are seriously threatened by the evolution of the modern globalizing world (Lacirignola 2016). Personally, I helped the shepherds from Kelmend to organize themselves as a professional group fighting as a civil society for their rights in respect of their traditional culture, and we have established links with indiviuals in the south of France. That was primordial for them, as they are feeling isolated and forgotten, and we are working together towards cooperative projects. Together we set up some development priorities, and I must confess that the shepherds in Kelmend are working in terrible conditions, which need to be improved. Nothing is organized and they do not have the education to do it alone. They need expertise and training. But the shepherds are convinced that they have to work in close relationship and collaboration with all the actors sharing the same territories—e.g., ecotourism, environmental protection, etc. Also, they are aware of the creation of professional shepherd schools in France which have induced a renewal of transhumant pastoralism. Despite the fact that they attach the utmost importance to learning from generation to generation within the family, they recognize the importance of supplementing it with scientific training for their young people, which could be done in professional schools that still have to be created. The shepherds are convinced that establishing professional training with a French-Albanian diploma will encourage youth to stay in the Albanian Alps and choose to become shepherds.

Recently in Kelmend, the shepherds have created the House of Shepherds and Transhumance (*shtepia e bariut dhe bjeshkimit*) linked to La Maison de la Transhumance in Salon de Provence in the south of France. This will create opportunities for them to connect with the local, regional, and central powers in order to work in close relationship with the civil society for an evaluation of the situation. Also, the perspective to join France, Spain, and some other countries asking UNESCO protection of the transhumance as a nonmaterial cultural heritage is presently giving new perspectives and a new visibility to this phenomenon and our region. This institution not only brings together transhumant shepherds from the Albanian Alps, but it also hopes to have regional repercussions in the Balkans in the medium term. Its long-term vocation is to become an interpretation center for pastoral cultures and an observatory of transhumance throughout the region.

Kelmend, a cradle of transhumant pastoralism, is presently convinced of the urgent necessity to associate with interdisciplinary researchers who will help them by giving scientific methods for enhancing development projects. The shepherds from Kelmend therefore began to mobilize to University of Shkodër and the agriculture university in Tirana. As an anthropologist, I joined them, offering my expertise, being convinced about the necessity of multidisciplinary research. But in the future we would like these research teams to be linked with international universities with high-level scientific standards within European projects.

But let us not forget our youth. Unfortunately, school programs are not giving any room for this pastoral culture. The children of the shepherds are often discriminated against; many of them do not have real access to school. In view of the school programs which are out of step with what the children of the shepherds experience in their everyday life, new forms of formal and nonformal education remain to be found. No sustainable development can really happen without it. In the past few years, we had tried a few nonformal school initiatives during summer holidays around nature. Now the children of the shepherds of certain pastures are beginning to get involved in writing a book on their pastoral life. The young people, who until now organized festive moments of dances and sports games between pastures, are preparing to invite the young people of the city next summer for a moment of conviviality and sharing, a pasture festival. And we all hope together that in the future they will be authorized to create cultural places seeking to bear witness to this pastoral ecoculture (for example, through a pastoral ecomuseum). Therefore, young shepherds are presently taking plenty of photos of their pastoral life.

Conclusion

I would like to say that in so many countries transhumance dynamics has ceased to exist. In some countries, such as France, transhumance pastoralism is experiencing a resurgence (Msika, Lebaudy, and Caraguel 2015). In some countries, transhumant pastoralism has never disappeared, in spite of all the historical wounds that have happened. We must learn from each other, meeting and exchanging information between the different countries. But please, do not forget that wherever we are, transhumance pastoralism is a global phenomenon, "un fait social et culturel total," according to Marcel Mauss, which includes so many facts which are all in interaction (interconnection). None of them makes sense without the others. We are presently thinking about what should be heritagization of transhumance, as nonmaterial heritage within UNESCO. We must define all the relevant factors and integrate them in this globalism phenomenon, in order to be protected. The shepherds from Kelmend have convinced me that we scientists, whether anthropologists from the agriculture field or elsewhere, must work together defining what is the transhumance dynamic and conduct research to help these pastoral populations. A documentary film has been made in Kelmend and the filmmaker called the shepherds, "the heroes of the High pastures." I would like to add that we need to be careful about folklorization of the transhumance, which could negate what is, in reality, a sacred experience. And this is what the forgotten shepherds from Albania with their herds, are reminding all of us. "Mire se vini në bjeshkët e Kelmendit": welcome to the high pastures of Kelmend.

Acknowledgments

The author would like to thank Mhairi Forbes for the revision of the English text.

Martine Wolff is a pastoral anthropologist, living and working with the transhumant shepherds from Northern Albania, and a PhD candidate on traditional transhumant pastoralism in Northern Albania and its heritagization in a comparative dimension with other countries from Western Balkans. Collaborating with professionals and researchers in the fields of agriculture, ecology/biology, and human geography, she undertakes research that also supports the shepherds standing as a civil society and orienting themselves towards a sustainable development in respect with their own traditional autochthonous culture. She has supported the roadmap towards the creation of the House of Shepherd and Transhumance, a center for interpretation of the transhumant pastoral culture and an open space for dialogue and confrontation, oriented towards ecotourism and protection of nature.

References

Bardhoshi, Nebi Lelaj Olsi. 2018. *Etnografi në diktaturë dija shteti dhe holokausti ynë*. Tiranë: Akademia e shkencave e Shqipërisë, akademia e studimeve albanologjike.
Breda, Charlotte, Marie Derridder, and Pierre-Joseph Laurent. 2013. *La modernité insécurisée. Anthropologie des conséquences de la mondialisation*. Louvain-la-Neuve: L'Harmattan.
Brisebarre, Anne-Marie. 2007. *Bergers et transhumances*. Romagnat: De Borée.
———. 2013. *Chemins de transhumance. Histoire de bêtes et de bergers du voyage*. Neuchâtel: Delachaux et Niestlé.
Brisebarre, Anne-Marie, Guillaume Lebaudy, and Pablo Vidal Gonzalez. 2018. *Où pâturer, le pastoralisme entre crises et adaptations*. Avignon: Cardere.
Charbonnier, Quentin, and Thomas Romagny. 2012. *Pastoralismes d'Europe: rendez-vous avec la modernité!* Die: Association française du pastoralisme.
Del'homme, Bernard. 2016. "Agriculture dans l'ouest des Balkans des potentialités réelles menacées par l'absence de politiques." CIHEAM watch letter 36. Retrieved 22 February 2022 from https://www.ciheam.org/uploads/attachments/301/021_Delhomme.pdf.
Descola, Philippe. 2005. *Par-delà nature et culture*. Paris: Gallimard.
———. 2018. *Les natures en question*. Paris: Odile Jacob.
Durham, Edith. (1909) 2013. *Shqipëria e Epërme*. Tiranë: ed. IDK.
———. 1928. *Disa dëshmi të fiseve ballkanike, dokumente ligjesh dhe të kulturës së Ballkanit*. Tirana.
Fabre, Patrick, Gilbert Molénat, and Jean-Claude Duclos. 2002. *Transhumance. Relique du passé ou pratique d'avenir? Etats des lieux d'un savoir-faire méditerranéen en devenir*. Salon de Provence: Cheminements.
Lacirignola, Cosino. 2016. *Crises et conflits en méditerranée, agriculture comme résilience*. Paris: L'Harmattan.

Mema, Denis, Sulltana Aliaj, and Alfred Matoshi. 2019. *Përtej Kufijve, raport hulumtues analitik mbi migracionin—rasti i Shqipërisë*. Tiranë: ed. GEER.

Msika, Bruno, Guillaume Lebaudy, and Bruno Caraguel. 2015. *L'alpage au pluriel, quel avenir pour l'alpage, ce lieu commun bien singulier?* Paris: Cardère.

Neziri, Zymer Ujkan. 2012–13. *Kelmendi në shekuj*. Simpozium shkencor. Shkodër—Prishtine: Komuna Malësi e Madhe.

Nori, Michele. 2016. "Shifting Transhumances: Migrations patterns in Mediterranean Pastoralism." *CIHEAM. Watch Letter 36—Crise et résilience en Méditerranée*. Retrieved 15 May 2020 from www.iamb.it/share/integra"les_lib/"les/WL36.pdf.

Palaj, At Bernardin. 2018. *Bota e maleve shqiptare, Dokumente historike, zakone, doke dhe tradita* biblioteka françeskane. Shkodër: AT Gjergj Fishta Kolana e shkrimtarëve françeskanë.

Progni, Kolë. 2000. *Malësia e Kelmendit* monografi Shkoder: Camaj Pipa.

Rama, Luan. 2020. *Bujtës të largët. Mbresa të udhëtarëve francez gjatë shekullit XIX në Shqipëri*. Tiranë: Ed Klean.

Tirta, Mark. 2016. *Etnokulture shqiptare ligji dokesore*. Tiranë: Akademia e shkencave e shqipërisë Mirgeeralb.

CHAPTER 6

Revisiting Transhumance from Stilfs, South Tyrol, Italy
The Everyday Diverse Economy of a Forgotten Alternative Food Network

Annalisa Colombino and Jeffrey John Powers

Introduction

This chapter discusses the pastoral practice of transhumance through the lens of the idea of an alternative food network.[1] The literature on alternative food networks in geography and agro-food studies is abundant. However, it has neglected to explore how transhumance is a lively agricultural practice that engenders and partakes in food networks that may be radical in their alternativeness. The chapter starts with introducing how alternative food networks emerged in the last thirty-five years or so as important foodways through which food is produced and consumed. It moves on to discuss transhumance as an agricultural practice and to focus, specifically, on summer transhumance—a form of extensive agriculture that is present especially in the European Alps and that produces cheeses primarily for local consumption. The chapter then draws on ethnographic data to offer an account of the everyday practices and routines of humans and animals working on the summer pastures right above the village of Stilfs/Stelvio in South Tyrol, northeast Italy. It concludes with a short discussion that points to how alpine transhumance may be seen as engendering radically alternative food networks which enrich what Gibson-Graham have notably called the "diverse economy" (Gibson-Graham 1997, 2014; Gibson-Graham and Dombroski 2020; see also Healey 2020); namely, the economy is conceptualized according to interconnections between a multiplicity of practices of production, exchange, and consumption, which are more than just profit-oriented, and which mainstream scholarship has long tended to overlook by framing them as "alternative" and, thus, marginal.

The Rise of Alternative Food Networks

Alternative Food Networks (AFN) have been increasingly present in society as both a production and consumption option and have been well represented in the academic literature. Agreed-upon definitions are difficult to come by, but a point of departure in defining AFN is by pointing to what they are not: AFN are not industrialized food systems. Angela Tregear, for example, identifies AFN as

> forms of food provisioning with characteristics deemed to be different from, perhaps counteractive to, mainstream modes which dominate in developed countries . . . [and which are] heavily reliant on industrialized methods of food production and processing, global sources and means of supply, corporate modes of financing and governance, and [have] an imperative towards operational efficiency. (2011: 419)

AFN are different from conventional, industrialized food systems as they involve short supply chains, sustainable forms of agriculture, and focus on producing and circulating "quality" food; food whose qualities are deemed to be good not only in organoleptic terms but also in terms of how that food was produced: locally, in a socially fair and environmentally sustainable way, according to (invented) traditions, etc. AFN are frequently described as emerging as a response to the global, industrialized food system considered responsible for detaching consumers from producers (Venn et al. 2006: 248) and, subsequently, for fostering an obliviousness about the ingredients used to process food and their geographical provenance. It is in fact frequently argued that people are increasingly removed from food production because farming has become a large-scale, specialized activity carried out by few big companies. AFN stem partly from a desire to resist or change this trend (Holloway et al. 2007). Their agenda is one which aims at reconnecting producers, consumers, and the origins of the food they respectively produce and eat (see Bruckner, Colombino, and Ermann 2019).

One of the ways in which this reconnection takes place is via "spatial proximity between farmers and consumers" (Jarosz 2008: 231), as AFN's distribution channels consist of direct sell, local farmers' markets, and small, local shops, for example. The narratives around AFN tend to emphasize, often in a nostalgic way, the role of proximity in assuring the quality of the food eaten in the past. Consumers, it is argued, once knew the producers and could observe how food was made. Trust in the quality of food was then a by-product of the local nature of food production (Renting, Marsden, and Banks 2003). With the advent of a more global, industrialized food system rising after World War II, food quality assurance changed and became institutionalized: governmental or semi-governmental agencies took the role of controlling the safety and quality of food. While consumers initially seemed to accept and rely on these formalized controls, today it appears that for the public this trust has been broken (Renting et al. 2003).

Mistrust in industrially produced and institutionally controlled food emerged as a response to the "food scares" (such as bovine spongiform encephalopathy, foot and mouth disease, the use of lethal doses of methanol in wine-making, etc.), which have been occurring since the late 1970s (see DuPuis and Goodman 2005; Goodman 2004; Knowles, Moody, and McEachern 2007). Consumers then started to buy food produced in other-than-industrial ways; namely, food produced and circulated within AFN as, for example, food produced locally by small farms, sometimes practicing organic agriculture that avoids the use of chemical fertilizers and pesticides. Mistrust in industrial food, combined with growing public concerns over additional issues such as animal welfare and sustainability have resulted in the establishment of AFN as an option for consumers' food provisioning (DuPuis and Goodman 2005: 360).

More generally, according to Venn et al. (2006), in order to be characterized as alternatives, food networks need to have four main characteristics. First, they must connect consumers, producers, and food in an economic space, which re-embeds food production and consumption. Second, their distribution channels and supply chains should be unconventional; that is, detached from industrial supply-and-demand distribution and corporately controlled food chains. Third, they should adopt principles of social embeddedness; namely, they should be linked with a specific geographical location, evolve around a sense of community and work through relationship of trust. Finally, AFN should focus on "quality" food and contribute to preserving local heritage.

AFN is then a broad descriptor and scholars apply it to a wide range of different foods and venues that range from community gardens and food cooperatives, allotment groups and a wide range of food self-provisioning practices whereby those who grow food also eat it (see Jehlička and Smith 2001; Smith and Jehlička 2013). Community supported agriculture, where the risks of production are shared by consumer-producer partnerships, and direct sales initiatives such as farmers' markets, farm gate sales, adoption/rental schemes (e.g., the project "Adopt-a-Sheep" in Abruzzo, Italy, see Cox et al. 2011), mobile food shops, box schemes, and producer cooperatives are comprised under the AFN umbrella term. AFN also include more profit-oriented venues such as specialist retailers, where sales are more direct than in conventional supermarkets, and which commercialize high value-added, specialty foods and which are often targeted by tourists or gastronomes (see Venn et al. 2006; see also Colombino 2018). What falls under the guise of AFN is thus extensive. Scholars in geography and in agro-food studies have abundantly explored AFN.[2] There is in fact a plethora of publications that, in the last thirty years or so, with very few exceptions, has primarily debated the production, circulation, and consumption of vegetables and fruit within AFN. Very few studies have explored animal-based products such as cheese, meat, and eggs in AFNs (exceptions include, for example, Baritaux et al. 2016; Bruckner et al. 2019; Colombino and Giaccaria 2013, 2016; Forney 2016; Miele 2011; Stassart and Whatmore 2003). Importantly for the aim of this chapter, transhumance has been rarely explored as an agricultural practice that specifically partakes in alternative food networks (but see Buller's [2008] discussion of the benefits of pasture-based systems in France, and Holloway and colleagues reference to transhumance in Italy in their works which analyze "internet-mediated food production and consumption" to suggest that the glocal "economy of care"—articulated in the Abruzzo's agritourism they analyze—may inspire sustainable farmland management strategies elsewhere; Holloway 2002; Cox et al. 2011). More generally, and importantly, transhumance has been neglected in recent geographical accounts of the Global North, perhaps because in the Anglo-American geographical debate, transhumance is thought to be an agricultural, pastoral practice that pertains to the past and to the Global South (cf. Urbanik 2012: 104–5). Yet, as this chapter and other contributions in this volume highlight, transhumance is a form of animal husbandry and extensive agriculture currently practiced in different parts of the world. Importantly, as we point out in the conclusion, it engenders food networks which may be radically alternative to conventional ones, and which contribute to nourishing the conceptualization and empirical study of the diverse economy.

Pastoralism, Transhumance, and *Almwirtschaft*

Transhumance is a form of pastoralism; namely, an agricultural practice in which domesticated animals play a primary role. Pastoralism "involves herds of ungulates, which, depending on location, can include cattle, yak, sheep, goats, horses, donkeys, reindeer, camels, llama and guanaco, as well as a number of non-ungulate species" (McGahey and Davies 2014: 1). Pastoralism is both a land-use strategy and a form of animal husbandry (Reitmaier et al. 2018), in which managed herd movements are key to producing animal-based products such as milk, cheese, meat, and also wool, for example. Estimates vary largely, but it is thought that pastoralism is practiced in over 75 percent of all nations on earth, by up to 500 million people and involving over a billion animals (McGahey and Davies 2014; see also Dong et al. 2011). It is an important form of subsistence, particularly in low-income countries where people rely on animal keeping for their family's income and nutrition. Although scholars tend to neglect the role of pastoralism in the Global North as most of contemporary studies focus primarily on the Global South (see Turner et al. 2016; Urbanik 2012), this agricultural practice plays an important role in more affluent countries where the "centrality of keeping livestock—and of meat or milk consumption—to traditional cultures and identities" is of great importance (Garnett et al. 2017: 10). This is particularly evident for the pastoral practice of transhumance, which has been recently rediscovered as an object of academic enquire and empirical investigation (Nori and Gemini 2011; Nori and de Marchi 2015; Hartel, Plieninger, and Varga 2015);[3] of theoretical reflection in the social sciences and humanities (Colombino and Palladino 2019; Palladino 2017; Philippopoulos-Mihalopoulos 2012); and also of public interest, as its recognition as UNESCO immaterial cultural heritage in 2019 suggests (see Bindi, Chapter 7, in this volume).

More specifically, transhumance refers to the regular seasonal movement of animals between two or more locations, where, in many instances, the distance between these places would be too great for daily return (see European Commission 2009). More precisely, transhumance can be seen as "a system of livestock farming which rests on the utilization of pastoral resources in complementary zones which, by themselves, can only support livestock for part of the year" (Clearly 1987: 107). These complementary zones are primarily the common lands, and/or rural areas which are left unproductive for parts of the year. These pastures, in which the animals graze, are productive for certain seasons only, or even time-periods within a season. The animals therefore move from pasture to pasture, sometimes covering great distances based on the availability of grass or forage, which differ accordingly to several variables, such as location, season, weather, altitude, and specific vegetation profiles, etc.

Transhumance is widespread throughout the world and it is practiced in different ways according to the history and geography of the regions and countries where this form of animal husbandry exists. Scholars of different academic traditions classify transhumance in various ways, which primarily depend on the time of the year in which it is practiced, on the distance covered by the herds, and on the direction that movement is accomplished by the animals. Transhumance may be thus characterized as long-distance, horizontal, winter transhumance, such as the one practiced in Romania (see Juler 2014) and also in France (see von Sturler 2013), or as short-distance, vertical, summer transhumance, as the one explored within this chapter and which is common in several European alpine regions (see Jeschke and Mandl 2012).

There is uncertainty over when exactly the practice of transhumance in Europe is thought to have started.[4] There is evidence that dates it back at least ten thousand years in some regions. Whereas in alpine regions of Europe the estimates put the date around 6000 BP (Zoller 1960; Bätzing 1996). It is hypothesized that areas closed to the European Alps were settled since 7000 BP. Looking more specifically at the Po Valley of Italy, for example, it is hypothesized that forage was not available in summer due to the long dry summers that are characteristic of this region. These conditions therefore necessitated the movement of domesticated animals to the higher adjoining alpine regions to take advantage of the mountain pastures. Although there is uncertainty in general regarding when vertical transhumance began, it is interesting to note that for some scholars the oldest European mummy, known as Ötzi (or Similaum Man), discovered at the border between Italy and Austria, in a location 40 km away from the area discussed later in this chapter, was possibly a high-altitude shepherd involved in transhumance (see Carroll 2000; Ruff et al. 2006). In the area of Vinschgau (Val Venosta), South Tyrol, where Ötzi was found, it seems that alpine grazing by domesticated animals began in the Middle Bronze Age (around 3300–3550 BP; see Festi, Putzer, and Oeggl 2014).

This form of transhumance common in the Alps is called vertical transhumance or alpine transhumance in English, which corresponds to the Italian *transumanza* and to the German *Almwirtschaft*.[5] The expression specifically refers to the seasonal movement of animals to mountain pastures at high elevations—generally at or above the tree line—for the summer and then back down to the valley during the other months of the year. It differs from horizontal transhumance primarily because the distance the livestock moves is generally shorter. Yet, importantly, the significance of the movement is one of altitude.

Vertical transhumance is one which is specifically geared towards mountain farming as it enables a very efficient use of forage resources. Farming in alpine regions was possible only with high levels of adaption to "seasonal changes of climate and vegetation [and] vertical transhumance evolved as one of these adaptations" (Sal, Herzog, and Austad 2004: 192). During winter, farmers had to keep their animals inside barns, where they would be fed with hay which needed to be produced in the summer. However, the combination of a shortened summer growing season and narrow mountain valleys meant that there were instances in which the hay or pasture resources from the valley were not enough. This resulted in the system of vertical transhumance where the farmers could take advantage of high-altitude mountain pastures during the summer leaving the valleys free to produce hay for the winter (Sal et al. 2004; Grasseni 2007; Laiolo et al. 2004; Zendri, Sturaro, and Ramanzin 2013).

Characteristic of vertical transhumance used to produce cheese is what can be seen as a "graduated" farm organization: animals are brought to different altitudinal pastures at different times of the year and, importantly, different animals have access to different altitudes. For example, lactating cows are given access to the best grass at a lower level to maximize milk production; non-lactating cows graze the pastures at a higher level;[6] and goats or sheep move to the highest level to graze the land, that is, to pastures with insufficient nutritional availability for cows and also which would be physically dangerous or impossible for cows to reach. This system not only uses pastures and grasses that are only available at certain times of the year and at certain altitudes but also, and importantly, frees up lower pastures for haymaking in the summer to feed the animals throughout the winter. This vertical movement or vertical ordering is a key component of alpine, summer transhumance.

The type of vertical transhumance practiced most commonly in the Alps today is a dairy vertical transhumance, where goats, sheep, and, more commonly, cows are brought from their valley homes to the mountain pastures in the summer months to take advantage of the grasslands to feed the animals and to produce local kinds of cheese and other dairy products.

The ways in which cheese and dairy products are manufactured is very similar to the way they were made in the past in the alpine European valleys. During the summer months, in fact, cheese was made in the *Alm* (*malga* in Italian); namely, a "chalet" with the facilities to make cheese and to host the workers during the season. The reason why cheese was made up in the mountains is because, as we explain below, it was not feasible for a single dairy farmer to make cheese. As Barbara Orland argues in her historical account of transhumance in Austria and Switzerland, to be able to manufacture "10 kilograms of cheese even in the best months—May to July—the milk of 10 to 15 cows was still necessary" (2004: 340). In the past, most valley meadows were common lands and everyone in the community had the right to graze their animals there. Because of this right, villages and communities in the Alps had limits on cattle or barriers to prevent that too many people owned large numbers of animals: "having 6 to 8 cows in the stall was already a sign of prosperity" (Orland 2004: 340). Single farmers could not make cheese by themselves due to a lack of resources, stemming from practical ones such as the equipment needed, the knowledge necessary, or the amount of milk necessary to make a wheel of cheese that would age for a long period of time. Therefore, to maximize resources and productivity, communities started to hire workers and put their animals together for the summer on the alpine pastures, which had areas with the necessary infrastructure to make cheese, i.e., the *Alm*. Importantly, the workers called in to make the cheese and care for the animals freed the owners of cattle from a labor standpoint so that they were able to make hay from the lower pastures for the winter.

In the past, but also nowadays, this type of farming practice was considered necessary to produce high-quality cheese and also to keep animals healthy. For example, it was claimed that the mountainous climate was generally much healthier for cattle and gave them considerably greater resistance to disease, and that the high percentage of ethereal oils contained in the vegetation eaten by the cattle improved their salivation and digestion (Orland 2004). Cheese and butter produced on the *Alm* is in fact considered unique. Because of the conditions of alpine living—such as the altitude, greatly increased walking distances compared to barn confinement, and in general much less or no supplementary or commercial feed—animals are slower to fatten than during the same length of time in the valley, and milk output decreases. As a result, at higher elevations the milk contains more fat (Mathieu 1992: 97). Additionally, alpine products were, and are, "considered to be tastier and healthier because of herbs found only there, containing high percentages of ethereal oils" (Orland 2004: 333). For example, butter and cheese produced from cows in alpine pastures can contain an orangish hue due to increased levels of beta carotene, which is the effect of the grasses, flowers, and herbs eaten by the animals in alpine meadows.

The practices of vertical transhumance and cheese making in the *Alm* continue today in the Alps: they exist as renovated cultural foundations for mountain communities and provide farming solutions and a valuable product in a challenging environment.

In the next section, we turn to discuss the Stilfser Alm, located in South Tyrol, above the village of Stilfs (Stelvio, in Italian), in the autonomous province of Bozen (Bolzano), Italy. After introducing the history of this specific Alm and offering a brief description of the transhumance of the animals to the pastures, we provide an account of the everyday practices and routines of humans and animals working together on the Alm to produce and distribute cheese.[7]

The Stilfser Alm

Stilfser Alm dates back to 1322 and is perhaps one of the oldest Alms in Vinschgau. The current Alm is actually the second location, the original building was located approximately four hundred meters up the moun-

Figure 6.1. The Stilfser Alm from the Upper Alm, 2019. © Annalisa Colombino

tain. It is thought that an avalanche destroyed the original building, yet the remnants of the original Alm in the form of stonewall ruins are still present. The current Alm was rebuilt in the sixteenth century at its current location.

The Upper Alm (Obere Alm), located approximately three hundred meters away, was built in the seventeenth century. In the past both Alms (Upper and Lower) had cows and made cheese. Today the Upper Alm has been turned into a restaurant and guesthouse and has no cows. It has overnight accommodations, but its main business is to serve meals to hikers and mountain bikers. When the Upper Alm did have cows, both Alms would milk cows owned by different farmers living in the community of Stilfs and make cheese. However, they would operate completely independently. The river Tramentan—originating from the various small streams and springs surrounding the Alm and running through the village of Stilfs—was used as a reference point for the community to determine whose cattle would be taken to which Alm. If the farmers lived on the left or south side of the Tramentan their cows went to the Upper Alm, if they lived on the right or north side then the cattle had to go the Lower Alm (i.e., the present, working Alm we discuss in this chapter). Both Alms received water in 1953. In 1972 the first milk machine was installed, before this all cows were hand milked. Up until the 1980s, the butter churn was not run on electricity but by water that was diverted through the house and to the butter churn which was located inside. Around 1990, a water turbine was constructed approximately three hundred meters downhill of the house and barn for electricity (before this a generator was used) on one of the streams that in the past was used to divide cows into the Upper and Lower Alm. Both the Upper and Lower Alm were on this closed system powered by the turbine until 2013 when they joined the community electricity grid. Operations of the water turbine were taken over by the community power company, the turbine is still in operation today supplying a percentage of the power for the wider Stilfs electrical grid. Due to a decrease in the total number of cows beginning in 1976, the Upper Alm no longer functioned as a cheese-making Alm and all animals were sent to the Lower Alm. In 2006, a section of the Alm barn was converted to house goats and make cheese from their milk.[8]

The management of the Alm is the responsibility of an Alm Meister (director of the Alm) who is appointed by the local community. Since 2014, the Alm Meister has been Ernst Pingerra, a lifelong resident of Stilfs who, as a child, spent summers herding the goats of the community in the surroundings of the Alm, and who has put cows on the Alm for over twenty years. As a productive unit, the Alm may be described as a form of cooperative. As in the past, the herd is not owned by a single farmer, but each farmer owns a few animals. The majority of cow owners have from one or two up to five cows grazing the pastures. Normally, they pay a certain amount of money for the summer per cow, whose milk is measured once a week and they are given the corresponding percentage of cheese that their individual cows produce. The same system is used for the goats.

Figure 6.2. Laura, the *Sennerin* of Stilfer Alm, making cheese in the summer, 2016. © Jeffrey John Powers

The workers on the Alm are not necessarily farm owners or have cows but are hired by the farmers from the village to take care of their animals. In summer 2016, there were two shepherds for the goats, one shepherd for the lactating cows and the horses and one for the "dry" cows, one cheese maker, and two dogs. The workers follow the directions of the Alm Meister. For example, when it comes to decide the areas where the milk cows graze, the shepherd follows unspoken constraints of general rules put in place by the Alm Meister on the basis of what has been done in the past.

The "actual" transhumance (i.e., the movement of the herd from down the valley to the Alm and then backwards) in Stilfs takes place in June and September. The movement of the animals towards the Alm in June is called, in the local dialect, *Auf-fohrm* (literally, going-up): the animals who are owned by farmers who live close by walk up with the owners to the Alm (around 50 percent of the entire herd); those who are located farther away are nowadays brought up to the pastures in trailers. The most important moment for the transhumance, in terms of heritage commemoration, is the celebration of the end of the season up in the pastures, in September. In Vinschgau, this event is called *O-fohrn*. This dialectical expression (*Almabtrieb*, in high German) means literally "drive down" and refers to returning the animals back to their original farms. In Stilfs, the *O-fohrn* involves the animals and the Alm's workers, who walk down with the herd and drop the individual animals off at their home farms. The way down towards the village follows a curving path, which passes each (or almost each) farm, where the owners wait for the herd to come past and separate the animals out from the group. For this occasion, the animals (the cows, in particular) can be adorned with flowers and large bells around their necks. Furthermore, it would be customary for the cow who has produced the highest quantity of milk during the season to be adorned with the largest of the bells and with more flowers, compared to the others. The *O-fohrn* ends once the village of Stilfs is reached by the group, where the farmers—those who live farther from the Alm and the path followed by the herders—pick up their animals and put them in trailers to drive them back to their farms. Around fifty per cent of the Alms in the Vinschgau valley would celebrate their *Almfest* (literally, the celebration of the Alm) in the village at the end of the cattle drive. However, the Stilfser Alm organizes its specific celebration on August the sixth, the day of St. Rochus, its patron saint.[9]

The infrastructure of the Alm consists of a chalet with accommodations, including a kitchen, for the workers, as well as the cheese-making facilities. These include the cheese-making room (*Sennerei* in German) with the cheese pot, a milk storage room (*Milchkeller*) with the cooling tanks and a butter churn, and a cellar for the cheese (*Käsekeller*). There is also a barn which can accommodate around sixty cows and which has an additional section for the goats, including a milk stand and indoor and outdoor areas for them to rest. Next to the goats' area, there is an outdoor fenced space for the pigs and also an indoor structure for them to sleep, which holds the two large whey tanks. The reason why pigs are on the Alm is to feed them with the whey, the by-product of cheese making, which cannot be disposed elsewhere. Approximately three hundred meters away from this Alm, there is what is referred to as the Upper Alm (*Obere Alm*). This used to be a cheese-making barn and housing, but it has been recently turned into a restaurant and guesthouse. It has overnight accommodation, but the main business is to prepare and serve meals, specifically lunch, to hikers and mountain bikers.

In general, the season runs from early to mid-June until mid-September, the specific date depends on the weather and corresponding amount of grass available. In summer 2016, the animals were brought to the Alm on 7 June and went back on 9 September. The team of four people, of which Jeffrey was part, had to look after sixty-two milk cows, sixty-eight goats, twenty pigs, and twelve horses.[10] As mentioned before, the animals themselves came from individual farmers.

A normal day at the Alm begins at four-thirty in the morning: two team members collect the cows from the night pasture and "hang" them or tie them in their individual spots in the barn. At the same time, the other two members of the team begin to work in the *Sennerei* as they take the previous days cheeses out of the molds, clean these molds, and prepare the large cooling tanks for that day's milk. The herding and the preparation of the cheese-making room take until about five or six in the morning. Then, one team member also has to move the goats from their night pasture, which is located in a small area next to the barn and, differently from the cows, rarely changes. The remaining team members would assemble the milking machines and begin milking.

The milking of cows and goats begins at approximately six in the morning, with two people milking the cows and the other two milking goats.[11] The milking takes around two and a half hours for the cows and three hours for the goats, to one and a half hours for the cows and one hour for the goats. The length of milking changes because the animals give less and less milk as the summer progresses and therefore it takes less time to milk. The goats are milked in a wooden milking stand constructed by one of the farmers, which allows the team members to milk eight goats at one time using three milking machines. The cows do not have a milk stand but are tied in a specific, individual spot, with thirty-one cows on either side of the barn. Before the milking, the cows are fed in their places: the milkers have a wheelbarrow of "noodles" (the local term for "dry feed"; *Nudeln* in German or *Nudl* in the local dialect) and go down the barn, giving each cow a specific amount. Two people milk the cows using two milk machines each. At approximately eight in the morning the milking for both animals is finished and the team members split into separate tasks: two shepherds move the cows and the goats on the pastures, and the *Sennerin* (female cheese-maker) starts to prepare the cheese-making process.

Depending on where the cows are supposed to be that day, the shepherd pushes them with the help of a dog to a specific area, which can be located from very close to the barn up to one hour of walking uphill (the longest push being a move of approximately two kilometers and a gain of about three-hundred meters in altitude). The goats are also led out of the barn by a shepherd. In contrast to the cows who are pushed, the goats follow the shepherd who leads the herd. The goats have no set pastures or areas they are supposed to be. The only rule is that they need to be moved above the pastures used by the cows. Normally, by eight-thirty or nine the cows and goats are moved in the area they are supposed to graze, and the shepherds move back to the *Sennerei*. One helps the cheese-maker and the other shepherd helps the fourth team member with brushing and cleaning the previously made cheese in the cellar. Including a short break to eat breakfast, the cheese making (Figure 6.2) and brushing processes take approximately three to four hours; that is, up to noon or one o'clock. At this point, any "extra" or non-routine tasks can be completed (additional cleaning in the *Sennerei* and milk cellar; fencing, and general maintenance and repair). This is also the time for regular work tasks such as the cleaning of cheese boards; the checking and giving salt to the horses; calling the veterinarian and farmers; additional cleaning of the barn and paperwork (the recording of the daily cheese-making process, amounts of milk and cheese and butter made, accounting of the cheese given to farmers or sold, etc.). These additional tasks vary but on average the entire team finishes them by 2 p.m. to take a break. At approximately 3:30 p.m., depending on where the cows and goats are located, one or all four team members collect the animals. The animals are herded back to the barn to be milked and fed, and then put back in the night-pastures. The cows are put in one of three night-pastures, which are rotated according to grass availability and additional factors, such as the state of the fencing, for example. Finally, basic tasks need to be accomplished in the *Sennerei*, milk cellar, and cheese cellar: the cheeses have to be flipped, the weights taken off the cheese molds, cooling turned on, and the milk cellar cleaned. The working day finishes at approximately 8 p.m.

Over the course of the 2016 summer, three different types of cow-milk cheese and only one kind of goat cheese were produced. The cows' cheese is referred to by local names in combination with the type of bacteria used. The first type of cow cheese is referred to as "Alm DIP," DIP being an acronym for "direct in production" and refers to specific freeze-dried bacteria culture used to make the cheese. The second, *Säuerwecker*, can be translated as "awakening of the acidity," and refers to the bacteria culture used to start the cheese-making process. Finally, the third type of cheese is called *Schweizer Kultur*, which again refers to the specific the "Swiss [bacteria] culture" used.

The choice of bacteria, and subsequently the kinds of cheese produced, depends on what was produced in the past and on the knowledge and experience of the cheese maker: the owners of the animals expect the cheese to be ready soon and that the cheese maker will start with producing a cheese that ages quickly. The *Sennerin* then has some flexibility in deciding which specific cheeses should be produced, for how long, and at which stage of the season.

For the 2016 summer, the *Sennerin* decided to begin making "direct-in-production" cheese in the first two days of the season. This was a result of concern by the cheese maker of the risk of unwanted bacteria both on the cows and in their milk. The cows were coming from many different barns operating with different standards. The *Sennerin* decided to use first the strongest bacteria culture—that is, direct-in-production—in order to combat any unwanted bacteria in the milk or on the animal in the first two days. The beginning period of cheese production is the most at-risk regarding unwanted bacteria, which can affect or possibly ruin the cheese. After this period, the team on the Alm has more control over where exactly the cows are grazing and this corresponds to a having more confidence in the milk being produced by the animals. On the third day, the team began making *Säurewecker* (produced until 17 July), because the farmers want cheese as soon as possible and *Säurewecker* is the quickest to age (it takes approximately one month in the cellar before it is ready to eat).

Beginning in August, the cheese of the Swiss type was made. This is the longest aging cheese, needing a minimum of three months and having an ability to age for much longer. This gives the farmers the best chance at having cheese for the majority of the year. As an experiment, the *Sennerin* decided to make, for two days in late August, a combination of the Swiss culture bacteria with a special culture of bacteria from milk mold; a little less than 150 kilograms of this cheese were produced. This cheese differed in that it required no cleaning or brushing, and it also had a more white and dry rind.[12]

For the entire summer, direct-in-production bacteria was used for producing goat cheese. In the past seasons, there had been problems with the goat cheese and the *Sennerin* surmised that the problem lied in unwanted bacteria in the milk. As mentioned before, direct start is the strongest bacteria, and it was used to correct or combat this problem. Finally, there were small quantities of soft cheese (less than a hundred kilograms of mascarpone, ricotta, and Topfen) that were produced and eaten by the team on the Alm and distributed to the farmers on a very small scale.

As for the distribution of the cheese, some is sold on the Alm itself to hikers and mountain bikers who are vacationing in the area, and to community members who walk up to the Alm from the village. Farmers also sell the cheese privately amongst friends and personal connections. Several farmers also run guest houses on their *Hofs* (farms), where they provide sleeping accommodation and also food: the cheese is then served or sold to the guests. In 2016, one farmer made an agreement with a local Spar grocery store to sell his cheese over the summer, and one member of the Alm team sold small amounts of cheese through personal connections in Graz and Vienna. Finally, a significant proportion of the total amount of cheese made is not commercially sold, but it is either eaten by the farmers themselves or given to friends and family.

Conclusion

In their extensive work, Gibson-Graham (1997, 2014, Gibson-Graham and Dombroski 2020) formulate an understanding of the economy as "diverse"; that is, as performed through manyfold and varied practices of production, exchange, and consumption which, in turn, enact an economic space of difference and experimentation, which is no longer dominated by purely profit-oriented capitalist enterprises. In so doing, they propose a profound critique of the common conceptualization of the economy as comprising two main, oppositional and distinct forces; namely, conventional, mainstream capitalism versus alternative, and presumably marginal, economic practices. In Gibson-Graham's theorization, the economy emerges as a plethora of different practices of production, circulation, exchange, and consumption, which are not only those of Capitalism with a capital C—a monolithic, nearly transcendental, economic force that dominates and exploits subordinate human and nonhuman subjects and matters. Such practices of economic diversity encompass human and nonhuman unpaid labor in households and farms (Barron and Hess 2020) and practices of food self-provisioning, for example (Grasseni 2020; Jehlička 2021 for a series of diverse economies' studies see Gibson-Graham and Dombroski's 2020 edited volume). Thinking of the economy as diverse enables the possibility of recognizing how, in different places, there are modes of production, circulation, and consumption, which are enacted to achieve aims that are not uniquely concerned with the extraction of monetary profit and capital accumulation. In the diverse economy, other relations, regimes, and registers of value are at stake, which, in turn, may contribute to fostering social and environmental benefits, unlike capitalist practices. These other-than-conventional economic formations do not exist nor emerge as distinct from conventional capitalist modes of production: often invisible, diverse economies' practices intertwine and coexist with the workings of conventional and more visible capitalist ventures. By adopting ethnographic research tools able to bring to light what theory may fail to grasp (Gibson-Graham 2014), the diverse economy approach is able to include, rather than marginalize and exclude, modes of existence and related economic practices, which the myth of alternative economies as marginal (and therefore not worthwhile of falling under the lens of mainstream academic enquiry and policymaking) tends to obscure. The burgeoning literature on alternative, diverse economies has been demonstrating the existence of a wide range of practices that are widespread, rather than marginal and unimportant, and which support a variety of real (rather than idealistic) livelihoods around the globe.

Our, admittedly partial, discussion of Stilfser Alm's microcosm of alpine transhumance has attempted to offer a glimpse into some of the practices that contribute to engendering the diverse economy nourished by a food network which emerges as radically alternative, when compared to conventional food production and exchange. We have discussed how daily work on the Stilfser Alm is organized and keeps summer transhumance alive and meaningful in the Vinschgau Valley. Our account suggests that alpine, vertical transhumance generates an alternative food network, which partakes in an economy which is diverse in the sense that its effects and aims are more than just profit driven.

Rather obviously, the Stilfser Alm is an extremely small and local production system. The animals, their milk, and the cheese produced come from Stilfs, a small village in South Tyrol. The cheese's identity and value are profoundly rooted in this locality: it is a very specific kind of cheese produced at a specific time of the year, made in a particular way that builds on local heritage, practices, and knowledges. The number of animals and workers involved, and the amount of cheese produced are modest. Over the course of the summer of 2016, sixty-two cows and sixty-eight goats produced respectively just over five thousand kilograms and around nine hundred kilograms of cheese. The Alm may be described as a very small, local cooperative, where individual animal owners and the community of Stilfs make decisions to determine the direction the Alm needs to take. Importantly, the direction of the Alm is constrained in a way that no industrialized food system would be: that is to say, by the local, cultural practices and desires of this specific community. Members of the local community visit regularly the Alm and have done so for generations. They personally know the farmers who own the animals on the Alm, as they are either their neighbors, close friends, or acquaintances. From the village, they can literally see the Alm's pastures and where the animals are grazing on a specific day. The Stilfser Alm thus can be seen as engendering a food network and a diverse economic formation that is socially embedded. A sense of community and of trust is omnipresent. The animals are not owned by the people running the Alm but are lent for the summer. This occurs in a rather high-risk environment: while making cheese is more profitable on the pastures in summer, it carries a higher risk compared to selling milk to the companies (which is what the famers would do for the other nine months of the year). Miscalculations by the cheese maker regarding timing, temperature, amounts of bacteria used to make cheese, contamination of any equipment, and the continued caring for the cheese over the course of months could result in failure of the cheese batch from that day and even for other days, thus resulting in a complete loss for the farmers. Furthermore, the environment itself is a more dangerous one for the animals: they are grazing at a high elevation over an extremely large area, which contains significant hazards such as steep slopes, rocks, and ravines. In addition, the animals who belong to different farms have diverse social orders, hierarchies and, therefore, there is instability in the power relations that needs to be established once they become part of a new single group up in the Alm. The danger that animals would be injured is a very real one, because of the morphology of the land but also because of confrontations between the animals in the new group. All of these risk factors necessitate a high level of trust between the farmers, the local community, and the Alm's team. We thus suggest that vertical, alpine transhumance partakes in alternative food networks and diverse economies that contribute to make agriculture socially and environmentally sustainable, compared to the unsustainability of current industrialized food systems. As an animal husbandry method, opposed to intensive and industrial animal production (see Porcher 2017), and as a type of extensive agriculture, alpine transhumance is a practice that does benefit the environment. In sharp contrast with conventional modes of rendering the land productive, alpine transhumance may be seen not only as an agricultural and economic practice that, in summer, minimizes costs as it maximizes the productivity of natural resources (which without transhumant herds would not be productive) to produce "quality" food. Also, and importantly, alpine transhumance acts as a tool for preserving biodiversity in the mountains and for containing the depopulation of these areas. In fact, farm animals in transhumance contribute to landscape and biodiversity preservation by grazing the land, which otherwise would be neglected. Moving across the landscape and dispersing seeds through their feces, farm animals contribute to maintaining biological and genetic diversity. In becoming food for predators, these domesticated animals prevent the disappearance of those wild animals who could not survive without the presence of transhumant livestock. Transhumance thus represents a mode of living together with animals (cf. Porcher 2017) and the environment which, in supporting lively and diverse economies of food specialties, keeps farmers on and across the land, which otherwise would be abandoned.

Transhumance, in its different historical and geographical manifestations, has a perhaps ironic relations with capitalist, economic expansion. From being an activity that historically has contributed to the spread of capitalism, transhumance has been more recently pushed to its margins (Chang and Koster 1994). Too often understood by governments as a backward, agricultural practice in sharp contrasts with the imperatives of modernization, transhumance has been discouraged in diverse parts of the world including Europe (see e.g., Juler 2014 on Romania), so much so that, along with other forms of pastoralism, it has sometimes been framed as an agricultural practice that does not any longer exist in the Global North (see, e.g., in geography Urbanik 2012). In a similar manner to food self-provisioning practices and other alternative food networks, notably explored by Jehlička and colleagues (e.g., Jehlička 2021; Jehlička et al. 2020; Fendrychová and Jehlička 2018), and as this book and the revival of interest in agricultural pastoralism in and beyond academia seem now to show, transhumance is widespread, rather than marginal. In a world in which the economies of conventional food production appear to be dominant, transhumance emerges as tenaciously resilient.

Acknowledgments

We would like to thank Laura Hanni, Sennerin of the Stilfser Alm, from 2015 to 2017, for her patience and knowledge of cheese making. We are grateful to Petr Jehlička, Paolo Palladino, and the anonymous reviewers for reading and commenting on the early draft of this chapter. Their comments helped us to revise the chapter and, hopefully, clarify our account of Stilfser Alm.

Annalisa Colombino is a human geographer currently working at the Department of Economics, Ca' Foscari, University of Venice, Italy. She obtained her PhD at the Open University, UK. Her current research interests lie at the intersection of critical food studies, human-animal relations, biopolitics, and the diverse economy.

Jeffrey John Powers completed his Master's degree in Sustainable Development from the University of Graz in 2019. For three summers, he worked with his wife Laura Hanni, the Sennerin at Stilfers Alm, as the shepherd and assistant cheese maker. He currently works for the United States Forest Service in Utah.

Notes

1. Far from being merely systems of food distribution, alternative food network (AFN) broadly refers to networks which comprise the production, circulation, and consumption of (usually local) food. AFN is thus an expression used to think about how these three empirical and analytical spheres should not be seen as distinct or linear, but as closely interrelated.
2. Despite a general positive view of AFN, geographers and agro-food scholars have explored AFN from a critical standpoint. Recently, critiques have emerged to point to how it is no longer possible to clearly separate the products produced and circulated as part of AFN from more conventional food networks such as those involved in supermarkets chains. This is because, nowadays, supermarkets also sell products associated with AFN (e.g., local and regional products) and put a premium price on them (See Goodman and Goodman 2009; Tregear 2011). The difficulty in clearly distinguishing AFN from conventional food networks is enhanced by the fact that multinational industrialized food companies have been buying smaller organic and "alternative" producers, for example, the takeover of Horizon and Cascadian Farms by the food giant Dean. From a socioeconomic perspective, some authors point to how AFN may be elitist and observe that some initiatives may maintain—rather than overturn—pre-existing inequalities between participants (Allen et al. 2003; Goodman 2004; DuPuis and Goodman 2005). Others point to how AFN may exhibit a conservative insularity and defensiveness rather than being open to progressive change (Winter 2003). Hinrichs (2000), for example, points out that local or short food supply chains do not necessarily translate into social justice. Saying that a food is local does not mean necessarily that there are not substantial exploitative relations at play. Local food systems may employ industrialized production techniques, exploit farm workers, and still produce organic food. Local food systems cannot be assumed to be uniformly "good" or progressive, based solely on a geographical basis (DuPuis and Goodman 2005; Winter 2003). Scholars have in fact pointed out that research on AFNs has frequently focused on consumers and has often ignored producers, thus neglecting to explore the social conditions of farmers and, especially, those of farm workers (Goodman 2004). There has also been

criticism on AFN regarding environmental impacts: some of the metrics used to determine how environmental or sustainable a system or product is, such as food miles, are not the best or an even accurate tools to judge appropriate environmental goals (Edwards-Jones et al. 2008; Oglethorpe 2009).
3. See also the extensive work being undertaken within the project, recently funded by the European Union, *Pastres: Pastoralism, Uncertainity and Resliance* led by Ian Scoones, Michele Nori, and Jeremy Lind (https://pastres.org).
4. For an historical and geographical account of transhumance in the Mediterranean see Braudel (1995). On diverse types of transhumance's entanglements with capitalism see, e.g., Shields (1992) and Chang and Koster (1994).
5. These terms in different languages are important to note for two main reasons: first, the case study explored later in the chapter is located in South Tyrol/Alto Adige, in Italy, where German is primarily spoke by its inhabitants; second, German-speaking academics would not classify *Almwirtschaft* as a form of transhumance (*Transhumanz* or *Wanderweidewirtschaft*) but as the "economy of the alpine hut" (its literal translation), which points to the specific economic activity related to pasturing and cheese production. See "Transhumanz" in Wikipedia, where *Almwirtschaft* is described as a mistaken form of transhumance: https://de.wikipedia.org/wiki/Transhumanz, retrieved 10 June 2020. See also https://de.wikipedia.org/wiki/Alm_(Bergweide), retrieved 11 July 2020.
6. Non-lactating cows are brought to the pastures to save money for their feeding.
7. Recent scholarship in human-animal studies and cognate fields have demonstrated how farm and other animals are also individuals who do work. See, notably, Jocelyne Porcher's extensive work (2014, 2017); see also Barua (2019), Coulter (2016), Lainé (2020). In this chapter, however, we maintain a humanist perspective on transhumance. A more-than-human/posthuman perspective on the economies of transhumance is at the core of Colombino's ongoing research (see Colombino and Palladino 2019).
8. The data on the history of the area of the two Alms was gleaned from an interview with Ernst Pingerra, the current Alm Meister (the director of the Alm). The data used for the account of the Alm are the outcome of participant observations conducted in the summer of 2016 primarily by Jeffrey as he was working as a cow shepherd and as an assistant to the cheese maker on the Alm. It must also be noted that Jeffrey worked on this Alm for four seasons. Annalisa conducted some ethnographic incursions in the same Alm and, from the vantage point of the Upper Alm, she observed the pastures and the Alm's human and animal workers for nearly two weeks in 2016, from 12 to 25 August. She conducted two semi-structured interviews with the Alm's workers and had several informal conversations with the shepherd responsible for the dry cows, with the Upper Alm's manager, and some of the valley's farmers who visited the Alm.
9. During the summer some individual animals go back to their farms if they are close to giving birth or if they are injured. More commonly, some cows leave the Alm when they are *galtvieh* (a young female cow who is pregnant but has

never delivered a calf before) and do not produce enough milk. In this case, they are moved higher in the pastures and are taken care of by another shepherd. When these cows are close to giving birth (about two weeks before) then they are brought back to their home farms by trailer.
10. The horses required little work or interaction and were located approximately one to two kilometers away from the Alm.
11. Approximately halfway through the season the schedule was changed to allow only one person to milk the goats and the other person would beginning making the butter.
12. For the 2016 season, direct start was also made in the last two weeks in July to add to the total direct start produced and give more of a variety to the farmers.

References

Allen, Patricia, Margaret FitzSimmons, Michael Goodman, and Keith Warner. 2003. "Shifting Plates in the Agrifood Landscape: The Tectonics of Alternative Agrifood Initiatives in California." *Journal of Rural Studies* 19(1): 61–75.

Baritaux, Virginie, Marie Houdart, Jean Pierre Boutonnet, Carole Chazoule, Christian Corniaux, Philippe Fleury, and Jean François Tourrand. 2016. "Ecological Embeddedness in Animal Food Systems (Re-) Localisation: A Comparative Analysis of Initiatives in France, Morocco and Senegal." *Journal of Rural Studies* 43: 13–26.

Barron, Elizabeth, and Hess, Jaqueline. 2020. "Non-Human Labour: The Work of Earth Others." In *The Handbook of Diverse Economies*, ed. J. K. Gibson-Graham and Kelly Dombroski, 163–69. Cheltenham: Edward Elgar Publishing. https://doi.org/10.4337/9781788119962.00026.

Barua, Maan. 2019. "Animating Capital: Work, Commodities, Circulation." *Progress in Human Geography* 43(4): 650–69.

Bätzing, Werner. 1996. *Landwirtschaft im Alpenraum—unverzichtbar, aber aber zukunftslos?* Berlin: Blackwell Wissenschafts-Verlag.

Braudel, Fernand. 1995. *The Mediterranean and the Mediterranean World in the Age of Philip II: Volume I.* Berkeley: University of California Press.

Bruckner, Heide K., Annalisa Colombino, and Ulrich Ermann. 2019. "Naturecultures and the Affective (Dis)Entanglements of Happy Meat." *Agriculture and Human Values* 36: 35–47.

Buller, Henry J. 2008. "Adding Value in Pasture Based Systems: Reflections on Britain and France." *Proceedings of the British Society of Animal Science*: 11–14.

Carroll, Rory. 2000. "Iceman is Defrosted for Gene Tests: New Techniques May Link Copper Age Shepherd to Present-Day Relatives." *The Guardian*, 26 September. Retrieved 25 July 2020 from https://www.theguardian.com/world/2000/sep/26/rorycarroll.

Chang, C., and Koster, H. A., eds. 1994. *Pastoralists at the Periphery: Herders in a Capitalist World.* Tucson: University of Arizona Press.

Cleary, Mark C. 1987. "Contemporary Transhumance in Languedoc and Provence." *Geografiska Annaler: Series B, Human Geography.* 69(2): 107–13.

Colombino, Annalisa. 2018. "Becoming Eataly: The Magic of the Mall, the Magic of the Brand." In *Branding the Nation, the Place, the Product*, ed. Ulrich Ermann and Klaus Hermankik, 67–90. London: Routledge.

Colombino, Annalisa, and Paolo Giaccaria. 2013. "Alternative Food Networks tra il locale e il globale. Il caso del Presidio della Razza Bovina Piemontese." *Rivista Geografica Italiana* 122: 225–40.

———. 2016. "Dead Liveness/Living Deadness: Thresholds of Non-Human Life and Death in Biocapitalism." *Environment and Planning D: Society and Space* 34(6): 1044–62.

Colombino, Annalisa, and Paolo Palladino. 2019. "In the Blink of an Eye: Human and Non-Human Animals, Movement and Bio-Political Existence." *Angelaki: Journal of the Theoretical Humanities* 25(5): 168–83.

Coulter, Kendra. 2016. *Animals, Work, and the Promise of Interspecies Solidarity*. New York: Palgrave Macmillan.

Cox, Rosie, Lewis Holloway, Laura Venn, Moya Kneafsey, and Elizabeth Dowler. 2011. "Adopting a Sheep in Abruzzo: Agritourism and the Preservation of Transhumance Farming in Central Italy." In *Tourism and Agriculture: New Geographies of Consumption, Production and Rural Restructuring*, ed. Rebecca Maria Torres and Janet Momsen, 151–60. London: Routledge.

Dong, Shikui, Lu Wen, Liu Shiliang, Zhang Xianfen, James Lassoie, Shaoliang Yi, Xiaoyan Li, Jimpeng Li, and Yuanyuan Li. 2011. "Vulnerability of Worldwide Pastoralism to Global Changes and Interdisciplinary Strategies for Sustainable Pastoralism." *Ecology and Society* 16(2): unpaginated.

DuPuis, E. Melanie, and David Goodman. 2005. "Should We Go 'Home' to Eat?: Toward a Reflexive Politics of Localism." *Journal of Rural Studies* 21(3): 359–71.

Edwards-Jones, Gareth, Llorenç Milà i Canals, Natalia Hounsome, Monica Truninger, Georgia Koerber, Barry Hounsome, and David. L. Jones. 2008. "Testing the Assertion That 'Local Food is Best': The Challenges of an Evidence-Based Approach." *Trends in Food Science and Technology* 19(5): 265–74.

European Commission. 2009. *New Insights Into Mountain Farming in the European Union*. Document: SEC(2009)1724. Retrieved 10 June 2020 from http://mountainlex.alpconv.org/images/documents/EU/CSWD_new_insights_mountain_farming.pdf.

Fendrychová, Lenka, and Petr Jehlička. 2018. "Revealing the Hidden Geography of Alternative Food Networks: The Travelling Concept of Farmers' Markets." *Geoforum* 95: 1–10.

Festi, Daniela, Andreas Putzer, and Klaus Oeggl. 2014. "Mid and Late Holocene Land-Use Changes in the Ötztal Alps, Territory of the Neolithic Iceman 'Ötzi.'" *Quaternary International* 353(1): 17–33.

Forney, Jérémie. 2016. "Enacting Swiss Cheese: About the Multiple Ontologies of Local Food." In *Biological Economies: Experimentation and the Politics of Agri-food Frontiers*, ed. Richard Le Heron, Hugh Campbell, Nick Lewis, and Michael Carolan, 67–81. London: Routledge.

Garnett, Tara, Cécile Godde, Adrian Muller, Elin Röös, Pete Smith, Imk de Boer, Erasmus zu Ermgassen, Mario Herrero, Corina van Middelaar, Christian Schader, and Hannah van Zanten. 2017. "Grazed and Confused? Ruminating on Cattle, Grazing Systems, Methane, Nitrous Oxide, the Soil Carbon Sequestration Question—And What It All Means for Greenhouse Gas Emissions." Food Climate Research Network, University of Oxford. Accessed on 10 June 2020 from https://library.wur.nl/WebQuery/wurpubs/fulltext/427016.

Gibson-Graham, J. K. 1997. "The end of capitalism (As We Knew It): A Feminist Critique of Political Economy." *Capital & Class* 21(2): 186–88.

———. 2014. "Rethinking the Economy with Thick Description and Weak Theory." *Current Anthropology* 55(S9): 147–53.

Gibson-Graham, J. K., and K. Dombroski, eds. 2020. *The Handbook of Diverse Economies*. Cheltenham: Edward Elgar Publishing.

Goodman, David. 2004. "Rural Europe Redux? Reflections on Alternative Agro-Food Networks and Paradigm Change." *Sociologia Ruralis* 44(1): 3–16.

Goodman, David, and Michael K. Goodman. 2009. "Alternative Food Networks." In *International Encyclopedia of Human Geography*, ed. Robert Kitchin and Nigel Thrift, 208–20. Oxford: Elsevier.

Grasseni, Cristina. 2007. "Managing Cows: An Ethnography of Breeding Practices and Uses of Reproductive Technology in Contemporary Dairy Farming in Lombardy (Italy)." *Studies in History and Philosophy of Science Part C: Studies in History and Philosophy of Biological and Biomedical Sciences* 38(2): 488–510.

———. 2020. "Direct Food Provisioning: Collective Food Procurement." *The Handbook of Diverse Economies*. Cheltenham: Edward Elgar Publishing. Digital.

Hartel, Tibor, Tobias Plieninger, and Anna Varga. 2015. "Wood Pastures in Europe." In *Wood-Pastures in Europe. Europe's Changing Woods and Forests: From Wildwood to Managed Landscapes*, ed. Keith Kirby, and Charles Watkins, 61–76. Nosworthy Way: CABI Press.

Healey, Stephen. 2020. "Alternative Economies." In Kobayashi, Audrey (ed.) *International Encyclopedia of Human Geography*, ed Audrey Kobayashi. Oxford, UK: Elsevier. https://doi.org/10.1016/B978-0-08-102295-5.10049-6.

Hinrichs, Clare C. 2000. "Embeddedness and Local Food Systems: Notes on Two Types of Direct Agricultural Market." *Journal of Rural Studies* 16(3): 295–303.

Holloway, Lewis. 2002. "Virtual Vegetables and Adopted Sheep: Ethical Relation, Authenticity and Internet: Mediated Food Production Technologies." *Area* 34(1): 70–81.

Holloway, Lewis, Moya Kneafsey, Laura Venn, Rosie Cox, Elizabeth Dowler, and Helena Tuomainen. 2007. "Possible Food Economies: A Methodological Framework for Exploring Food Production—Consumption Relationships." *Sociologia Ruralis* 47(1): 1–19.

Jarosz, Lucy. 2008. "The City in the Country: Growing Alternative Food Networks in Metropolitan Areas." *Journal of Rural Studies* 24(3): 231–44.

Jehlička, Petr. 2021. "Eastern Europe and the Geography of Knowledge Production: The Case of the Invisible Gardener." *Progress in Human Geography* 45(5): 1218–36.

Jehlička, Petr, Miķelis Grīviņš, Oane Visser, and Bálint Balázs. 2020. "Thinking Food Like an East European: A Critical Reflection on the Framing of Food Systems." *Journal of Rural Studies*. Retrieved 22 July 2020 from https://doi.org/10.1016/j.jrurstud.2020.04.015.

Jehlička, Petr, and Joe Smith. 2011. "An Unsustainable State: Contrasting Food Practices and State Policies in the Czech Republic." *Geoforum* 42(3): 362–72.

Jeschke, Hans Peter, and Peter Mandl. 2012. "Eine Zukunft fur die Landschaften Europas und die Europaische Landschaftskonvention." *Klagenfurter Geographische Schriften* 28: 185–325.

Juler, Cainerol. 2014. "După coada oilor: Long-Distance Transhumance and its Survival in Romania." *Pastoralism* 4(1): 1–17.

Knowles, Tim, Richard Moody, and Morven McEachern. 2007. "European Food Scares and Their Impact on EU Food Policy." *British Food Journal* 109(1): 43–67.

Lainé, Nicolas. 2020. *Living and Working with Giants: A Multispecies Ethnography of the Khamti and Elephants in Northeast India*. Paris: Muséum national d'Histoire naturelle.

Laiolo, Paola, Francesca Dondero, Enza Ciliento, and Antonio Rolando. 2004. "Consequences of Pastoral Abandonment for the Structure and Diversity of the Alpine Avifauna." *Journal of Applied Ecology* 41(2): 294–304.

Mathieu, Jon. 1992. *Eine Agrargeschichte der inneren Alpen, Graubunden, Tessin, Wallis 1500–1800*. Zurich: Chronos.

McGahey, Daniel, and Jonathan Davies. 2014. *Pastoralism and the Green Economy—A Natural Nexus?* Nairobi: IUCN and UNEP.

Miele, Mara. 2011. "The Taste of Happiness: Free-Range Chicken." *Environment and Planning A: Economy and Space* 43(9): 2076–90.

Nori, Michele, and Valentina de Marchi. 2015. "Pastorizia, biodiversità e la sfida dell'immigrazione: il caso del Triveneto." *Culture della sostenibilità* 8(15): 78–101.

Nori, Silvia, and Michele Gemini. 2011. "The Common Agricultural Policy Vis-à-Vis European Pastoralists: Principles and Practices." *Pastoralism: Research, Policy and Practice* 1(1): 1–8.

Oglethorpe, David. 2009. "Food Miles – the Economic, Environmental and Social Significance of the Focus on Local Food." *CAB Reviews: Perspectives in Agriculture, Veterinary Science, Nutrition and Natural Resources* 4(72): 1–11.

Orland, Barbara. 2004. "Alpine Milk: Dairy Farming as a Pre-Modern Strategy of Land Use." *Environment and History* 10(3): 327–64.

Palladino, Paolo. 2017. "Transhumance Revisited: On Mobility and Process Between Ethnography and History." *Journal of Historical Sociology* 31(2): 119–33.

Philippopoulos-Mihalopoulos, Andreas. 2012. "The Triveneto Transhumance: Law, Land, Movement." *Politica & Società* 3: 447–68.

Porcher, Jocelyne. 2014. "The Work of Animals: A Challenge for the Social Sciences." *Humanimalia* 6(1). Retrieved 1 October 2020 from http://www.depauw.edu/site/humanimalia/issue percent2011/porcher.html.

———. 2017. *The Ethics of Animal Labour. A Collaborative Utopia*. New York: Palgrave MacMillan.

Reitmaier, Thomas, Thomas Doppler, Alistair W. G. Pike, Sabine Deschler-Erb, Irka Hajdas, Christoph Walser, and Claudia Gerling. 2018. "Alpine Cattle Management During the Bronze Age at Ramosch-Mottata, Switzerland." *Quaternary International* 484: 19–31.

Renting, Hank, Terry K. Marsden, and Jo Banks. 2003. "Understanding Alternative Food Networks: Exploring the Role of Short Food Supply Chains in Rural Development." *Environment and Planning A: Economy and Space* 35(3): 393–411.
Ruff, Christopher B., Brigitte M. Holt, Vladimir Sládek, Margit Berner, William A. Murphy, Dieter Zur Nedden, Horst Seidler, and Wolfgang Recheis. 2006. "Body Size, Body Proportions, and Mobility in the Tyrolean 'Iceman.'" *Journal of Human Evolution* 51(1): 91–101.
Sal, Antonio Gómez, Felix Herzog, and Ian Austad, eds. 2004. *Transhumance and Biodiversity in European Mountains. Report of the EU-FP5 Project TRANSHUMOUNT*, IALE Publication Series 1. Alterra: Wageningen.
Shields, S. D. 1992. "Sheep, Nomads and Merchants in Nineteenth-Century Mosul: Creating Transformations in an Ottoman Society." *Journal of Social History* 25(4): 773–89.
Smith, Joe, and Petr Jehlička. 2013. "Quiet Sustainability: Fertile Lessons from Europe's Productive Gardeners." *Journal of Rural Studies* 32: 148–57.
Stassart, Pierre, and Sarah J. Whatmore. 2003. "Metabolising Risk: Food Scares and the Un/Re-Making of Belgian Beef." *Environment and Planning A: Economy and Space* 35(3): 449–62.
Tregear, Angela. 2011. "Progressing Knowledge in Alternative and Local Food Networks: Critical Reflections and a Research Agenda." *Journal of Rural Studies* 27(4):419–30.
Turner, Matthew D., John G. McPeak, Kramer Gillin, Erin Kitchell, and Niwaeli Kimambo. 2016. "Reconciling Flexibility and Tenure Security for Pastoral Resources: The Geography of Transhumance Networks in Eastern Senegal." *Human Ecology* 44(2): 199–215.
Urbanik, Julie. 2012. *Placing Animals: An Introduction to the Geography of Human-Animal Relations*. New York: Rowman & Littlefield.
Venn, Laura, Moya Kneafsey, Lewis Holloway, Rosie Cox, Elizabeth Dowler, and Helena Tuomainen. 2006. "Researching European 'Alternative' Food Networks: Some Methodological Considerations." *Area* 38(3): 248–58.
Von Sturler, Manuel, dir. 2013. *Winter Nomads*. Louise Productions. Documentary.
Winter, Michael. 2003. "Embeddedness, the New Food Economy and Defensive Localism." *Journal of Rural Studies* 19(1): 23–32.
Zendri, Francesco, Enrico Sturaro, and Maurizio Ramanzin. 2013. "Highland Summer Pastures Play a Fundamental Role for Dairy Systems in an Italian Alpine Region." *Agriculturae Conspectus Scientificus* 78(3): 295–99.
Zoller, Heinrich. 1960. "Die kulturbedingte Entwicklung der insubrischen Kastanienregion seit den Anfangen des Ackerbaus im Neolithikum." *Bericht des Geobotanischen Institutes der ETH Stiftung Rubel* 32: 263–79.

Part II
Discontinuities and Transformations

CHAPTER 7

Transhumance Is the New Black
Fragile Rangelands and Local Regeneration

Letizia Bindi

Introduction

New studies and policies have been conducted and subsequently set up in Italy to safeguard and promote: ancient pastoral tracks and landscapes; transhumance as a cultural system; and the relationships between pastoralism, agriculture, and rural communities. For centuries, Italy has hosted transversal transhumance from the inner, mountainous areas of the south-central Apennines (Abruzzo, Molise, Basilicata, Campania) to the coasts, and plains transhumance is also widespread in northern Alpine regions (Grasseni 2003; Viazzo 1989) as well as southern regions such as Abruzzo, Molise, and Apulia (Petrocelli 1999; Russo 2002). Vertical transhumance is also practiced in the central regions of Italy, as seen in Tuscany and Lazio (Trinchieri 1953; Metalli 1903) and the islands of Sicily and Sardinia. This ancient practice, deeply rooted since the Roman Empire, has influenced settlements and routes, local landscapes, and sociocultural structures (Ballacchino and Bindi 2017).

One of the fieldwork-based studies on this phenomenon focused particularly on the centuries-old system connecting the southern and inner regions of Italy to the planes of Apulia, where during the winter cattle were driven on foot as far as Foggia, where the sheep customs station was located from 1468 to 1806. Today the associated system of breeding and moving flocks has been substantially abandoned. Current ethnography thus focuses on "heritagization" processes, consequent conservation policies, and the exploitation of biocultural heritage for tourism purposes.

In the last few decades, Italian regions have been radically affected by the dismantling of transhumant sheep breeding, a productive practice that contributed to structuring the kinship relations, symbols and settlements of Mediterranean and Alpine communities over the centuries. This form of breeding, selecting and managing flocks and herds, effectively contributed to shaping the landscapes in which it was practiced. It also holds a very significant place in the family, social, and political structures of local communities and, finally, the practice of transhumance represented a sense of identity and belonging, despite the ambivalent connotations of these notions for the social sciences.

A second field involving traditional pastoralism is the heritagization process unfolding in Amatrice and the Laga and Gran Sasso mountains. The 2016 earthquake severely affected these areas, almost completely destroying the urban settlement of Amatrice as well as other small towns and villages in the area and, ultimately, interrupting the breeding activities that previously took place in the surrounding mountains. In recent years, however, there have been efforts to revive transhumance. Indeed, this revival was underway in neighboring areas before the earthquake and it remains effective in pushing communities towards new forms of regeneration.

A Local Issue

In the last ten years, there have been a number of regional initiatives (often conducted through the mediation of Local Action Groups (LAGs) or NGOs such as Legambiente, Italia Nostra, etc.) aimed at defining and protecting residual regional sites of transhumance and mapping these places.

The conservation and valorization of natural and cultural landscapes has been much less systematic. Nevertheless, scholars have noted that many agricultural and planning permits were still granted during this period despite the recommendations and planning constraints of the Regional Superintendence of the Ministry for Cultural Heritage. In 2011, the Regional Superintendence of the Ministry of Cultural Heritage reaffirmed the significance and cultural scope of this archaeological and environmental heritage, and asked for a regional *tratturi* (herd path) plan to map all the transhumance tracks throughout the region so as to establish buffers[1] around them and the built heritage in need of protection. The intention of this legal intervention was to assess the value of environmental biodiversity as well as its relative cultural and social expression. However, what followed was a process in which the regional institutions responsible for this policy area ended up considering a significant number of exemptions to the initial assessment. It was not until a few years ago that the biocultural value of *tratturi* was officially confirmed, and many issues still remain unresolved.

Legal and political quarrels and debates over *tratturi* uses and permissions make the "heritage field" controversial and challenging (Herzfeld 2004). The subsequent redefinition of agency and governance over local areas affects institutions at different levels; indeed, Cultural Heritage Ministry offices at the regional level, as well as various communities of practice, have found themselves grappling with this issue.

In the background, the 2004 Council of Europe's Landscape Convention has paid increasing attention to sustainable rural development policies, calling for the participation of the communities involved while providing expertise on safeguarding and enhancing rural areas and new policy-making strategies for developing marginal and peripheral areas (Papa 2013; Barlett 2016; Cejudo Garcia and Navarro Valverde 2020). Communities are thus redefining their relationship to the past (Herzfeld 1985) which is part of a wider discussion on the future of regional/transregional inland areas. The current debate on this issue is being carried forward by the National Strategy on Inner Areas (a governmental program set up to monitor and intervene on specific critical issues facing inland regions in Italy). This move reflects the current framework in which all of these concepts are being addressed and taken into consideration. The last element of debate we need to consider is the one around land use (the officially, but not effectively, protected *tratturi* lands). This is a very contentious territorial and legal issue. It is particularly signficant in a number of marginalized areas that have long been considered a sort of "no man's land," free to be utilized, basically without the application of already existing rules and permits despite their being formally/legally defined as public and common goods.

Those herders who still practice the transhumance in Molise, as well as in the other southern mostly mountainous areas of Lazio, Abruzzo, Campania, Basilicata, and Apulia, have decreased dramatically in last few decades, as the research and studies into the history and cultural significance of the transhumance increased.

Meanwhile, other problems and difficulties are emerging for the sector as well. One of the most severe of these is the increasing misuse of pastures in which they are allocated for animal-breeding activities on false premises, only to secure the funding provided by the CAP (EU Common Agricultural Policy), a phenomenon commonly called pasture *mafias* (Calandra 2019; Mencini 2021). This trend reveals the harsh contradictions systematically facing contemporary shepherds and herders and how hard it is for them to remain in their local areas and continue practicing the forms of breeding and production they inherited from previous generations and are trying to maintain. At the same time, it also points to the continuous alternation of aspects more closely connected to the safeguarding of a heritage of practices and knowledge forms and the concrete sustainability of traditional agropastoral production within the gears of the neoliberal food market.

There has recently been renewed interest in traditional breeding activities and pastoralism. Alongside such interest, cultural associations—particularly those engaged in slow tourism projects—as well as innovative, public-oriented small farms leaning towards more sustainable, high quality, and ecological goals have shown a strong commitment to developing national/regional development programs for inner areas. These projects link up with local promotional initiatives and people involved in high quality agrifood production (cheese, meat, herbs, phytopharmacological products, and so on) as well as Slow Movement proposals for healthy and experiential tourism.

This connubium of actors and projects provides the perfect metaphor for heritagization processes. Examining UNESCO applications, data collection and various promotional activities, we see an increasing shift towards conservation and the valorization of biocultural goods coupled with an evident focus on valorizing local areas. Some young and return herders are organizing various projects to revive traditional and nomadic pastoralism (e.g., short transhumant pathways for tourists to follow, educational farms, and experiential cheese-making workshops for tourists). Moreover, a new commitment to creating short supply chains goes hand in hand with growing attention on the quality of local products. Natural feeding methods and stock movements lead to a consistent improvement in milk, meat, and wool production while ensuring a better quality of life for animals.

Meanwhile, greater environmental sustainability and the appeal of pastoral landscapes have proven fertile grounds for the growth of experiential tourism in which visitors share in the spaces and practices typical of local communities (Palladino 2017) such as transhumance routes, processing raw materials (milk, wool, handcrafts), and nomadic/semi-nomadic regimes. This transformation has been recognized and shared by shepherds and herders: they have made it their own and integrated it into some of their behaviors and routines as well as their overall approach to production. Discourses are framing "bone lands" as resilient in contrast to the "pulp lands" of coastal areas, the dilemma of whether to leave or remain, the dichotomy between traditional and innovative: all of these elements constitute a sort of semantic carpet on which the people who graze animals make choices and carry out actions, choices and actions that are actually very concrete and pedestrian.

In addition to this new interest in extensive pastoralism shown by the National Strategy for Internal Areas, another project dedicated to this issue has been launched and sustained by the local groups and associations taking part in the APPIA Network for Pastoralism (www.retepastorizia.it). This network of practitioners, activists, and experts on extensive pastoralism has worked together with other groups and institutions, such as Riabitare l'Italia (an association aimed at fostering innovative ways of considering and planning Italy's inland areas) and CREA (the National Council for Research on Agriculture and analysis of the Agrarian Economy) to establish a National School of Pastoralism (SNAP) modeled after similar schools already active in France and Spain. This development also took place as part of programs for valorizing sustainable breeding and supporting so-called return shepherds in various European areas (École des Bergers du Domaine du Merle / Escuela de Pastores de Andalucia). (Lebaudy 2016; Ugarte et al. 2014).

In the Molise region, transhumance has deep historical roots and was once recognized as one of the area's main rural activities; today, however, it has decreased markedly. Its continuity and visibility have been ensured in the last decade by the Colantuono family from Frosolone, traditional herders who continue to practice the transhumance tradition of moving their cattle from the Molise mountains to the plains of Apulia every year. Meanwhile, other families of shepherds and herders are reactivating small to medium-sized transhumance routes. Antonio Innamorato has conducted an interesting case study in this area. He restored the ancient transhumance track between Campitello Matese and the archaeological site of Sepino—along the main, larger *tratturo* between Pescasseroli and Candela—which has attracted an increasing of number tourists since 2017 and received significant media coverage over the last three years.

These activities are highly ambivalent and interesting from an ethnographic point of view, particularly in relation to the process of reviving rural and pastoral practices in Italy's inner areas as a way of linking up heritage communities and researchers with the aim of uncovering and recovering memories of local transhumance. During these revivals, local actors often present contradictory tales of the past and ambivalent representations of what they consider to be their own cultural heritage. One key aspect of such work is that the presence of ethnographers in the field has implicitly encouraged shepherds to think about and acknowledge the cultural value of their practices. A second important element to take into consideration is the rise in funding for and legal frameworks focused on *tratturi* and transhumance conservation and valorization as a potential resource for new rural development.

The heritagization process underway around transhumance highlights the link between local practice, landscape conservation, and cultural heritage as well as the powerful associations surrounding these routes, from

Figure 7.1. Lu vic p'dent, Bojano, 2017. © Rossella De Rosa

religious and cultural to fitness and wellness associations. Today, these associations are increasingly evoked, sustained, and promoted at the national and European levels as a key element for fostering sustainable tourism, local development, and the empowerment of heritage communities, especially in inland areas of Europe (Council of Europe 2019).

A Long Way to Go

Since 2008, Colantuono's, a well-known cheese producer in the Frosolone area, has launched a campaign of transhumance revitalization. The Colantuono farm is located in an inner area of the Molise region, which has been supported over the last decade by local institutional funds and especially impacted by LEADER Strategy and the activities of an area LAG (Local Action Group). They move their cattle along the traditional "green highway," a 300-kilometer-long track from Frosolone in Molise to San Marco in Lamis, where their farm in Apulia is located, following in the footsteps of their ancestors.

Over the last decade, this newly revitalized transhumance has become a popular tourist attraction in which many local associations, public institutions, and private citizens take part. The abovementioned Local Action Group also oversaw the first attempt to submit a dossier to the UNESCO World Heritage Sites list. The submission was not accepted as written, but UNESCO did request a reformulation and, as a result, the application was shifted from a tangible heritage/landscape submission to an intangible cultural heritage one. This shift has had certain consequences and implications in terms of strategies for safeguarding and valorizing the heritage in question. First, it led the applicants to instead present an International Network application, including Austria and Greece in the ICH element submitted for inclusion in 2018. The dossier was presented by the Italian Ministry of Agriculture, Food, and Forests in collaboration with other national authorities for Austria and Greece (the Ministries of Culture, in these cases) and brought in multidisciplinary clusters of researchers as experts to establish the scope of transhumance as an intangible cultural heritage through significant studies of this phenomenon.

As a result, the National Ministry for Agricultural Policies supported the Local Development Agency and regional institution in shifting from a local/regional process of safeguarding and valorization to a national one and, finally, in ratifying the international network application. This process entailed registering *tratturi* in the national inventory of historical rural landscapes and, subsequently, providing support with the set of procedures required to present the UNESCO ICH List application. Through this process transhumance has been granted greater visibility and institutional attention, but it has also been packaged in a markedly hierarchical way through safeguarding measures and protocols and been framed as a consumable good in keeping with today's marketing-oriented logics for promoting pastoral routes.

Meanwhile associations, NGOs, and more informal movements and groups of citizens have also been involved in the implicit heritagization of the landscapes and cultures of ancient pastoralism in various areas. There are many associations and voluntary groups engaged in promoting local areas through forms of slow and experiential tourism, for example. These include activities such as taking tourists along drovers' routes from one farm to the next as well as small breeding units and other medium-sized enterprises. There are walking groups both large and small, horse riders, and cyclists as well many other people involved in public events and ceremonial occasions linked to transhumance in the various regions I have been monitoring through my multisited ethnography. Similarly, many local communities, even ones only marginally connected to transhumance and animal breeding, are working to preserve a significant number of practices such as food traditions and narratives (oral poetry, folk songs, and dances, and so on) and traditional beliefs despite the intense sociopolitical and biocultural changes they have experienced in the last few decades.

Through such activities, transhumance-as-an-event has undergone a process of commodification and rhetorical/mediatic reshaping. At the 2015 Milan Expo, an image of Carmelina Colantuono was used to symbolically represent the Molise region. This choice was highly significant in many ways. Carmelina is often represented in the local and national media as "the cowgirl of Molise" or a sort of "native American horse rider who takes care of her cattle along the pathway" (Bindi 2012). At the Expo, she became an iconic image representing a region nestled between rural pastoral society and a brand-new range of themed attractions revolving around ancient, revitalized, and heritagized practices. This representation has exalted what had been in the past marginal and invisible and made it visible on the national/international tourist destinations and foodscapes markets.

The 2015 Slow Food Convention for the Apennine Communities was organized in Castel del Giudice, a small village in Molise considered an excellent example of local area regeneration processes. At this convention, a group of young rural entrepreneurs presented a potential new model for rural economic development in the region, a model focused on new rural and breeding activities based on sustainability and typicality as well innovative, social, and inclusive ways of producing agri-food products (Goodman, Dupuis, and Goodman 2012). Many of the presenters were and are returning farmers, breeders, or herders. Although the trend of return farmers and herders remains quite elitist at the moment (Van der Ploeg 2008; Padiglione 2014), events such as this represent extremely interesting examples of the emergence of a "new ruralist" movement.

At the political level, there are various critical issues affecting the legal frameworks characterizing conservation/valorization policies. As mentioned above, many conservation laws are unclear or inconsistent, and the same is true of those relating to regional planning and systems of distribution.

At the same time, in recent years, ecological movements and heritage communities' concerns are growing up as well as the vigilance on uses and misuses of the lands by informal groups of private citizens.

In at least one case in Molise, an ecomuseum has been established that organizes experiential walks along the sheep tracks. This ecomuseum, called Itinerari Frentani, is modeled after other similar initiatives in Italy and other parts of Europe (such as the Pontebernardo Ecomuseum of pastoralism that includes walkways, training camps, and farms and sells natural, ethically produced food and other goods).

This question seems have become increasingly an object of crucial confrontation among different institutional levels, civil society, associations, and private citizens and the issue of consensus-building on the territory is not exempt from this theme. In this sense transhumance is transformed into a privileged context for evaluating institutional and power relationships at a local level, and into an ethnographic context to highlight the role anthropologists play at the local level, their engagement with and service to the communities with which they work.

A multidisciplinary approach can help us to understand the extent to which inland areas are affected by neo-endogenous development models as well as processes that redefine local identities and senses of belonging (Bender 2001; Yuval-Davis 2006; Mee and Wright 2009; Wright 2014) and that repurpose local areas as tourist destinations based on the marketability of a food heritage-scape (Bindi and Grasseni 2014) and "authentic/ genuine" products (handcrafts, events). This complex scenario proves even more important for economically and sometimes socially depressed areas. National governmental strategies for Europe's inner regions, including programs protecting transhumant tracks and recovering civic uses of the land and protected areas, constitute a challenging site in which to ethnographically investigate local development policies and the ways communities care for their own areas and landscapes. Meanwhile, given these complex factors, we as researchers must also reconsider regional/

national policies and supernational "heritage regimes" as well as the very definition of an embedded cultural asset.

Ultimately, intangible cultural heritage could represent a promising but also controversial opportunity for local development and the enhancement of sustainable tourism. It could be a real turning point for inner regions and a shift in the way a traditional practice such as transhumance interacts with the regional/national/global frameworks of local regeneration and development by rearticulating the link between the mountain and the plains, inner areas and the coasts. In other words, a return to the relationship of the past based not on hegemonic differentiations between different areas but rather a recognition that all are necessary and complementary.

A Pathway Toward Regeneration

In the area of Amatrice and the Park of Gran Sasso-Laga, the process of safeguarding and valorizing transhumance has grown to become a widespread discourse on mountains and fragile areas. In recent years, this discourse has become part of a reflection on the opportunities and potential for rethinking the country's backbone, a point made by many observers including and especially during the months of pandemic isolation. The mountain-herding-depopulation nexus plays an increasingly role in structuring the narrative on these places, mixing with concepts such as taking care, proximity, staying, and regenerating (De Rossi 2018; Teti 2016).

In this context of renewed attention to internal, mountainous, secluded, peripheral, and pastoral areas, Amatrice undoubtedly plays a role. The regeneration of wandering pasturage and revival of all the activities related to sustainable and eco-ethical livestock breeding has come to represent a significant site of local engagement through multidisciplinary cooperation with social scientists, economists, agronomists, zootechnics, jurists, and eco-systemic services experts. This approach has the potential to effectively address research concerns alongside institutional policies through participatory interaction with local communities. More recently, the mountains, along with fragile internal areas, have been reconsidered as sites for investing in the kind of short food supply chains and proximity food production evoked by "feeding the planet" advocates (as it was evoked in the 2015 World Expo in Milan) in intersection with the desires of communities and those calling for regeneration.

"Transhumance-as-an-event" can be seen as part of this rearticulated rural narrative on the past and present, encompassing as it does the memory of traditional practices and the forms of local knowledge evoked in

Figure 7.2. The route of transhumance, Amatrice, 2019. © Letizia Bindi

multiple Italian poems. "The *pecoraro* (shepherd) did not lack anything, yet there was an economy and a hierarchy among shepherds" recounts a poet-shepherd in a recital at the occasion of the last *Via della transumanza*, a festival and reenactment of historical transhumance held in Amatrice in September of 2019.

Oral poetry alludes to a rigid internal organization designed to ensure survival and mutual support, articulating an opposition between high and low, white-collar and rural blue-collar workers, between cooperative human/animal relationships and relationships that recognize animals' value exclusively in terms of economic purchasing power. Traditional poets still highlight the distinction between their world and a sanitized, medicalized, urban world as well as the public's lack of knowledge about mutton and lamb, products that have become less attractive for city-based consumers. Today, after months of pandemic-inspired speeches on health precautions, this theme appears extraordinarily topical and significant. In the abovementioned song presented by Mario, the "learned shepherd" (Ciaralli 1997), herders—erroneously and hegemonically represented from the outside as isolated, backward, and almost savage people—make a claim for the past and the present alike.

In a significant passage from one of these songs, the two singers challenging each other present the clash between a *laudator temporis acti* (one who praises past times) and a voice exalting the present and its virtues.

At the end of one of these exchanges, one of the two poets celebrates the return to bartering, that is, exchanging the products derived from agricultural and pastoral work. Through its registers and messages, oral and traditional poetry presents a case for supporting sustainability and equal exchange, in contrast with the inequalities of capitalist society. It would be mistaken to consider these poetical improvisations nothing more than a poetic cliché. Rather, they are conscious rearticulations of a local way of thinking that calls into question modernity and the very idea of local development. At the same time, they are both a *mise en forme* as well as, at times, an almost mystical approach to the pastoral way of life coupled with discourses on sustainability and the value of reconnecting human to animals, culture, and nature in a way that goes beyond strictly economic considerations and the possibly unpleasant aspects of this practice.

Local ways of narrating transhumance deal with the present and future of communities, their "conscience du lieu" and new "worlds of life" (Magnaghi 2017; Poli 2019). As a further demonstration of the value of this art of speaking, shepherd-poets speak to the present and help us to navigate it. At times, poets also adopt a more intimate register. For instance, Antonio Cannavicci does this when he proposes another way of dealing with trauma by rehabilitating local areas through an exercise of remembering the area's pastoral roots.

These narratives encompass principals, knowledge archives, transmission tools, and frames for interpreting the world. Their words reconstruct the inhabitability of the land and use language as a way of reshaping places. Indeed, Ingold and Vergunst have noted this when observing that the rhythm of writing-narration can be compared to the beat of footsteps: a pedestrian grammar for a well-planted reading of the world through the feet (Ingold and Vergunst 2008; Bindi 2020).

Walking with Animals after an Earthquake

Every year around the end of September, in proximity to the emblematic date of 29 September dedicated to honoring St. Michael as the protector of wanderers and walkers, people gather to celebrate a pastoral trail, with the flocks flowing along urban streets leading down from the mountain and heading towards the Pontine countryside (Metalli 1903; Trinchieri 1953).

In Amatrice, this "reenactment" of transhumance—an ambivalent practice located somewhere between memory and performance—took place even before 2016. *La via della transumanza* (route of transhumance) project gained greater momentum after 2011 as part of "Ecorutour," a component of the EU's LIFE program.² The program also involved other such projects organized in the same period in various regions of southwestern Europe such as: "Pasturismo" (Monllor and Soy 2015), conceived and created in Catalonia; and the Italian-French project *La Routo* (Lebaudy et al. 2012) connecting the abovementioned Ecomuseum of Pastoralism in Pontebernardo to the Domaine du Merle along a sheep track shepherds traditionally used to move their flocks.

From the beginning, *La via della transumanza* has been engaged in recovering the history and culture of local communities and promoting tourism. In this project, transhumance has been associated with a narrative of the land based on agricultural and pastoral identity, the abovementioned discourse of mountains, and sustainable development linked to biocultural heritage.

> Reenactment is, in fact, an activity of reinventing history fueled by a series of phenomena specific to contemporary culture: the growing importance of visual and media aspects, the acceptance of new forms of relative authenticity, the now nearly consolidated interconnection between leisure and culture in many tourist and cultural consumption practices, the preponderant role of experience and sensitivity, the growing need to focus, recover or invent new individual and collective identities so as to face the identity liquidity of contemporary society, and the strong need, from this point of view, for the past and roots. It is this latter that partly explains the renewed interest in history, archeology, and local identities, not only at a tourist level but also in televised dissemination and practices of consumption. (Melotti 2013: 147)

The idea, therefore, was to develop sustainable tourism mobility in the area of Amatrice, province of Rieti (but also Accumoli and Arquata), the Central Italian Apennines, and Gran Sasso and Monti della Laga National Park. For three years this initiative involved the local population, administration, and civic association members in a common project of revitalizing and rediscovering the promotional value of pastoralism in this area. Some tensions and misunderstandings arose between local shepherding/breeding families and public administrations, with the latter probably considering the community-based initiative to be excessively autonomous and free from political constraints. As in many other cases, it was families of shepherds and breeders who ensured the continuity and durability of the project after the initial startup period, and this increased frustration among project proponents as they experienced this disinterest and opposition on the part of the administration with great disappointment.

In 2017, after the inevitable period of discomfort and disorientation following the 2016 earthquake, a number of actors proposed the idea of resuming the practice of transhumance (descent from the high summer pastures). These actors included several local associations (Laga Insieme, and Appennino Solidale) as well as The Magnificent Lands of Centro project and CAI (Italian Alpine Club). A series of entrepreneurial subjects (primarily restaurateurs in the "area of taste," a brand-new part of the city totally reconstructed after the earthquake and presently occupied only by restaurants and a Congress Hall), but also several breeders with pastures in the area such as the Scialanga family and others, also lent their support alongside the municipal administration, determined to relaunch sustainable tourism in this area. These historical, heritage reenactments are often framed as leisure and tourism events located between infotainment, recreation, and play. They often convey an idea of history and memory as supposedly nonpolitical or post-political, unrelated to a political commitment to local areas and communities (Carnegie and Maccabe 2008).

At the same time, both before and after the earthquake, *La via della transumanza* in Amatrice definitely appears aimed at conveying or articulating "new forms of local resistance to globalization or new regionalisms or anti-national hyper-localisms" (Melotti 2013: 150), but also resilience in the face of depopulation (Cejudo Garcia and Navarro Valverde 2020 and the abandonment of cropland and pastureland (Bakudila 2017). Such projects to relaunch tourism-cultural promotion and the local economy as a whole in a lasting and sustainable way is therefore a way of "reinhabiting" depressed and isolated regions and places (De Rossi 2018; Teti 2016). These initiatives involve rebuilding an imaginary and giving meaning back to the land, reconnecting breeding practices, and reorienting production from a focus on agri-food chains to a focus on landscape, biodiversity. and intangible cultural heritage made up of trails and sheep-raising products (handcrafts, food and conviviality, soundscapes and oral histories). The "heritage turn" has allowed communities to look at extensive pastoralism not as an exclusive mode of production, but also as a cultural asset and potential tourism resource (Bindi 2020).

After the earthquake, relaunching the *La via della transumanza* project represented not only a revival but also a real sign of rebirth after the trauma suffered by the community. It has involved focusing on collective memories, family landscapes, and a world of shared practices and knowledge that, over the centuries, has allowed these settlements to maintain and perpetuate their own forms of life and internal organization.

The revitalization of transhumance in a post-disaster context like that of Amatrice must be read as a form of resilience: a perspective in which environments and communities are closely interconnected (Folke 2006), implying the need for a radically multidisciplinary approach. Multidisciplinarity is needed not only as an interpretative tool but also a means of supporting a governance of these processes necessarily characterized by adaptive and self-organizing skills (Adger 2000). "Socioecological resilience" is the process by which ecosystems adapt by trying to absorb shocks and disturbances, putting more emphasis on the element of resilience than on that of sustainability. This approach enhances the concrete possibilities available to a given biocultural context before the abrupt break caused by the earthquake or any other form of change (Cork 2010).

According to this perspective, the reenactment of transhumance beginning in 2017 should be read as a form of local-area regeneration involving families as well as a grouping of institutions, associations, private entrepreneurial actors, and the general public. The Scialanga family and its flock of Comisane sheep (not an autochthonous type of livestock) remain on their farm in Pratica di Mare, near Rome during the winter season; they return to the mountains in springtime along a route that takes several days of walking, although today the move is carried out largely with trucks. They voluntarily resumed this practice of returning to their grazing areas after the earthquake despite the logistical difficulties involved, including an old, damaged farmhouse, and agreed to do a section of *demonticazione* (descending with the flock from the high pasture on foot). Over time, this has become a strongly symbolic act and a sort of celebration for the local community. When I observed this event, on their arrival in the Amatrice town square in front of the recently rebuilt "House of the Mountain," a show of transhumance took place displaying a lively mix of colors and gestures, practitioners and visitors. Silvestro Scialanga, head of the flock, assisted by his sister Vittoria and other collaborators, walked solemnly, accompanied by people in traditional costumes and decorated sheep as well as photographers, journalists, and researchers.

An enthusiastic audience watched the parade, supporting and appreciating it. They took photographs and made short video recordings with their mobile phones, devices that have become people's main tool for asserting their own presence at an event and participating in ceremonial occasions and historical reenactments, essentially as spectators. At a certain point, the procession took the road towards the former school building, a prestigious and emblematic all-inclusive structure in Amatrice that had welcomed elementary, middle, and high school children over the decades. Locally, the school is familiarly known as Don Minozzi after the priest who originally founded it. Today, the building no longer exists. Nonetheless, Don Minozzi was at the center of an image also used for the poster publicizing the *La via della transumanza*; the image depicts an austere shepherd in traditional costume standing in front of the colonnade of this building as portrayed in an old postcard from the early twentieth century.

The symbolic association between past and present was condensed into two, superimposed images: the historical photo of the shepherd in front of

Don Minozzi and the reenacted scene featuring Silvestro Scialanga with his flock entering the battered historic center of Amatrice. Like a missing limb that the amputee still feels, we are once again reminded of the power of images when the memory captured by photography projects the image of the past onto the present. The past is like a thin veil laying atop the current state of things, allowing us to see the invisible and to imagine what is no longer or has not yet occurred. Between multiple uses of the past and the emotions triggered by heritage, between reenactment and commemoration, the shepherd in gaiters and a wide-brimmed hat becomes a guiding image, a condensation of the past and a prosect for the future in an ambivalent arena characterized by both a need to relaunch and spectacularize the event and the urgent need for the town to recover and regain its resilience.

At the same time, the multidisciplinary research (anthropology, agrarian economy, and zootechnics) carried out in this field has revealed people's explicit aspirations of reviving extensive pastoralism, both as a provider of eco-systemic services and as a way of bolstering a niche agri-food market oriented towards short supply chains, responsible consumption, and the valorization of local areas but also as a source of tourism development based on the paths once tread by wandering pastoralists. The governmental institutions involved, civic associations, and the producers themselves all aim to pursue an integrated approach. Despite these intentions, however, I found that the various political, economic, and association-based actors struggle to develop effective internal synergies.

The ancient idea of commonalities shared among the pastoral communities of these areas until at least the end of the nineteenth century has been threatened by the growing privatization of businesses and farming practices. In addition, traditional pastoralism has become less and less economically sustainable due to the need to buy pastures and the tendency to optimize farming costs through the stabling and consequent sedentarization of flocks. At the same time, herders face challenges in using pastures: once largely civic and common lands, these fields are increasingly privatized and broken up. Such parceling also tends to dismantle the network of proximity and cooperation that was historically embedded in the practices and uses associated with shared land. In Amatrice as in Molise more generally, there is clearly a clash between two different visions: a heritagizing move that evokes and supports the value of traditional and transhumant pastoralist practices as a potential driver of rebirth for areas and communities plagued by depopulation, economic degrowth, and social disaggregation, on one hand, and a productive, intensive concept of breeding on the other hand that is indifferent to the environmental and sociocultural impact of intensive farming practices. Indeed, such intensive production is increasingly aimed at meeting the demands of—and, in many cases the extortion exerted by—large-scale distribution channels, thereby further undermining transhumant shepherds and exacerbating their precarious conditions.

Resilient Pastoralism

Local systems manage to change by adapting to ongoing transformations; at the same time, they tend to generate enrichment and the mobilization of available resources, sometimes hidden, unexplored, or underutilized ones (Steiner and Markantoni 2013). Active participation in change—as in the case of the community engaged in reviving transhumance in the Monte Laga area—can therefore be read as a way of creating and enhancing intangible resources connected to social, symbolic, and cultural capital. This process begins from relationships between local actors and the biocultural landscape in question, from the imaginary of this practice understood as a revival but also as the potential for a new eco-systemic approach to reconstruction and a context in which to observe the capacity of local agency in terms of rebuilding and escaping the post-disaster crisis (Herzfeld 2001; Norris et al. 2008).

In this way, reenacted transhumance is conceived as a powerful form of sociocultural as well as environmental capital (Putnam 2000, 2007; Breda and Lai 2011) to be deployed in the post-disaster scenario. It empowers people's sense of belonging and participation in the community as a proactive way of responding to changes by creating social bonds, external networks, and relationships of trust among citizens and towards the local and super-local institutions supporting these kinds of initiatives. Regenerating the transhumance route plays a powerful symbolic role in that it encourages people to think about their belonging to a local area and sharing in a knowledge-practice system. Nonetheless, such an approach implies a holistic concept of community, a concept that has been discussed by the social sciences and which leads to a reconsideration of the very notion of resilience (Wilson 2012; Steiner and Markantoni 2013).

More generally, we have documented a flourishing range of initiatives, conferences, trails, and walks dedicated to protecting, rediscovering, and valorizing sheep tracks after transhumance was included in the UNESCO ICH List. This new degree of attention counterbalances the relative neglect characterizing the past as well as the fact that this phenomenon is often merely evoked and allocated funds are regularly diverted rather than actually being used for projects related to safeguarding and valorizing this important form of biocultural heritage.

In order to support agropastoral development, especially in fragile and depopulated areas, draft laws have recently been written, regional offices have been set up, and PSR funds and other local development measures, such as LEADER strategy funds—through the mediation of Local Action Groups—have been established. Meanwhile, on a different scale, a widespread discourse has been produced, framing transhumance as a keystone for promoting local areas traditionally characterized by the knowledge-practice system of wandering pastoralism.

Transhumance is frequently presented as an opportunity for overall development and for rethinking the tourist market, alongside the promotion of artisan products and related agri-food chains. Nevertheless, it remains to be seen if this process, passing through a network of municipalities and declaring transhumance as a transversal and participatory vocation, can potentially consolidate a strategic vision of safeguarding and valorization, ultimately enabling an integrated array of interventions and a brand-new cultural approach to this issue.

The theme of pastoral routes and transhumance is not at all backward or atavistic, but it needs to be framed outside of late modern "structural nostalgia" for societies' agropastoral roots (Herzfeld 2004). Moreover, the question is how to take care of the sheep tracks as well as herders' communities, documenting the agricultural permits that authorities have continued to grant despite the superintendency's rules protecting transhumant routes since 1939.

Similarly, it is necessary to reconstruct the system of mobility that allowed men and flocks as well as ideas, relationships, craftsmanship, words, attachments, rituals, and biocultural knowledge to travel. Extensive breeding also contributes to the depopulation of the inner and fragile areas of the Apennine ridge, all of which are fighting their increasing marginalization and the negative representations of shepherds that are nowadays tinged with new connotations and expectations.

At the same time, the theme of transhumant sheep farming raises very urgent issues linked to the today's most pressing concerns, such as the conservation of biodiversity and the landscapes connected to these paths, the artisan practices associated with this traditional activity, and the increasing attention to animal welfare as well opportunities for a new form of experiential and slow tourism. Multispecies heritage has recently become established as a field of research considering more-than-human communities (Morris 2014: 51) and the interrelationships and "contact zones" (Haraway 2008: 244) "where human and animal lives biologically, culturally and politically intertwine" (Aisher and Damodaran 2016). These studies question human/animal co-being (Davis, Maurstad, and Cowles 2013) in a fiercely critical perspective that forges connections between the social sciences and animal studies as an alternative way of dealing with natural and land resource management, ideas about development, ways of inhabiting places and, obviously, interspecies ethics. Processes of protecting and valorizing sheep tracks as well as transhumance constitute a highly political arena, one that encompasses aims of environmental and landscape protection, place-based knowledge about conserving the land in connection with a sense of identity. These aims can be reached through a participatory decision-making process that ensures shared governance of land use as well as local public-private entrepreneurial activities. Today, therefore, both sheep track and transhumant experiences can be considered a predominantly cultural journey through which communities can become fully involved in heritagizing and planning for local development.

Transhumance at the Heritage Turn

After the UNESCO nomination in December of 2019, transhumance has been continuously evoked as an antidote to depopulation and the progressive loss of soil and biodiversity, presented as a kind of panacea for crisis and a crucial element in the definition of local identity, supported by experts and policy makers.

Pastoralism is an excellent multidisciplinary field for study, encompassing as it does biocultural heritage and ecology, rural economy, and geography as well as biodiversity and agrarian and landscape studies. Through such research, it is possible to assess environmental sustainability, ecological approaches, and participation in decision-making processes and the governance of territories as a strategy for inland and peripheral areas fighting depopulation and marginality in relation to global processes, the loss of biodiversity and cultural diversity in keeping with the agendas of the major global agencies.

At the same time, increasing animal welfare activism and concerns about healthy/natural and even ethically produced food is giving rise to new attention towards and investments in extensive pastoralism. More and more often, urban consumers are committed to animal welfare and environmental safeguards and tend to boycott products derived from intensive and industrial farming practices. At the same time, they are interested in buying natural milk/meat/wool products and in sustaining low-carbon-emissions extensive breeding (transhumance and wandering pasturage) (Soussana et al. 2007), and they voice an ethical commitment to the respect and valorization of local cultural communities. Similarly, legal debates over common lands and natural resource uses must be considered a way of "improving governance of the pastoral lands" (Davies et al. 2016;

Behnke and Freudenberger 2013). During and after the lockdown and isolation imposed by the COVID-19 pandemic, there has been a greater recognition of the value of mountain footpaths and the chance to directly experience areas and landscapes along with a focus on sustainable, extensive, and animal welfare-oriented ways of farming, at least in European countries. This shift has also deeply influenced the representation of pastoral routes and trails. In this period, "at a distance" field studies and virtual interactions with locals revealed their experiences of social distancing and lockdown in the temporary residences, prefabricated houses, and empty streets of desolate towns. For farmers and breeders, the pandemic has further undermined the sustainability of their activities.

The UNESCO listing is often brandished as a seal of excellence rather than a pact for safeguarding and valorization; indeed, it is often treated as a sort of positive appraisal and a privilege for areas that usually lay outside the major economic and tourism circuits. In many cases, participatory processes are evoked rather than practiced. Herders and shepherds are often externally driven, sometimes inspired by the exceptional spectacularization of transhumance-as-an-event and the mirage of economic growth and support from external programs and funds. This can make them victims of facile marketing operations and/or the illusion of incontrovertible development processes.

Meanwhile, the discourse on internal areas has progressively shifted to a resilience-oriented "poetics of staying," a new rhetoric of "small is good," and a narrative of "belongingness" (Müller 2021). The social distancing imposed by the pandemic—certainly not looked-for, but still not to be wasted—put us face to face with the question of what this pandemic might imply for fragile areas, small villages, isolated parts of the countryside, and pastoral movements (Boeri 2020; Tantillo 2020; Zane 2020). The problematic aspects of the urban life system and the limits inherent in large concentrations of people became particularly clear (Moreno 2016; Hidalgo 2020) while the difficulties of coexistence and the wide-open character of the most depopulated areas displayed all its positive aspects in a time of social distancing, limited and exclusively internal tourism, and slow mobility. Inner and mountainous areas have been revealed to be hospitable contexts, beyond arcadian, picturesque tropes of the secluded, lonely mountain. People have begun to look to rural homes as new and alternative residences to recover from the city and to live in greater balance with nature, conceptualizing/creating a new life involving more time and more care. The transhumant mode of breeding and more generally extensive ways of livestock farming are presently recognized as more suitable and sustainable than before, in contrast to the unsustainability of intensive herding with its excessive pollution and land use.

The heritage turn—coupled with a critique of reckless, unsustainable post-capitalist economic growth—has finally enabled communities as well as policy makers to profoundly reconsider transhumance and extensive farming in terms of a multifunctional approach bringing together respectful animal breeding, artisanal cheese/meat/wool processing, biodiversity conservation, ecosystem services, landscape conservation, handicrafts revitalization, and slow and experiential tourism.

On one hand, the reenactment of transhumance implies a reconsideration of people's symbolic relationship with familiar landscapes and a strongly embodied sense of a human-animal community, suggesting a nostalgic and somehow neoatavistic, conservative, and arcadian representation of traditional breeding. On the other hand, traditional and extensive pastoralism calibrated to the climate crisis, changes in local land management and post-disaster transformations supports resilience discourses and practices as well as post-capitalist and participative, neo-endogenous processes of reconstruction and local development that are truly embedded in local rurality, giving them a common sense of belonging as heritage keepers as well as a renewed emotional commitment to the landscape and a reevaluated perception of their *savoir faire*. Transhumance thus allows for a rearticulation of the past/present relationship, a chance to reconcile with ancient modes of living and producing that were previously associated with backwardness and negativity but are now being seen as ecologically and ethically sustainable.

Policies of biocultural heritage conservation represent the real challenge and political field in which to critically understand new rhetorics and representations of rurality as well as new processes of decision-making around local development. The reshaped and narrated pastoral field becomes a stage in which the tensions of an increasingly competitive agri-food market and various policies of local/rural development surface clearly. This attests to the relevance of heritage rhetorics in the sense of belongingness, the local branding of food heritagization, symbolic and social capital for defining rural potential, and local identity in the multifaceted heritage framework.

Acknowledgments

The author would like to thank Penny Barron and Angelina Zontine for having proofread this chapter.

Letizia Bindi has been Professor of Cultural and Social Anthropology at several Italian universities and a visiting scholar at several European universities. She is a member of the major national and international societies of cultural and social anthropology. She is presently a professor at the University of Molise, Italy, where she directs BIOCULT, the research center on biocultural heritage and local development. The main ongoing projects she coordinates include: the Erasmus + Capacity Building Project—EARTH (Education, Agriculture, Resources for Territories and Heritage) and the Italo-Argentinian Project—TraPP (Trashumancia y pastoralismo como elementos del patrimonio Inmaterial).

Notes

1. Buffers are "areas of respect" located on both sides of the sheep track. It is advised to establish such buffers to protect the track from infrastructure and buildings so as to safeguard the visibility of the route and the maintenance of the landscape.
2. http://www.transumanzaamatrice.it/Transdoc/presentaztransum.pdf.

References

Adger, N. W. 2000. "Social and Ecological Resilience: Are They Related?" *Progress in Human Geography* 24(3): 347–64.

Aisher, Alex, and Vinita Damodaran. 2016. "Introduction: Human-Nature Interactions Through a Multispecies Lens." *Conservation & Society* 14(4): 293–304.

Bakudila, Anselme. 2017. "La Pastorizia Nelle Aree Appenniniche: Analisi Dell'ultimo Decennio." Stati Generali Delle Comunità Appenniniche. *Slow Food*. Retrieved 18 July 2020 from http://www.slowfood.it/wp-content/uploads/2018/07/LA-PASTORIZIA-NELLE-AREE-APPENNINICHE.pdf.

Ballacchino, Katia, and Letizia Bindi. 2017. *Cammini Di Uomini, Cammini Di Animali. Transumanze, Pastoralismi E Patrimoni Bioculturali*. Campobasso: Il Bene Comune Edizioni.

Barlett, Peggy F. 2016. *Agricultural Decision Making: Anthropological Contributions to Rural Development*. Orlando: Academic Press. First published in 1980.

Behnke, Roy, and Mark Freudenberger. 2013. *Pastoral Land Rights and Resource Governance, Overview and Recommendations for Managing Conflicts and Strengthening Pastoralists' Rights*. Retrieved 21 July 2020 from http://www.usaidlandtenure.net/sites/default/files/USAID_land_tenure_pastoral_land_rights_and_resource_governance_brief_0.pdf.

Bender, Barbara. 2001. "Landscapes On-The-Move." *Journal of Social Archaeology* 1(1): 75–89.

Bindi, Letizia. 2012. "Manger Avec Les Yeux. Alimentation, Representations De La Localité Et Scenarios Translocaux." In *Patrimoine Et Valorisation des Territoires*, ed. Laurent Fournier, Denis Crozat, Charles Bernié-Boissard, and Charles Chastagner, 53–64. Paris: L'Harmattan.

———. 2020. "Take a Walk on the Shepherds Side. Transhumance and Intangible Cultural Heritage." In *A Literary Anthropology of Migration and Belonging*, ed. Michelle Tisdel, and Cecilie Fagerlid, 22–53. New York: Palgrave Macmillan.

Bindi, Letizia, and Cristina Grasseni. 2014. "Media Heritagization of Food." *Arxius De Ciencias Sociales* 30: 58–72.

Boeri, Stefano. 2020. "Via Dalle Città, Nei Vecchi Borghi C'è Il Nostro Futuro." *La Repubblica*, 20 April. Retrieved 21 April 2020 from http//rep.repubblica.it/pwa/intervista/2020/04/20/news.

Breda, Nadia, and Franco Lai. 2011. *Antropologia del "Terzo Paesaggio." Il mestiere dell'antropologo, Antropologia dell'Ambiente*. Roma: CISU.

Calandra, Lina. 2019. "Pascoli E Criminalità in Abruzzo. Quando La Ricerca Si Fa Denuncia (L'Aquila, 30 Giugno 2019)." *Semestrale Di Studi E Ricerche Di Geografia*. XXXI(2): 183–87.

Carnegie, Edward, and Stephen Mccabe. 2008. "Reenactment Events and Tourism: Meaning, Authenticity and Identity." *Current Issues in Tourism* 11: 349–68.

Cejudo Garcia, Eugenio, and Francisco Navarro Valverde. 2020. *Neoendogenous Development in European Rural Areas: Results and Lessons*. New York: Springer.

Ciaralli, Mario. 1997. *Il Paese Dei Dotti. Cornillo Nuovo La Storia, Le Immagini*. Roma: Editrice Aurelia.

Cork, Steven. 2010. *Resilience and Transformation: Preparing Australia for Uncertain Futures*. Sidney: Csiro Publishing.

Council of Europe. 2019. "Cultural Routes of the Council of Europe Programme—Activity Report." Retrieved 31 July 2020 from https://rm.coe.int/cultural-routes-of-the-chttps://rm.coe.int/cultural-routes-of-the-council-of-europe-activity-report-2019/16809e5074 percent20ouncil-of-europe-activity-report-2019/16809e07ae.

Davies, John, Pedro Herrera, Jabier Ruiz-Mirazco, Jennifer Mohammed-Katerere, and Emmanuel Nuesiri. 2016. *Improving Governance of Pastoral Lands. Governance of Tenure Technical Guide 6*. Rome: FAO.

Davis, Dona, Anita Maurstad, and Sarah Cowles. 2013. "Co-Being and Intra-Action in Horse-Human Relationships: a Multispecies Ethnography of Be(Com)Ing Human and Be(Com)Ing Horse." *Social Anthropology* 21(3): 322–35.

De Rossi, Antonio, ed. 2018. *Riabitare l'Italia. Le aree interne tra abbandoni e riconquiste*. Roma: Donzelli.

Folke, Carl. 2006. "Resilience: The Emergence of a Perspective for Social-Ecological Systems Analyses." *Global Environmental Change* 16(3): 253–67.

Goodman, David, Melanie E. Dupuis, and Michael K. Goodman. 2012. *Alternative Food Networks: Knowledge, Practice, and Politics*. New York: Routledge.

Grasseni, Cristina. 2003. *Lo Sguardo Della Mano*. Bergamo: Bergamo University Press.

Haraway, Donna. 2008. *When Species Meet*. Minneapolis: University of Minnesota Press.

Herzfeld, Michael. 1985. *The Poetics of Manhood: Contest and Identity in a Cretan Mountain Village*. Princeton: Princeton University Press.

———. 2001. *Anthropology: Theoretical Practice in Culture and Society*. New York: Blackwell.

———. 2004. *The Body Impolitic: Artisan and Artifice in the Global Hierarchy of Value*. Chicago: University of Chicago Press.

Hidalgo, Anne. 2020. *Le Paris Du Quart D'heure*. Retrieved 21 June 2020 from https://annehidalgo2020.com/wp-content/uploads/2020/01/dossier-de-presse-le-paris-du-quart-dheure.pdf.

Ingold, Timothy, and John L. Vergunst, eds. 2008. *Ways of Walking: Ethnography and Practice on Foot*. London: Routledge.

Lebaudy, Guillaume. 2016. *Les métamorphos du bon berger. Mutations, Mobilités, mutations et fabrique de la culture pastorale du sud de la France*. Avignon: Cardère.

Lebaudy, Guillaume, Patrick Fabre, Stefano Martini, and Maria Elena Rosso. 2012. *La Routo. Sulle Vie Della Transumanza Tra Le Alpi E Il Mare*. Ecomuseo Della Pastorizia/Maison De La Transhumance. Cuneo: Nerosubianco Edizioni.

Magnaghi, Alberto. 2017. *La Conscience Du Lieu*. Paris: Eterotopie France.

Mee, Kathleen J., and Sarah Wright. 2009. "Geographies of Belonging." *Environment and Planning A: Economy and Space* 41: 772–79.

Melotti, Marxiano. 2013. "Turismo Culturale E Festival Di Rievocazione Storica. Il Re-Enactment Come Strategia Identitaria E Di Marketing Urbano." In *Mobilità Turistica Tra Crisi E Mutamento*, ed. Romina Deriu, 144–54. Milano: Franco Angeli.

Mencini, Giannandrea. 2021. *Pascoli Di Carta. Le Mani Sulla Montagna*. Vittorio Veneto: Kellerman Editore.

Metalli, Ercole. 1903. *Usi E Costumi Della Campagna Romana*. Roma: Nuove Edizioni Romane.

Monllor, Neus, and Emma Soy. 2015. *El Pasturisme: Un Producte Turístic Nou Per a Les Comarques Gironines? Anàlisi De Recursos I Propostes De Futur*. Retrieved 25 July 2020 from https://premisg.costabrava.org/wpcontent/uploads/2017/11/02_ybettebarbaza_pasturisme_issu.pdf.

Moreno, Carlos. 2016. *La Ville Du Quart D'heure: Pour Un Nouveau Chrono-Urbanisme*, Retrieved 21 June 2020 from https://www.latribune.fr/regions/smart-cities/la-tribune-de-carlos-moreno/la-ville-du-quart-d-heure-pour-un-nouveau-chrono-urbanisme-604358.html.

Morris, Brian. 2014. *Anthropology, Ecology, and Anarchism. A Brian Morris Reader*. Oakland: PM Press.

Müller, Oliver. 2021. "Making Landscapes of (Be)Longing. Territorialization in the Context of the Eu Development Program Leader in North Rhine-Westphalia." *European Countryside* 13(1): 1–21.

Norris, Fran H., Susan P. Stevens, Betty Pfefferbaum, Karen F. Wyche, and Rose L. Pfefferbaum. 2008. "Community Resilience as a Metaphor, Theory, Set of Capacities, and Strategy for Disaster Readiness." *American Journal of Community Psychology* 41: 127–50.

Padiglione, Vincenzo, ed. 2014. "Etnografie del contemporaneo II: Il Post-agricolo e l'Antropologia." Special issue, *Antropologia Museale* XII (34/36).

Palladino, Paolo. 2017. "Transhumance Revisited: On Mobility and Process Between Ethnography and History." *Journal of Historical Sociology* 31(2): 119–33.

Papa, Cristina. 2013. "Sviluppo Rurale E Costruzione Della Qualità. Politiche Globali E Pratiche Locali." *Voci* X: 153–62.

Petrocelli, Edilio, ed. 1999. *La Civiltà Della Transumanza*. Isernia: Cosmo Iannone.
Poli, Daniela. 2019. *Rappresentare Mondi Di Vita: Radici Storiche E Prospettive Per Il Progetto Di Territorio*. Milano: Mimesis.
Putnam, Robert D. 2000. *Bowling Alone: The Collapse and Revival of American Community*. New York: Simon and Schuster.
———. 2007. E Pluribus Unum: Diversity and Community in the Twenty-First Century. The 2006 Johan Skytte Prize Lecture. *Scandinavian Political Studies* 30: 137–74.
Russo, Saverio. 2002. *Tra Abruzzo E Apulia: La Transumanza Dopo La Dogana*. Milano: Franco Angeli.
Soussana, Jean-François, Vincent Allard, Kim Pilegaard, and Per Ambus. 2007. "Full Accounting of the Greenhouse Gas (CO2, N2O, CH4) Budget of Nine European Grassland Sites." *Agriculture Ecosystems & Environment* 121(1): 121–34.
Steiner, Artur, and Marianna Markantoni. 2013. "Unpacking Community Resilience through Capacity for Change." *Community Development Journal* 49(3): 407–25.
Tantillo, Filippo. 2020. "Il Paese Remoto, Dopo La Pandemia." *Dialoghi Mediterranei* 43/2020.
———. 2016. *Quel Che Resta. L'Italia Dei Paesi, Tra Abbandoni E Ritorni*. Roma: Donzelli.
Trinchieri, Roberto. 1953. *Vita Dei Pastori Nella Campagna Romana*. Roma: F.Lli Palombi Editori.
Ugarte, Enrique, Isabel Casasús, Ian Doria Otaegi, Miguel Barado, Maria Ruiz, and Roberto Ruiz. 2014. "Las Escuelas De Pastores: Iniciativas Innovadoras Para La Potenciacion Del Sector Ovino, El Ejemplo Del País Vasco Y Cataluña." In *CYTED—Ciencia Y Tecnología Para El Desarrollo* (Programa Iberoamericano). Zaragoza: Cita.
Van Der Ploeg, J. Douwe. 2008. *The New Peasantries: Struggles for Autonomy and Sustainability in an Era of Empire and Globalization*. London: Sterling.
Viazzo, Pier Paolo. 1989. *Upland Communities: Environment, Population and Social Structure in the Alps since the Sixteenth Century*. Cambridge, UK: Cambridge University Press.
Wilson, Geoff. 2012. "Community Resilience, Globalization, and Transitional Pathways of Decision-Making." *Geoforum* 43(6): 1218–31.
Wright, Stephen. 2014. "More-Than-Human, Emergent Belongings: A Weak Theory Approach." *Progress in Human Geography* 39(4): 391–41.
Yuval-Davis, Nira. 2006. "The Politics of Belonging." *Patterns of Prejudice* 40: 197–214.
Zane, Massimiliano. 2020. "Paesaggio, turismo e pandemia: un "valore" da maneggiare con cura." *ArtTribune*. Retrieved 18 April 2020 from https://www.artribune.com/turismo/2020/04/paesaggio-pandemia/.

CHAPTER 8

Continuities and Disruptions in Transhumance Practices in the Silesian Beskids (Poland)
The Case of Koniaków Village

Katarzyna Marcol and Maciej Kurcz

Transhumance is a phenomenon difficult to define, covering many different practices, which are shaped by ecology and culture. It is a variety of pastoralism, which is supplemented by land cultivation and trade. According to Schuyler Jones (2005), it includes: rural settlement, land cultivation, and periodic migration of people and animals between mountain pastures. In 2004, Piotr Kohut, a resident of Koniaków, managed to bring back to life a seemingly completely extinct institution of the pastoral community in the area of the Silesian Beskids—the so-called *sałasz*. *Sałasz* is a type of farming specific to the Carpathian area, which consists in setting up shepherding companies in which a few or more sheep owners join their flocks, placing them under the supervision of a *baca* (chief shepherd) who is responsible for grazing of animals and shares proportionately the benefits obtained from the milk. This communal grazing, which was associated with certain cultural behaviors and the way of life, was a key determinant of Carpathian pastoralism for many centuries, but economic and political factors disturbed its functioning. After many years of interruption, the sheep of local *gazdas* (owners, breeders) were merged together into one flock again, which was tended by a collectively elected *baca*—a head of the shepherding team—until the end of the pastoral season. This event can be considered as the actual beginning of shepherding restitution in the Cieszyn Silesia region.[1]

Pastoralism in the Silesian Beskids today takes on the form of a nostalgic story about the cultural heritage and history of the region, and in particular about an ideal order existing "before time." For breeders, academics, or the region lovers, pastoralism has a "structurally nostalgic" character (to use Michael Herzfeld's phraseology). It is a kind of myth about prosperity and well-being, about a state of perfect harmony between man and the ecosystem, but also between the state and the citizen, which (not irretrievably at all) was lost in the second half of the nineteenth century. For many, the golden period in the history of the Silesian Beskids ended with the imperial patent of 1853 and the gradual disappearance of *sałasz* pastoralism. Since then, pastoralism in this area has been something constantly returned to—in the period after Poland regained its independence in 1918, after the end of World War II or finally after the fall of Communism. It is a project that has never fully come true—constantly disrupted, more in the sphere of wishes than reality. In other words: the transhumant pastoralism in the Silesian Beskids, at least since the second half of the nineteenth century, has been an object of continuation, but also of disruption—both factors have been equally cocreating different variants of this phenomenon in the area, leading simultaneously to its ups and downs. As Frances Pine (2002: 98) noted, in the post-socialist reality, almost every discussion about the change and continuation is disturbed by a certain regularity. Namely, many social or economic processes that seem to be quite new, at a closer look, turn out to refer to former structures, and those that would seem to be continuations in fact appear as completely new creations. The institution of *baca*, *sałasz*, or *redyk*—so characteristic of the Silesian Beskids' pastoralism—seem to a large extent be a part of this very logic.

The COVID-19 pandemic was not originally considered in our research plans. We have decided, however, to extend them by this very subject, and despite the fact that at the time of this study the end of the pandemic still cannot be seen. We think, however, that it will undoubtedly become another disturbance in the series of disturbances in the process of restoring pastoralism in the conditions of post-socialist Poland. This is a phenomenon that affects the changes in the landscape of animal husbandry like none before. Today's shepherds are not only forced to comply with the requirements of work in the epidemic regime, but they also have to thoroughly rethink their activities: work structure, trailing routes, or distribution systems. With the pandemic, the way of thinking about shepherding is also changing, which in turn has a huge impact on the dynamics of this activity in the economic structure of the inhabitants of the region.

For this reason, among other things, the perspective of "continuation and disruption"—particularly popular when discussing the turbulences of the world economy or neoliberal globalization—has become the center of our investigations into modern pastoralism. Among other things, it raises the following questions: What kind of dynamics can we observe in transhumance in peripheries of Eastern Europe that can be grasped through the continuity and disruption perspective? What kind of relationships can (new) pastoralism establish with territories, animals, knowledge production, or gender? What are the relationships between pastoralism and the processes of heritagization? To what extent is public support not properly used? To what extent do the new discourses about more sustainable/responsible development, animal rights, or climate change influence herding practices?

The Shepherd's Centre in Koniaków as a Case Study

The transformations of the Carpathian pastoralism in the post-socialist period, as yet, evoke moderate interest of researchers (Costantin2003; Sendyka and Makovicky 2018). In Poland, for years the attention of scholars has actually focused on one area only—the so-called Podhale, that is, the region closest to the Tatra Mountains, while little is known about pastoralism in the Low Beskids or Silesia Beskids. That is why we have decided to analyze a case study of *sałasz* farming in Koniaków in the Silesian Beskids. In the village of Koniaków, the highest elevated village in the Silesian Beskids, inhabited by about 3,500 people, pastoralism is developing dynamically under the watchful eye of *baca* Piotr Kohut, who is responsible for summer grazing of one of the largest flocks of sheep in the region of about 1,200 sheep. Grazing takes place on Hala Ochodzita, Hala Barania, Magurka Radziechowska, Podgrapy, and Kamesznica, that is, within an area covering about fifty square kilometers. Activities connected with *sałasz* farming are carried out by the members of the Kohut family, especially Piotr's wife, Maria, and include running a Shepherd's Centre, where activities are carried out with the tradition of sheep grazing being the main common denominator. The Shepherd's Centre in Koniaków is a complex which consists of buildings inspired by the architecture of traditional Beskids buildings: *na szańcach* (a mountain hut), *gazdówka* (a shepherd's hut, with utility rooms where sheep and goats are kept), and the Highlander Shop. In the period of summer grazing—when sheep are on pastures—the local shepherd's hut hosts demonstrations of how to make sheep milk cheese, such as *bundz, bryndza, redykołka*, and *oscypek*. Seasonally, apart from sheep milk cheese, tourists can also buy lamb meat. All year round, there is an exhibition of shepherding utensils, and a photo exhibition showing life on pastures.

The Shepherd's Centre offers an interactive educational path on the processing of sheep fleece and wool with the use of old and modern methods. Visitors can participate in lectures on traditional sheep grazing in the Carpathians, its importance for the preservation of cultural heritage and

Figure 8.1. The Shepherd's Centre, Koniaków, 2019. © Katarzyna Marcol

biodiversity, and taste traditional regional cheeses or enjoy a demonstration of how milk is processed into *bundz*. Maria Kohut is the president of the Transhumance Pastoralism Foundation, whose main objective is to support activities related to traditional pastoralism in the Carpathians and the Balkans by, among other things, promoting sheep products, organizing events, conferences, seminars, workshops, training, festivals, and concerts referring to the cultural and material heritage of the Carpathian and Balkan communities. The intentions of the foundation also focus on activities to protect the environment and natural heritage of these regions, as well as to support *bacas, juhases, gazdas*, and sheep breeders. The Shepherd's Centre and the foundation are the organizers of events devoted to traditional pastoralism, which are very popular among the region's inhabitants and tourists: mixing of the sheep (gathering animals belonging to many owners into a single flock) and *redyk* (the first spring trailing of the sheep onto mountain pastures) which start the pastoral season at the beginning of May, as well as the Shepherds' Fair and the Tastes of the Carpathians culinary workshops, promoting folk crafts and traditional Carpathian regional products, especially sheep milk cheese.

The purpose of this study is to define the sociocultural situation in Koniaków under the influence of pastoralism which is being reborn, from the

Figure 8.2. An exhibition of shepherding utensils in the Shepherd's Centre, Koniaków, 2016. © Katarzyna Marcol

perspective of continuation and disruption of this process. This analysis will be used, on the one hand, to show the ways of negotiating social relationships in a local community, and on the other hand, emancipation of shepherds as a separate social group. We will provide examples of actions taken to revive awareness of pastoralism in the local community, giving it meaning through reference to collective memory and establishment of cultural heritage of the Wallachians as the basis of a common identity. In our analysis it is also important to illustrate the individual motivations of the members of the Kohut family to take action for the development of traditional pastoralism, often in spite of adversities like, for example the ones COVID-19 pandemic has recently brought. Listening to the stories of *baca* and his wife has allowed us to observe the way they assign meaning to their work, establish the sense of daily activities, and create self-identification based on pastoral traditions. Because the narratives contain the way of interpreting the reality, they give meaning to activities taken up (Taylor 2001: 95). Understanding one's own experience and giving it a form is possible by referring to known and accepted cultural codes, which makes the human being a self-interpreting creature (Taylor 1985: 54; Ricoeur 2003: 297–354). The narrative is constitutive not only for creating the reality, events, or experiences, but also for shaping the identification of the narrator who acts and experiences (Carr 1986: 126; Carr 1991: 45–99).

The case study is a qualitative approach that uses in-depth data collection procedures of a diversified character (Creswell 2007: 73). The main research techniques were unstructured interviews conducted in Koniaków and other places of the Silesia Beskids in 2018–20. Beskids' shepherds were the respondents, and the interviews concerned the daily lives of their families, current problems related to farming, the meaning of pastoral culture in keeping the region's traditions, influence of economic factors on the changes of modern culture, and relationships in the local community. Talking about the shepherds we mean both *bacas* themselves (main shepherds running a shepherding company), *juhases* (shepherds grazing sheep on mountain pastures), and *honielniks* (boys helping with sheep herding), and whole multigeneration families for whom shepherding is, on the one hand, a source of income, and on the other, a way of life (in talks with members of shepherds' families a phrase kept appearing that work sets the rhythm of their life and it is not possible to separate work from free time).

The research also included observational studies during visits to cultural events and educational workshops devoted to pastoralism in various towns of the Silesian Beskids, including Koniaków, Istebna, and Wisła in the years 2016–19. These were events connected with spring and autumn *redyks* (i.e., trailing of the sheep onto the mountain pastures and back from the pastures to a farm for winter), museum exhibitions and workshops held at the Shepherd's Centre in Koniaków, concerning, among other things, wool processing, or sheep milk cheese production. We also participated in workshops being a part of the "Cultural Ecology"[2] project devoted to shepherding and running a multifunctional pastoral farm, taking into account organizational, legal, social, and economic aspects, as well as in educational meetings devoted to activities aimed at preserving the natural and cultural heritage of mountain areas.

This study includes also an analysis of how texts are a part of the contemporary public discourse in terms of its cognitive aspects, making it possible to read a certain repertoire of sociocultural beliefs (Van Dijk 2001: 26). In this case these beliefs will be related to the construction of cultural memory and the self-identification of contemporary shepherds and Beskids highlanders. An important role in shaping the image of the past is also played by statements from and actions taken by social actors, i.e., people involved in activities for their group, who are *bacas* and Beskids *sałaszniks* enjoying social respect, such as, Józef Michałek of Istebna, Jan Kędzior of Wisła, and Piotr Kohut of Koniaków. In this chapter we are interested both in the selection of information and the use of language, and the images and symbols accompanying it, that evoke certain visions of the world that influence the process of modeling collective memory, shaping, as a consequence, our own community identity. Our research approach to media sources and public messages (workshops, lectures, promotional materials) assumes their constructivist role in establishing social life, as these messages strongly affect the recipients' beliefs about the world and have an impact on their self-image (they influence self-identification). During the research we also analyzed documents from EU, international, and local government programs supporting the revitalization of pastoral economy in Europe, including the Silesian Beskids, which, on the one hand, made the objectives and priorities of the subsidized activities visible and, on the other hand, made it possible to observe the effects of the awarded grants.

History of Shepherding in the Silesian Beskids— Continuation and Disruptions

The sałasz and pastoral farming in the Silesian Beskids was the basis for the existence of local highlanders at least until the mid-nineteenth century (Popiołek 1939; Kopczyńska-Jaworska 1950–51). This was the migration route of high mountain shepherds, known as Wallachians, who grazed sheep moving along the ridges of the Carpathians from the southeast part of Europe, from the Balkan, Wallachian, Moldavian, and Transylvanian areas as early as at the turn of the thirteenth and fourteenth centuries (Kocój 2015: 276–77; Kocój 2018a: 55). In sixteenth-century documents from Cieszyn, the Wallachians are recorded as settlers who brought the ability to survive in difficult mountain conditions and to farm based on sheep pastoralism and milk and wool processing to the area of the Silesian Beskids (Spyra 2007: 39–41). They entered into relations with the local population, especially in winter, when they sought refuge in lower villages, which resulted, on the one hand, in certain elements of the Wallachian culture and economy being taken over by the Beskids settlers and, on the other hand, in migrants settling in local villages (Jawor 2000: 39–41). Therefore, over time, the Wallachians passed their culture on to the Indigenous peoples, and with migration many customs and rituals related to sałasz farming and pastoralism were adopted here, especially the already mentioned community economy (Kocój 2018a: 55–131).

The seventeenth century and the first half of the eighteenth century are considered to be the "golden period of sałasz farming in Silesia" (Kiereś 2010: 27). Sheep and dairy products were to be the guarantee of prosperity and stability of the local population. Thanks to the favor of the rulers of the Duchy of Teschen, the highlanders increased their flocks and pastures at the expense of the trees, paying high taxes to the Piasts of Cieszyn (until 1653) and later to the Habsburgs. The situation changed with a new economic policy, which tightened the regulations concerning the ban on tree cutting in order to preserve forest resources, necessary for developing heavy industry. Wood from the Beskids forests became a scarce commodity as it was used to burn the smelter furnaces in which the local ore deposits were processed. Thus, the forests became an extremely valuable asset, and the sałasz farming was an obstacle to maximize profits of the Teschen Chamber. However, the decline of the Beskids sałasz farming was eventually sealed by the issue of a patent by Emperor Franz Joseph I in 1853, under which the Teschen Chamber was recognized as the sole owner of the forest areas and did not have to respect previous agreements with the highlanders as regards to access to pastures, which effectively prevented sheep from being grazed on collective pastures (Spyra 2007: 143). Financial burdens and restrictions on access to pastures imposed upon the highlanders by successive authorities of the Teschen Chamber made shepherds move over time to mixed pastoral and agricultural farming. The change in the lifestyle of the inhabitants of the Beskids villages was later affected not only by the development of the wood economy and metallurgical industry but also by the flourishing of tourism, initiated at the turn of the nineteenth and twentieth centuries.

In the twentieth century, a further gradual and systematic decline of the pastoral economy continued. After World War I and the collapse of Austria-Hungary, the Beskids highlanders, who found themselves on the Polish side of the border with Czechoslovakia, took legal actions to regain the sałasz farming areas taken from them, which were owned by the Polish state after the war. The contingent of highlanders who wanted to restore the shepherding in the Silesian Beskids accepted the new Polish state with high hopes. Although the Polish Parliament accepted the sałasz shepherds' request and in 1921 passed a revision law that established the equivalents for lost property, and the established sałasz companies were slowly restoring pastures for grazing, World War II put an end to these efforts (Kiereś 2019: 59–71). After World War II, all attempts to revitalize the traditional pastoralism were rejected by the communist authorities, aiming to limit the freedom of economic activity and to liquidate dairy cooperatives (Kocój 2018b: 86). The scale of regression in pastoralism is demonstrated by statistics concerning the number of sheep grazed in the Silesian Beskids: in 1910 there were 4,411 sheep; in 1937 there were 2,310; and in 1947, there were 1,660 (Program Aktywizacji Gospodarczej 2007: 18). The economic changes that followed the collapse of communism in

1989 in Poland led to an even greater decline in the sheep population and to almost complete disappearance of the profession of baca and shepherd (Program Aktywizacji Gospodarczej 2007: 5).

The situation improved with Poland's accession to the European Union in 2004 and with the economic stabilization of Poland within the democratic and capitalist system. Membership in the EU made it possible to benefit from programs and subsidies for the revitalization of sheep farming and pastoralism from an economic, natural, and sociocultural perspective. An example is the Interreg V-A Poland-Slovakia 2014–20 program (financed by the European Regional Development Fund), under which the cross-border Wallachian Culture Trail was created, i.e., an integrated cultural trail to popularize pastoral traditions on the Polish-Slovakian border. In 2012–16, in turn, a project supporting knowledge about the protection and sustainable development of the Carpathians, written in the "Carpathian Convention," was implemented (an international agreement concerning a single mountain region, established on the basis of treaty rules of international law), using the funds of the Swiss program of cooperation with the new EU Member States: Carpathians Unite—Mechanism of Consultation and Cooperation for Implementation of the Carpathian Convention. The main activities carried out by Poland within this framework were: creating a sustainable tourism strategy and a code of good practice for spatial planning; supporting the development of pastoral management (training of shepherds, construction of shepherds' huts, purchase of sheep); promoting knowledge about the nature and culture of the Carpathians; developing a draft national action plan for the implementation of the provisions of the protocol on the "Protection and Sustainable Use of Biological and Landscape Diversity to the Carpathian Convention in the Polish Part of the Carpathians"; and developing a draft protocol on "Cultural Heritage to the Carpathian Framework Convention." In addition, since 2007, in Silesia funds has been awarded for activities related to pastoralism from the budget of the Self-Government of the Silesian Voivodship under the Sheep Plus: Program for Economic Activation and Preservation of Cultural Heritage of the Beskids and the Kraków-Czestochowa Upland. The latest edition of the program, planned until 2020, sets two main objectives, the first of which concerns activities to preserve, protect, and restore elements of nature on the basis of pastoral economy, and the second, to nurture the cultural identity associated with pastoralism, promoting the traditions of folk culture, the development of crafts, and the processing of products of sheep and goat origin. As a result of the implementation of this program, in accordance with the Regulation of the Minister of Labor and Social Policy of 27 April 2010 on the classification of professions and specialties for the needs of the labor market and the scope of its application, two new professions were registered: baca and juhas.

The programs supporting the preservation of pastoralism have resulted, on the one hand, in a greater dynamics of the economic activity (the number of sheep grazed in the mountain meadows has increased, thus increasing the availability of sheep products, enriching the range of local products) and, on the other hand, they have increased the awareness of the local community and people from outside the region about the natural and cultural values of mountain areas (Wojewódzki Program Aktywizacji 2015: 6–7). The visitors to the Beskids and journalists increasingly enjoy events such as: miyszani owiec (mixing of the sheep) and trailing them to the mountain pastures; the autumn rozsod, which is the return of sheep from the pastures and giving them back to their gazdas (owners); as well as the tasting of regional products, performances of highlander groups and bands, folk handicraft fairs. Sheep milk products are increasingly sought after by consumers, and Brinda podhalańska and oscypek cheese, produced in the Beskids among others, have been entered by the European Commission in the register of Protected Designations of Origin and Protected Geographical Indications. Piotr Kohut, as one of the few sheep milk cheese producers in Poland, is authorized to produce these local products.

What is most important for our study, however, is that the revival of pastoral traditions is accompanied by a process of restoring memory of the Wallachian heritage in the local community, and together with it, a sense of identity based on the continuation of this heritage by the Beskids shepherds. The collective memory, lying at the foundations of the rebuilt identity, is stimulated by institutions (mainly NGOs) which raise funds to promote the cultural heritage of the region. Thus, we are dealing with contemporary cultural memory, which is expressed by the group's conscious attitude to the past embedded in a specific cultural space, and its carriers are people involved in the activities of institutions supporting memory and thus affecting group identity (Assmann 2016: 68–71).

Continuity? Beskids Shepherds as Heirs to the Wallachian Tradition

The revitalization of pastoral traditions stimulated by international and local government programs has resulted in a rebirth of memory of the past, especially about the pastoral roots of the local community. As Jan Assmann (2016: 85–86) notes, it is the power—understood in this text as the policy of the state and the European Union as regards setting financial priorities—

that is a strong stimulator of memory. Funds transferred to "nurture and maintain traditions, customs and other elements of folk culture associated with pastoralism" and to "preserve the identity, individuality and heritage of the inhabitants of the region" (as defined in the documents of the Sheep Plus program until 2020) have contributed to the process of searching for sources of identity not only among shepherds but also among the members of the local community. This applies in particular to people who—although not directly associated with pastoralism—are under the influence of factors shaping memory and collective identity based on pastoral traditions, among others, who are participants of cultural events, workshops, lectures on cultural heritage, members of folk bands and ensembles, folk artists presenting and selling their products during handicraft fairs organized during events connected with pastoralism.

The creation and processing of identity determinants that are important for the group in the process of distinguishing one's own (shepherds, highlanders) from others (tourists, people from outside the local community) takes place in relation to collective memory (Halbwachs 1969: 217–61). The identity function of memory is mentioned expressis verbis by Barbara Szacka. She defines collective memory of the past as images of the past of one's own group, constructed by individuals from the information they remember coming from various sources and reaching through various channels. Then this information is selected and transformed according to one's own view of the world and of socially generated cultural standards. Thanks to these cultural standards, shared by members of a given community, the images of the past are codified, which, in a consequence, allows members to speak about the collective memory of history of one's own group. The collective memory understood in this way is not static, but changeable and dynamic, and additionally it is a field of clashes and mixing of the past images constructed from various perspectives (Szacka 2006: 47–54).

The basis of contemporary construction of memory of the pastoral tradition is the so-called Wallachian heritage. For the members of the local community, it is a very strong self-identification factor. Wallachians are considered by contemporary shepherds as ancestors, to whom they owe not only the way of farming and organization of social life but above all their own culture and identity, as well as some sort of a transnational connector with other communities involved in sheep grazing and living in the Carpathians. A mediator between tradition and modernity, between ancestors and contemporary inhabitants of the Beskids Mountains is the figure of baca—the chief shepherd, and a cultural guide, too. He is the one who is predestined by tradition and community to carry out activities that bring back the memory of the whole shepherding culture and make the preservation of the shepherding ethos a reality. According to tradition, baca was the most important person in sałasz farming, and most frequently an older, experienced person became baca. He was responsible for processing sheep milk into cheese and for its quality (he was punished for mismanagement), he allocated grazing areas to shepherds and helped with milking. At the beginning of the season, he received sheep from the miszaniks, at the end of the season he had to return the same number or compensate the owner for each missing animal (Štika 2005: 15–20). The restoration of the sałasz economy in the Silesian Beskids resulted in giving baca another "new" function—he has become a guardian of tradition and the duty to preserve the cultural heritage rests upon him. In the statements of Beskids' bacas and shepherds, often a reference to the Wallachian tradition appears, as a source of their own identity and of all Carpathian highlanders. The words of Piotr Kohut can be an example here. He identifies the modern pastoral and sałasz economy in reference to the "Wallachian heritage," speaking of the "Carpathian community." Piotr Kohut was one of the participants of the Carpathian Trailing of the Sheep: Transhumance 2013, i.e., the shepherds' trek with a flock of three hundred sheep, with dogs, donkeys, and horses through the areas of Romania, Ukraine, Poland, Slovakia, and the Czech Republic. During the trailing, in the period from 11 May to 14 September 2013, shepherds with animals covered the distance of about twelve hundred kilometers. The project participants stressed from the beginning that their effort was a tribute paid to the Wallachian ancestors, and the journey is a testimony to the shepherds' identity, the maintenance of which is the duty of modern highlanders. The Watra Podhale Information Service published an announcement of the Carpathian Trailing of the Sheep together with a statement of Piotr Kohut, who explained the aim of the project as follows:

> With our journey through the Carpathians, we pay homage to our Wallachian ancestors, pointing to an identity that should be protected and guarded—above all by ourselves. The dignity of the shepherd results from his awareness of his own identity on the way. Let us not be deceived by the illusory impression of a civilization leap, the presence offering prosperity and longevity in exchange for undermining our spirituality. The world of modern technologies should be used by us as much as possible, but let us not forget, however, that our strength lies in tradition and pastoral roots. ("Ruszył Redyk Karpacki 2013" 2013)

This statement shows that the processes of memory transmission and remembering take place on two planes: social and historical. Robert Traba, discussing Jan Assmann's theory, points out exactly to the aspect of combining culture, which is common for a group at a certain time, and history, which is a sense of link with ancestors:

> The transmission of memory and remembering takes place in specific conditions of cultural development. It creates this culture and, at the same time, makes individuals build bonds between them that enable them to identify themselves as a group with a common culture.... On a social level, a sense of cultural belonging is created between the members of the group due to living at the same time. On the historical level, however, a sense of relationship is created with previous generations defined as our ancestors. (Traba 2016: 14)

The community of memory is therefore based on the assumption of continuity between the past and the present. The group constituting itself as a community of memory creates the awareness of identity in time, which means that the remembered facts are selected and defined by virtue of their relevance, similarity, and continuity. Such a procedure is used in order to maintain the group's durability, which is the supreme value requiring the changes to be blurred and perceiving history as unchangeable continuation. If a group realized its own transformation, it would cease to exist as such and would give rise to a new group (Assmann 2016: 56).

Showing the continuity of the Wallachian heritage, and thus organizing the image of the world as a rational consequence of events, without inconsistencies and diversity, is made possible due to mythical thinking. The myth of Wallachians as the ancestors of today's highlanders is formed by repeated stories about the past, being an introduction to almost all workshops, lectures, or exhibitions devoted to pastoralism in the Beskids (e.g., organized at the Andrzej Podżorski Beskids Museum in Wisła, the Shepherd's Centre in Koniaków or the Těšín Museum in Český Těšín). Repeated and standardized texts about Wallachians, who walked along the peaks of the Carpathian Mountains with a flock of sheep and brought their laws and shepherds' habits to the Beskids, are a source of connotations and an interpretative matrix, bringing associations with highlanders as heirs to the Wallachian tradition to mind. Myths, as Jan Assmann notes, are typical figures of cultural memory as they constitute a remembered, founding story which is told in order to explain the present from the perspective of the primeval beginnings. The mythical narrative serves here both social integration and the creation of a collectively shared image of the world and a common system of values. Myth, as Roland Barthes (2008: 239–80) points out, gives things a foundation of nature and eternity and gives them clarity which does not result from explanations, but from the fact that today's shepherds are heirs of the Wallachians, and the Wallachian law, being the basis of the sałasz economy, not only established the rules of economic functioning, but is also connected with the code of ethics of shepherds resulting from the harmonious coexistence of shepherds with nature.

Collective memory functions as long as it is transmitted in the process of communication, so the transmission of knowledge about the past is a condition for the memory to exist. Memory lives in and through communication, and if the transmission is interrupted, changed, or the frame of reference disappears, then forgetting occurs (Assmann 2016: 53). The revival of the pastoral economy in the twenty-first century made it possible to return to the memory of the cultural heritage of the ancestors, which as a result of sociopolitical processes from the mid-nineteenth century to the first decade of the twenty-first century was forgotten by the community and ceased to be a factor shaping collective identity. The memory of shepherding did not disappear altogether, but functioned within a communicative memory, passed on orally from generation to generation in families where sheep breeding has been a daily activity for many generations. Nevertheless, it did not have a dimension shaping the identity of the local community, as it did not go beyond the family circle of narration and did not take on an institutional form. Community forgetting can have various foundations and can be the result of different social processes. It often results from the fading of social memory, located in individual memories, when the generation that kept the knowledge about the past in its memory passes away. The reason for forgetting can also be the concentration of attention on the present and modification of an attitude towards the past, in which generational changes play an important role. Finally, forgetting can be a result of deliberate political actions, carried out by various centers of power and with the use of various methods of exerting pressure (Connerton 2008, 2012: 87–93). It seems that in the case of the Beskids highlanders, each of these factors contributed in part to the process of forgetting the cultural heritage of the shepherds and to moving away from the identity shaped on its basis. Cultivation of this heritage has not been fostered by the authorities (from the Habsburgs' management at the turn of the eighteenth and nineteenth century, to governments during Communism and the period of political transformation in the twentieth century), whose efforts have led to the successive collapse of the pastoral economy and the marginalization of the professions of baca and shepherd, resulting in the disappearance of the reference frameworks for collective memory. At the same time, there was gradual disappearance of memory as a result of the interruption of intergenerational transmission and disappearance of the collective ritual practices that would maintain this memory. With the revitalization of pastoral traditions, images of the past of the Beskids highlanders came to life, which refer to pastoralism as the source of their own identity. As Paul Connerton (2012: 87–93) notes, the images of the past and knowledge of it are stored and maintained only in the process of communication, which takes place through language as a carrier of symbolic content and through collective ceremonies or ritual practices. It is due to the specific discourse and ritual practices currently undertaken by the Beskids shepherds that the memory of the past is shaped, which affects their own identification and the identity of the local community.

However, collective memory, like identity, is not given in its form once and forever, but undergoes transformations under the influence of external factors. The dynamics of memory results from the fact that it is an outcome of everyday discourse, constantly constructed by it from scratch. Thus, collective memory is a social construct variable in time, not a natural feature of the group, and it is closely related to the present and responds to the cultural and social needs of the community.

Disruptions Resulting from Reviving Pastoralism

The revival of traditional sheep grazing in the Silesian Beskids is not without disturbances at the level of social relations. As the Beskids bacas note, pastoralism is met with incomprehension on the part of those villagers who do not benefit from either sheep breeding or grazing. The problems vary, like for example from an unpleasant smell to violation of land ownership rights during transhumance. These issues give rise to special emotions among those who profit from tourism—i.e., the decisive majority of the rural community—for whom only the right direction for the village to develop is to transform it into something like a health resort and completely get rid of the agricultural activity. What is interesting is that in no way does it prevent the number of sheep and goat micro-breeders from growing. In the case of three villages called Trójwieś (Jaworzynka, Istebna, Koniaków) one can even talk about how trendy it is to have these animals in one's livestock and—what is worth emphasizing—without a clear objective.

That is why the shepherds call themselves a minority among the local highlanders, because of the system of values rooted in the pastoral ethos, which is supposed to make them different from the inhabitants of the cities (the so-called lords), but also from their closest neighbors who, nevertheless, lead the lives of ordinary farmers or entrepreneurs in the tourist industry. This is why local shepherds often refer to themselves as an exclusive, separate group—a kind of ethnic minority or perhaps we should say economic and ethnic. According to the theory of classical anthropology, the identity associated with a place or group is always relational and associated with maintaining differences, sometimes referred to as social boundaries (Barth 1969). In the case of Polish shepherds—or, more broadly, even highlanders—in the second half of the twentieth century the bureaucratic and economic structures of the state became an emanation of the "foreign" (Makovicky 2014).

At the beginning of the twenty-first century, regulations and projects of the European Commission became a new axis of conflict, which for some of the inhabitants of mountainous regions stood in opposition to the previous national prerogatives of the state. This is also linked to the emergence of a new understanding of peripherality—a category often referred to by shepherds from the Silesian Beskids—in which peripherality is understood not as an exclusionary force but as a building force, constituting a kind of value that should become a carrier of identification in the local or regional (pan-Carpathian) dimension. The famous Carpathian transhumance of 2013 referred to this kind of imaginary community. "The peripheries take over the role of the center, at the moment when the center becomes aware of the need for the periphery to function. When there was a problem in the village, when someone was sick, one went to baca for advice. Then he was in the center. He often knew better what was going on in the village," that is what one of bacas of Istebna said in one of the interviews about the uniqueness and the role of shepherds.

One of the basic problems every baca faces is to find the right personnel to work as shepherds (juhases). In the past they were usually young, unmarried men (Reinfuss 1959). Today, as in other pastoral communities in Europe, people involved in grazing are generally older men who, more importantly, have no one to replace them. Juhases' work is no longer attractive to the young generation, which is no longer able to reconcile it with the modern rhythm of life: school, work, or private life. Above all, however, it is not a job that has an economic value (Sendyka and Makovicky 2008: 5). Fortunately, in the first decade of the twenty-first century, the gap in the structure of shepherds of the Silesian Beskids managed to be filled by temporary workers from Ukraine. Interestingly, these are not accidental people. They come from the region of the Eastern Carpathians (the so-called Hutsul region), where shepherding traditions are still alive. It is worth adding that this region of Ukraine was within the borders of the Second Polish Republic before World War II—for this reason, these people can enjoy the status of familiarity in the Beskids (although they have little in common with Polishness today). Lack of problems in mutual communication is emphasized, for example. Ukrainians from the Hutsul region communicate in a dialect, which is supposedly understandable also for the Beskids highlanders. Since 2011, the Kohut family has been cooperating with one and the same group of Ukrainian shepherds. Piotr Kohut met them during his peregrinations in the Eastern Carpathians. Two of them, aged about sixty-five, have extensive experience in working with sheep; the others, although they come from the same region, had to learn the job. However, they also live every day on farms with smaller or larger livestock. "So, there was no need to teach them the responsibility for animals and the routine of working on the farm," argued Maria Kohut. They choose their own new co-workers (including Ukrainians) and train them to work as shepherds. The employment of workers from Ukraine enforced legal and organizational changes. The Kohut family house in Koniaków remains the logistic center; while in the field, the shepherds live in *koliba*s or campers. They feed themselves, receiving additional funds for this purpose. They also receive snuff rations. In addition, they have a passenger car and an off-road vehicle at their disposal. The Kohuts also provide them with a router for the internet. They are legally employed, and social security contributions and other required taxes are paid for them. They receive their wages in cash after the end of the season, usually on the day of departure. They are also entitled to milk products and wool without restrictions. They work in shifts—coming twice a year, in the spring-summer or autumn-winter season. Some also change after two or three months to carry out the necessary livestock-related works at home. Those who do not have agricultural farms usually stay for the whole season. Ukrainian workers also help in the winter season and take care of animals during lambing.

Another important factor negatively affecting the revitalization of traditional pastoralism is the fragmentation of land between numerous owners and the accompanying strong sense of private ownership. The postsocialist period is marked by a significant increase in the volume of trade in land, the transformation of land use plans by conversions of agricultural land and pastures into plots of land for development and their subsequent sale by owners to incoming people from large cities who use them for recreation or commercial purposes, running hotels and boarding houses. The fragmentation of large areas into smaller plots of land is also caused by the division of land between children and grandchildren so that they can run their own farms. These processes make the shepherding areas shrink immensely, and the preserved meadows and pastures are the property of people who do not always see the need to make them available for grazing. As the shepherds do not have their own land on which to graze their sheep, the shepherd is forced to negotiate the conditions for using the pastures individually with many owners (in 2020 the number of landowners on which Piotr Kohut grazed his animals reached eight hundred). This is not the case in Romania, where the mountain pastures are mostly publicly owned and managed by local authorities and can be used after paying a tax, or in the French Central Pyrenees, where access to the grazing land is legally established by municipal syndicates (Sutcliffe et al. 2013: 62–63;

Constantin 2003: 68). However, Polish legislation does not provide for such possibilities even if the pasture is wasteland. The situation is further hindered by the EU law on so-called direct payments. These are granted to persons who actually use the land and should therefore be collected by bacas for grazing sheep on pastures, while owners do not want to give up collecting payments and therefore do not agree to the entry of the flock. Shepherds, wishing to run the sałasz economy, must make individual arrangements with each landowner and establish rules of settlement with each one of them in accordance with the principle that "Łod dogwory moc zależy" (a lot depends on the agreement). That is why the preservation of the "free sheep grazing" in mountain areas is considered by the Beskids shepherds to be absolutely fundamental not only for the continuation of pastoral practices but also for the preservation of ecological balance. So far, this demand has not received any reaction from the authorities.

Shepherding in the Narratives of Piotr and Maria

Ideology, religion, and values are other important issues we have decided to devote a little more attention to during our research. In historical literature, transhumance was generally reduced to the spatial mobility of only one gender or age group—in contrast to nomadism in which all members of the community were in movement. In the case of contemporary transhumance pastoralism, the issue seems a bit more complicated. Especially in order to understand contemporary forms of pastoralism, it is necessary to listen to the narratives of the different people involved. Pastoralism today (but also yesterday) is not only about people (usually men) involved in grazing: shepherds, gazdas, or juhases—as in the case of the Carpathian arc. This institution also includes their families, whose individual members play almost equal roles in the production process, and pastoralism is a factor determining their lives regardless of the season. This is especially true today when pastoral activity often has to be combined with other, more profitable activities. In the reality of post-socialist Poland, the following polarization became the rule: baca is a sheep breeder, his wife is a businesswoman who sells the products of the shepherding farm. This arrangement is fostered by legal regulations which, for example, prohibit a shepherd from selling cheese directly to shops (Sendyka and Makovicky 2018: 11–12). This is not a completely new situation. In the past, the division of labor by gender was also kept—a man produced, a woman, in turn, was responsible for the sale and distribution of dairy products to the local markets. The woman's work was not a separate activity but was covered by a variety of family roles and responsibilities—simply it was what belonged to the woman's duties as a wife, sister, or daughter of a farmer (Pine 2002: 101). Today, we are dealing with a modification of this model, to be more precise, with the intensification of the role of a woman in a shepherding farm, who takes on her shoulders much more than she used to, and her activity is formally separated (most often in the form of her own business) and focused on the private sector. All this makes her a person at least as important in the whole system as baca himself, and she converts the whole activity into a kind of family business. A good example of this phenomenon is the Kohut shepherding family.

The Kohuts, as we have mentioned, live on pastoralism and related activities. This in turn irresistibly brings business activity to mind. For example, one of the recent projects was the construction of a guesthouse for tourists. The couple already run a shop where they sell souvenirs and food products such as cheese, cold cuts, or meat. On the premises of the farm—called a Shepherd's Centre—there is also a living museum. I would not call the conditions they live in as spartan. Their traditional style villa is one of the most impressive in the village. But the Kohuts would never call themselves entrepreneurs. The term business was rather reluctantly referred to. The reason—as they said—is that income is uncertain, and the business requires constant financial support. There seem to be other reasons too.

For both, being engaged in sheep breeding is more than a profitable activity. In our conversations, the topic of economics was always pushed into the background. The main theme was usually culture. And so, according to Piotr and Maria Kohut, the essence of their activity is tradition—to be more precise, immersion in the archaic pastoral practices and customs of the Western Carpathians. Interestingly, they both come from two different micro-regions: Piotr from Koniaków in the Silesian Beskids, and Maria from Zakopane at the foot of the Tatras. They did not take over the knowledge and shepherding skills from their ancestors, in direct intergenerational transmission, but they acquired it through their own intellectual search or baca courses. Both of them belong to the enthusiasts of Carpathian shepherds' cultural heritage. The practices and customs they collect are incorporated into—as they call it themselves—traditional, cultural grazing, in which they actively participate. The event in which the cultural drives of the couple become visible is the ritual of mixing of the sheep. It is organized in Koniaków at the beginning of May to officially start the pastoral season in the Silesian Beskids. The date of the ritual is given on the internet as a tourist attraction of the region; the website of Istebna commune informs about it. It is an all-day event filled with numerous rituals and magical rites, such as burning incense, blessing sheep, burning bonfires, all with the participation of a large audience and to the

Figure 8.3. The ritual of mixing of the sheep, Koniaków, 2017. © Katarzyna Marcol

accompaniment of highlander music. The focal point is the mixing of the sheep and trailing them three times around a fir tree. Each year, the centuries-old customs are added.

All this, according to the Kohuts, is to revive the tradition, so that it is not just an open-air museum spectacle. Their role in this process seems extremely important. They also have a clearly defined opinion on this subject. For example, in their opinion, the cultural heritage of the Beskids shepherds should stem from both written sources and oral traditions, and should be open to people from outside, to tourists. Piotr Kohut, however, has one reservation: "The mixing of the sheep is our (pastoral) holiday. Tourists are important, but not the most important." Piotr Kohut is improving the tradition all the time—he sees himself as a promoter-restorer of archaic customs. He also stresses the social dimension of his work—the need to be an authority, a man of impeccable reputation. In his case, one can even talk about the ethos of the Beskids baca: a man as hardworking as he is honest and trustworthy. Without this, cooperation with local breeders and landowners would not be possible.

Another matter—I do not know if not even more important—shepherding is a kind of lifestyle for both of them. And so freedom is important for both spouses—but each understands it a bit differently. For the man it means the possibility of full devotion to the shepherding lifestyle, a kind of escapism, effected through active participation in seasonal grazing. The man dons a woolen coat, puts on a hat and fills a pipe. So equipped, he sets off on his journey, becomes a wanderer, even a pilgrim—as Piotr Kohut himself puts it. This journey has a strong symbolic, religious or as an ethnologist would say: liminal meaning. He finds peace of mind in the pasture, says Piotr Kohut in one of his interviews. He likes this unhurried rhythm of life when he can patiently perform his duties. Looking calmly as time passes through his fingers and contemplating immeasurable landscapes, so different at different times of day. According to Piotr, the shepherd puts himself under the protection of God, and he is also an executor of His orders: he farms the land, takes care of the animals, slaughters them with his own hands, following the kosher principles, which, in his opinion, is the most humane. The same applies to milking. Piotr even uses the term "tactile pastoralism" on this occasion. He always milks with his own hands, which for him is the essence of shepherding—just like the closeness between a man and a woman in marriage. This allows him to create a special relationship with the animal. "Every sheep passes through our hands. We know everything that happens to her. When milking [manually] we know how she is, we can plan . . . we have a touch, there are sheep, that if someone milks her badly, she will not go to him," Piotr Kohut explains. Therefore, being in the pasture has a deep, existential sense for him—it transforms him, at the same time bringing him closer to the mystery of life. For him, wandering with sheep is a symbol of life. According to him, life is a road. It has its beginning and end. It has everything that is important: love, responsibility, diligence. Finally, Cain and Abel, a global pastoral myth about the polarization of human societies, appear in his narratives. This makes him feel that he is part of something larger. In the stories of Piotr Kohut, echoes of the sacredness of the institution of *baca* in folk culture can be heard. "Every *baca* has his own magic, but keeps its secrets to protect the sheep and the people," write Maria and Piotr Kohut (2018: 242). However, not only quasi-liturgical practices are meant such as wearing amulets, burning incense, or having the flock blessed, but understanding, full of mysticism, of the functions of *baca*.

> Above all, it is to protect against all evil. The responsibility is taken not only for people, but also for animals, nature, dogs and various property. Mixing of the sheep protects. That's why the incense herbs and blessing three times. In the past everything that was supernatural allowed *bacas* to produce the energy that protected the flock during that mixing. And they had to watch. At the beginning, because you get to some things, after another mixing and another, [judging by] various behaviors, I know what the season will be. I

can already predict certain things, even during the mixing. Because some things happened to me, here and there, and it happened so later. Maybe this is my imagination? But some things are becoming clearer to me. (Kohut and Kohut 2018: 243)

And some other time: "Nature talks a lot, only very slowly, sometimes for 10 years, you have to learn to read signs. Just like with a tree that is withering. Earlier, it calls for help. This is a language. This observation must be learned too. It is very similar with animals. First of all, the animal understands a lot."

For Maria Kohut, in turn, shepherding is more a matter of what is inside, the possibility of building an intimate world based on feelings and emotions, a parallel between work and family. A woman is characterized by a certain ambivalence towards standard business procedures, such as, for example, using loans for companies or transferring ownership rights to a company. Each thing, however, matters to her, has an emotional value and is connected with some sentimental detail. Many objects used in the Centre have been in her or his family for generations. Above all, she appreciates the possibility to decide and the self-sufficiency through multiplication of activities (production of cheese, meat, wool, artisan souvenirs, hotel activities). All these fields are, above all, her specialties. They all require creativity from a woman. What to do nowadays with wool—a product, as the interlocutor said—is "unsellable." Family is also important in all this, raising children. The Shepherd's Centre is both a workplace and a home. Different quality spaces intersect there. Shepherding has also been elevated to something almost sacred, full of mysticism. Thanks to the children, the eschatological element becomes visible in it. Children observe the mystery of death, learn the relationship between humans and animal, develop their own attitude towards carnivorousness, and finally learn humanitarianism in a half-Christian spirit (the idea of sacrifice, the idea of divine order) and half-archaic or even animistic spirit (the idea of mediation, equality in the world of nature). Why do we kill animals, can we guarantee them a dignified death?

The Kohuts—like many parents today—do not want their children to do what they do. The activity is too difficult and uncertain, and is connected with a label of outsiderness, peripherality. Consciously or not, they understand that their children's world will be diametrically different. This fact gives the shepherding an even deeper dimension—a breeding ground of values, of what is true, of what is unchangeable, of a certain post-capitalist, but also post-traditional philosophy of life.

To sum up: tradition, determination, creativity, sensualism, and freedom are the key words of the Kohut's pastoralism. They should be read in a relational, non-formal way, as each spouse looks at sheep farming in a bit different manner. The new neoliberal economic system, which has existed in Poland since the 1990s, is responsible for replacing the local work and labor arrangements. This seems to be particularly evident with regard to gender. New dichotomies—private-public, internal-external—have become elements of these new dynamics. The renegotiation of work, entitlements, or production are new forms of relations of the human being and market in rural areas. Finally, these elements are linked to the accelerating proliferation of the Western culture, often also in the opposition to it, in terms of economic life or consumption (Pine 2002: 98).

Disruptions during the COVID-19 Pandemic

During the COVID-19 pandemic there was no formal ban on grazing, and during the lockdown the sheep were still wintered at the breeders. However, the business activities of the Kohut family became dormant, primarily the operations of the shepherd's center in Koniaków (it was closed to tourists and residents until 4 May). Not only was the animation activity (lectures, workshops) discontinued, but also the sale of dairy products was stopped. The Kohuts decided also that in 2020 there would be no direct sale of dairy products at the baca's hut. Finally, there was also no mixing of the sheep organized every year during the long May weekend (1–3 May). The biggest problem of the initial phase of the pastoral season was not the lack of tourists, but the labor supply shortage. Closed borders prevented Ukrainian shepherds from coming, who, as we have reported, have been the core of the shepherding team for some time. There was hardly anyone who could replace them—not only because this activity is no longer popular among the local people, but what is important, during the pandemic an additional need appeared for a large part of the population to take care of children and provide them with help during distant learning. For many weeks this paralyzed, or at least made all kinds of livestock-related work or professional life of the inhabitants of the three villages much more difficult. Eventually, a team of Ukrainian shepherds arrived with some delay and in a reduced composition. It was decided that two oldest shepherds from Ukraine would not participate in this year's season (there were fears concerning their health). Four Ukrainian shepherds managed to arrive on 5 May and after a two-week quarantine they started working. The flock belonging to the Kohuts set off from the farm on 2 May, and within the following days they were joined by sheep of other breeders. Officially, the pastoral season was inaugurated on 9 May without the participation of tourists, "within the group of the closest family, friends and newspapers," as Maria Kohut reported on Facebook. During the quarantine of the shepherds, gazdas (flock owners) were working with the sheep. "This is community grazing, so everyone understood the situation and knew that if they wanted to give a sheep for grazing for the whole season, it was necessary to help during these two or three weeks," that is how Maria Kohut explained the whole situation.

Changes occurred in other fields too. In Koniaków and other neighboring regions, greater fragmentation of flocks occurred—division into separate sałasz flocks and putting them under the care of other bacas. Thus, at the beginning of the season, joint grazing of lambs and ewe lambs was organized in the Żywiec Beskids. Due to the new sanitary regime, the trailing route was also reduced, and the flocks were directed to more deserted pastures. Finally, in the case of grazing the Kohuts' flocks, their number was reduced by about 150 animals. In the 2020 season, 650 sheep were milked, while in 2018 about 1,500.

The answer to the question to what extent the pandemic brought about disruptions in the pastoral season and whether the disruptions will make permanent changes is obviously difficult at this point. We will be able to say this only after a long period of time, when the consequences of what happened are fully known. Looking at shepherding in the age of pandemic we can suppose however, that it will result in some changes, and in any case, as it is evident from the talks with the shepherds, it provoked some discussion among breeders and shepherds about how transhumance pastoralism should change.

Attention was paid to the increasing role of the family—reduction in business activity and employment of external staff. Particularly aware of this fact was Maria Kohut, who had to take on her shoulders, among other things, the role of driver (distributing the farm's products to customers). In her case it was all the more necessary as the gainful activity she was customarily engaged in—running the shepherd's shop—"got frozen." During the pandemic crisis, cooperation between breeders (gazdas), who had to engage more in joint grazing, turned out to be extremely important. Another issue was the relationship of shepherding and tourism. The question that had kept appearing for some time was in which direction should pastoral activity go, should it serve the most faithful reconstruction of cultural grazing of sheep, or should it rather meet the expectations of the commune and tourists halfway, join in the promotion of tourism in the region? An event on which this dispute focused was the aforementioned "mixing of the sheep" festival, or more precisely, its date. It is obvious for all those living on revenue from tourism that this holiday should become an inherent part of the long May weekend, during which a lot of tourists go to the mountains. For Piotr Kohut it is a factor of secondary importance—according to him only the welfare of the flock, for which he is responsible, should count. It is possible, or even necessary, to turn even more towards tradition. The pandemic crisis showed that this is a feasible postulate.

The pandemic also revealed smoldering animosities within the rural community. In the opinion of the inhabitants, the shepherds became a kind of privileged group, and thus their economic distance increased even further. At a time when people were losing their jobs and income from tourism, shepherds were allowed to carry out their activities as before. "Isn't that unfair?" asked one of the interlocutors running an agritourism farm in the area. The pandemic will probably result in even greater dynamization of the social structure in the rural areas of the Śląski Beskids.

Conclusion

Unchangeability and continuity are the most desirable human experiences, deeply inscribed in the cultural patterns of many communities (Becker 1999: 5). Mythical images of the Wallachian ancestors and the heritage they left behind, which became active with the revitalization of the sałasz pastoralism in the Beskids, and which became part of the cultural memory of the shepherds, give a sense of continuity, and with it, a sense of work done. However, should the practices described above be regarded as continuations or, on the contrary, are they completely new acts? The work of baca is tantamount to a vocation. It is no longer the fruit of intergenerational transmission. It is based on culture-building actions, constant development of a range of memories, and refers to belonging, based on an imaginary and in a way created from scratch, geography. At the same time, institutionally supported collective memory is a guarantee of the preservation of the pastoral identity, since the traditional intergenerational message no longer serves as a means of conveying knowledge about the past and no longer plays a decisive role in shaping identification. The collective identity, that is, the image of one's own group, is created through the identification of its members, and its permanence depends on the extent it motivates the thinking and actions of people belonging to the community. However, the willingness to maintain continuity, which is a guarantee of identification, is very often put to the test by various types of interference. These distractions—or, if you prefer, a continuum of distractions—have an important place in the structure of memory of people engaged in shepherding, and thus are a component of group identity. It is the memory of the man-shepherd struggle with adversities, the source of which was most frequently seen in the activities of a distant political center. In this story shepherding is a heroic passion. And its history consists of constant attempts to return to the sources.

European governance structures prove to be important for the rebirth of pastoralism in the Silesian Beskids. They influence, for example, the policy of local authorities on the issue of restitution of the shepherding subculture. Of course, they are also accelerators of local culture and identity. However, the perception of the influence of EU institutions by social actors is far from unambiguous. Some see the impulses coming from Brussels as the source of another bundle of distractions as a part of the myth of a distant and hostile center.

The basic way to consider distractions and continuity is the narrative that human beings tell about themselves. They reflect their experiences in the way the human being wants to see them and in the way they want others to see them (Becker 1999: 25). The stories of the shepherds reveal the experience of the relationships with nature, animals, and people. Sheep are not so much a passion or hobby for the Kohuts—in their narratives—as they are the foundation of their self-identification. The shepherds' relationship with animals anchors them in the ecosystem, in the calendar of nature, which, in turn, is connected with the Christian calendar. It strengthens their sense of holism and universality. Nature gives a sense of permanence and invariability even in the event of a sudden pandemic like COVID-19. The pastures became green, the sheep have to go out to graze and their milk has to be processed into cheese, which in turn has to be sold. Isn't this proof that the pastoral lifestyle is right?

Katarzyna Marcol is Assistant Professor at the Institute of Cultural Sciences, University of Silesia in Katowice and an anthropologist, folklorist, and literary scholar. The main field of her research is linguistic anthropology: she studies how language shapes communication, social identity, and group membership; how it organizes beliefs and ideologies; how it develops a common cultural representation of the social world. Her areas of interest include ethnic groups at the borderlands and multiethnic societies; traditional folklore and its adaptation in contemporary culture; and cultural ecology, especially in relation to Carpathian pastoralism and ecotourism.

Maciej Kurcz is Associate Professor at the University of Silesia in Katowice. His research interests lie in culture dynamics in modern Africa both in the context of rural societies and urban centers. His publications include *Urban Now: A Human in the Face of Borderliness and Urbanization in Juba, South Sudan* (2021), based on ethnographic research in South Sudan.

Currently, he is working on two ethno-archeological projects *Soba—the Heart of Kingdom of Alwa* and *The "Good Shepherd" of Maseeda*.

Notes

1. In the Beskids we are currently dealing with three types of sheep grazing: seasonal transhumance, when the sheep are taken out of their farms at the end of April (traditionally after St. Adalbert and St. George, i.e., after 23 April) and move with the shepherds for six days to pastures located in different mountain locations or places near the mountains within different voivodships, and for the winter they return for five days to their own farms around mid-October (traditionally on St. Michael's day); *redyk*, where the herds move under the watchful eye of shepherds between mountain pastures in different localities within a relatively short distance from their own farms, staying there from the end of April to mid-October; *sałasz*, where grazing takes place during the summer season in one village, on one or two pastures. In any case, it is communal grazing (the flocks are made up of sheep belonging to many owners) in areas not owned by the shepherds (they are usually owned by different private individuals), and the main benefit of grazing is to obtain milk for the production and sale of cheese and other dairy products.
2. The "Cultural Ecology" project was carried out by the Faculty of Ethnology and Educational Sciences of the University of Silesia in 2015–16, supported by Norwegian and EEA funds, coming from Iceland, Liechtenstein, and Norway.

References

Assmann, Jan. 2016. *Pamięć kulturowa. Pismo, zapamiętywanie i polityczna tożsamość w cywilizacjach starożytnych*. Translated by A. Kryczyńska-Pham. Warszawa: Wydawnictwa Uniwersytetu Warszawskiego.

Barth, Fredrik, ed. 1969. *Ethnic Groups and Boundaries: The Social Organization of Culture Difference*. London: George Allen and Unwin.

Barthes, Roland. 2008. *Mitologie*. Translated by A. Dziadek. Warszawa: Wydawnictwo Aletheia.

Becker, Gay. 1999. *Disrupted Lives. How People Create Meaning in a Chaotic World*. Berkeley: University of California Press.

Carr, David. 1986. "Narrative and the Real World: An Argument for Continuity." *History and Theory* 25(2): 117–31.

———. 1991. *Time, Narrative, and History*. Bloomington: Indiana University Press.

Connerton, Paul. 2008. "Seven Types of Forgetting." Memory Studies 1: 59–71.

———. 2012. *Jak społeczeństwa pamiętają*. Translated by M. Napiórkowski. Warszawa: Wydawnictwa Uniwersytetu Warszawskiego.

Constantin, Marin. 2003. "Capitalism and Transhumance: A Comparison of Three Pastoral Market Types in Europe (1950–2000)." *New Europe College Yearbook* 11: 55–116.

Creswell, John W. 2007. *Qualitative Inquiry and Research Design. Choosing Among Five Approaches*. Thousand Oaks: Sage Publications.

Halbwachs, Maurice. 1969. *Społeczne ramy pamięci*. Translated by M. Król. Warszawa: Państwowe Wydawnictwo Naukowe.

Jawor, Grzegorz. 2000. *Osady prawa wołoskiego i ich mieszkańcy na Rusi Czerwonej w późnym średniowieczu*. Lublin: Wydawnictwo Uniwersytetu Marii Curie-Skłodowskiej.

Jones, Schuyler. 2005. "Transhumance Re-Examined." *Journal of the Royal Anthropological Institute* 11: 357–59.

Kiereś, Małgorzata. 2010. "O gospodarce sałaszniczo-pasterskiej w etnograficznej pigułce." In *Owce w Beskidach czyli Owca Plus po góralsku*, ed. Józef Michałek, 22–36. Istebna: Porozumienie Karpackie, Województwo Śląskie.

———. 2019. "Beskidzkie sałasznictwo na przestrzeni wieków w świetle badań archiwalnych i terenowych." In *Beskidzkie sałasznictwo na przestrzeni wieków*, ed. Józef Michałek, 21–75. Cieszyn: Fundacja Pasterstwo Transhumancyjne.

Kocój, Ewa. 2015. "Artefakty przeszłości jako ślady pamięci. Dziedzictwo kulturowe Aromanów (Wołochów) na Bałkanach." *Prace Etnograficzne* 4: 271–305.

———. 2018a. "Ginący świat pasterzy karpackich. Życie codzienne pasterzy wołoskich na szałasach na pograniczu polsko-słowackim w XXI wieku jako dziedzictwo kulturowe regionu Karpat (wybór zagadnień)." In *Bacowie i Wałasi. Kultura pasterska na pograniczu polsko-słowackim*, ed. Ewa Kocój and Józef Michałek, 55–131. Cieszyn: Stowarzyszenie Wspierania Inicjatyw Gospodarczych Delta Partner.

———. 2018b. "Powroty do tematów pasterskich. Zwyczaje i wierzenia związane z rozpoczęciem sezonu pasterskiego na pograniczu polsko-słowackim w XXI wieku." *Etnografia Polska* 62 (1–2): 85–106.

Kohut, Maria, and Piotr Kohut. 2018. "Rozważania o wołoskich pasterzach i sztuce ręcznie wytwarzanych serów." In *Bacowie i Wałasi. Kultura pasterska na pograniczu polsko-słowackim*, ed. Ewa Kocój and Józef Michałek, 222–48. Cieszyn: Stowarzyszenie Wspierania Inicjatyw Gospodarczych Delta Partner.

Kopczyńska-Jaworska, Bronisława. 1950–51. "Gospodarka pasterska w Beskidzie Śląskim," *Prace i Materiały Etnograficzne* 8–9: 155–322.

Makovicky, Nicolette. 2014. "Old Minorities in New Europe. Enterprising Citizienship at the Polish-Czech Border." In *Neoliberalism, Personhood, and Postsocialism: Enterprising Selves in Changing Economies*, ed. Nicolette Makovicky, 163–87. London: Routledge.

Reinfuss, Roman. 1959. "Problem Karpat w badaniach kultury ludowej." *Polska Sztuka Ludowa* 13 (1/2): 3–11.

Pine, Frances T. 2002. "Retreat to the Household? Gender Domains in Postsocialistic Poland." In *Postsocialism: Ideals, Ideologies and Practices in Eurasia*, ed. Chris M. Hann, 95–114. New York: Routledge.

Popiołek, Franciszek. 1939. *Historia osadnictwa w Beskidzie Śląskim*. Katowice: Wydawnictwa Instytutu Śląskiego.

Program Aktywizacji Gospodarczej oraz Zachowania Dziedzictwa Kulturowego Beskidów i Jury Krakowsko-Częstochowskiej [Programme for economic activation and protection of cultural heritage of the Beskids and Kraków-Częstochowa

Upland]. 2007. Katowice. Retrieved 13 June 2020 from Self-Government of the Silesian Voivodship/ Sheep Plus Programme, https://www.slaskie.pl/content/wojewodzki-program--owca-plus-do-roku-2027.

Ricoeur, Paul. 2003. *O sobie samym jako innym*. Translated by B. Chełstowski. Warszawa: Wydawnictwo Naukowe PWN.

"Ruszył Redyk Karpacki 2013." 2013. *Podhalański Serwis Informacyjny*, 14 May. Retrieved 2 June 2020 from https://www.watra.pl/rotbav/wiadomosci/2013/05/14/ruszyl-redyk-karpacki-2013.

Schuyler, Jones. 2005. "Transhumance Re-examined." *Journal of Royal Anthropological Institute* 11(2): 357–59.

Sendyka, Pawel, and Nicolette Makovicky. 2018. "Transhumant Pastoralism in Poland: Contemporary Challenges." *Pastoralism: Research, Policy and Practice* 8(5). Retrieved 4 July 2020 from https://link.springer.com/article/10.1186/s13570-017-0112-2.

Spyra, Janusz. 2007. *Wisła. Dzieje beskidzkiej wsi do 1918 roku. Monografia Wisły*, Vol. 2. Wisła: Urząd Miasta w Wiśle.

Sutcliffe, Laura M. E., Inge Paulini, Gwyn Jones, Rainer Marggraf, and Nathaniel Page. 2013. "Pastoral Commons Use in Romania and the Role of the Common Agricultural Policy." *International Journal of the Commons* 7(1): 58–72.

Štika, Jaroslav. 2005. "Ovčoř na leto i na zimu." In *Sałasznictwo w Beskidach*, ed. Leszek Richter and Jan Szymik,15–20. Czeski Cieszyn: Sekcja Ludoznawcza ZG PZKO.

Szacka, Barbara. 2006. *Czas przeszły, pamięć, mit*. Warszawa: Wydawnictwo Naukowe Scholar.

Taylor, Charles. 1985. *Philosophical Papers*: Volume 2, *Philosophy and the Human Sciences*. Cambridge, UK: Cambridge University Press.

———. 2001. *Źródła podmiotowości. Narodziny tożsamości nowoczesnej*. Translated by M. Gruszczyński. Warszawa: Wydawnictwo Naukowe PWN.

Traba, Robert. 2016. "Pamięć kulturowa—pamięć komunikatywna. Teoria i praktyka badawcza Jana Assmanna. Wstęp do wydania polskiego." In *Pamięć kulturowa. Pismo, zapamiętywanie i polityczna tożsamość w cywilizacjach starożytnych*, ed. Jan Assmann, trans. A. Kryczyńska-Pham, 11–25. Warszawa: Wydawnictwa Uniwersytetu Warszawskiego.

Van Dijk, Teun A. 2001. "Badania nad dyskursem." In *Dyskurs jako struktura i proces*, ed. Teun A. van Dijk, trans. G. Grochowski, 9–44. Warszawa: Wydawnictwo Naukowe PWN.

Wojewódzki Program Aktywizacji Gospodarczej oraz Zachowania Dziedzictwa Kulturowego Beskidów i Jury Krakowsko-Częstochowskiej—Owca Plus do roku 2020 [Voivodship programme for economic activation and protection of cultural heritage of the Beskids and Kraków-Częstochowa Upland—Sheep plus until 2020]. 2015. Katowice. Retrieved 13 June 2020 from Self-Government of the Silesian Voivodship/ Sheep Plus Programme, https://www.slaskie.pl/content/wojewodzki-program--owca-plus-do-roku-2020.

CHAPTER 9

Contemporary Transformation of the Pastoral System in the Romanian Carpathian
A Case Study from Maramures Region

Cosmin Marius Ivașcu and Anamaria Iuga

Introduction

Traditional rural households in Romania used to be self-sufficient, until the end of World War II, when Romanian peasants relied mostly on the products of their own household, rather than on products they could buy. This was mostly a result of traditional animal husbandry, which provided the families with all the necessary products of animal origin (milk, cheese, wool, meat) that a household would need for its survival and development (Netting 1993). In the mountain regions of the Carpathian Mountains, cattle and sheep breeding is still quite common, decisively influencing the socioeconomic and cultural system of local cultures. When agriculture and small-scale animal husbandry were the main occupations of the inhabitants of these regions (until the mid-twentieth century), their life revolved around the needs of the animals which supported their livelihoods. For example, in some areas, livestock was of such importance that new families would build a barn first and their house second (information from Șurdești village, Maramureș region, northern Romania, 2013).

Moreover, pastoral calendars are of great importance in the communities, the year being divided according to the main work that must be done for taking care of the livestock. A dominant activity is finding good pastures for the animals (sheep, goats, but in some cases also cattle), activity reflected in four different types of pastoral practices, as framed by the ethnographer R. Vuia (1964). The first type is the local agricultural pastoralism, which was the most widespread pastoral practice not only in Romania but all over the Carpathians and the Balkan area (Vuia 1980). It is considered one of the most ancient forms of pastoralism because it implies year-round grazing on the village territory and manuring the fields with the help of the animals (Vuia 1980). Secondly, there is agricultural pastoralism with the sheepfold in the mountain pastures during summer. Thirdly, there is the pastoral practice that entails year-round pendulation (Romanian term for short-distance transhumance) between village territory and mountain pastures: during summer, grazing is done on mountain pastures; during autumn, winter, and spring grazing is done on the meadows situated within village territory. The fourth practice is the long-distance transhumance, when sheep flocks are travelling long distances starting in autumn to reach the winter pastures in the lowlands.

In the present study, we focus on the changes that occurred in the pastoral system in northern Romania (Maramureș region), where a mixed type of pastoral practice has been developing in the last twenty to thirty years, fusing the first and second types of pastoralism described by R. Vuia, due to socioeconomic and political changes. Among the drivers that generate change in the pastoral system all over the world, uncertainty (environment and resource uncertainty, economic uncertainty, and administration uncertainty, see Nori and Scoones 2019) is considered an element of great importance to livestock management. Most of it is also true in our case study, based on an interdisciplinary approach, using social science and natural science methodologies, and conducted in three villages (Botiza, Ieud, and Șurdești, in the Maramureș region of Romania), in the past sixteen years. Nonetheless, most transformations are mainly linked to social changes—the focus of our study—such as the massive emigration of locals to temporary or permanent jobs in agriculture in Western Europe, along with the aging population.

A Short History of Pastoral Practices in Maramureș

Maramureș is a historical and cultural region in northern Transylvania (Romania), situated in the largest depression of the Eastern Carpathians; until the year 1918, it also covered some parts of Zakarpatia area, situated in Ukraine today. This chapter is focused on Maramureș, a region situated in Romania, surrounded by mountains and hills on all sides, many of the mountain peaks are above two thousand meters, the highest being Pietrosul Rodnei (2,303 m); the lowest altitude is found near the Tisa River (214 m). The climate is temperate continental, with excessive precipitation and harsh, long winters. Maramureș is rich in forests, almost 60 percent of its area covered by broadleaf and coniferous forests. The lush grasslands with remarkable biodiversity (see Johansen et al. 2019; Wehn et al. 2019; Dahlström, Iuga, and Lennartsson 2013) are found here, spread all over the landscape, and are proof of the region's rich heritage in animal breeding and agriculture. Human presence in this region dates back to the Neolithic period (Popa 1997), being known that the communities of that time relied mostly on agriculture and animal breeding; hunting, fishing, and fruit gathering were less important. Sheep and goat breeders have been present here since Antiquity, starting with the Dacian culture, also found on the territory of Maramureș. Ancient sources even mention Dacians being skilled in fodder production and keeping their animals in stables during wintertime (Crișan 2007). Yet, the first documents that name the historical region of Maramureș and its sociopolitical organization are from the thirteenth century, depicting an independent voivodeship, inhabited mostly by small Romanian nobility (see Popa 1997).

The fourteenth-century documents related to the political and economic life of Maramureș mention several times the existence of arable fields, forests, rivers, hay meadows, enclosed meadows, sheepfolds, pastures, and mountain pastures used for grazing, among other geographic and cultural units (Mihaly 2009). For example, Ieud village is mentioned in the year 1435 by a document that uses as landmarks nine sheepfolds spread over a territory of 130 km^2, representing the border of the village, among other natural elements as rivers and mountain peaks. According to Popa (1997), the number of sheep in a single village during this medieval time could be around two to three thousand animals. The same fourteenth-century documents, when mentioning the possessions of certain villages, use specific phrases like *descensum in alpibus* (descent in the alps/mountains), *descensum vel caulam ovium* (descent or sheep fold) or *loci estivales* (places for spending the summer) (Mihaly 2009). This terminology is a certain proof that, back then, the pendulation of sheep flocks for summer grazing in the mountains or outside the village territory was already common. This is evidence that, in the fourteenth century, the system of pastoral pendulation ("short-distance transhumance") was already in use in the region, at least for sheep husbandry.

This system of "short-distance transhumance" is more accurately described in seventeenth-century documents, which mention many more mountains with high alpine pastures grazed by sheep flocks during summer. An interesting social phenomenon involving the minor Romanian noblemen was happening then. The number of their flocks increased and were used to compensate for the compulsory military conscription (see Ardelean 2012). Furthermore, documents mention several conflicts and trials with the locals in the neighboring areas (mainly the Bistrița-Năsăud region), on the rights to use the alpine pastures, conflicts that had a late reconciliation in the mid-nineteenth century (Ardelean 2012).

A drastic change occurred in the year 1919, after the union of Transylvania and almost half of Maramureș (up to the Tisa River, which served as the border with Czechoslovakia at that time, see Filipașcu 1940) with Romania (1 December 1918), when more than a hundred mountains used for summer grazing in northern part of historical Maramureș region were no longer available. Therefore, the number of flocks and herds decreased considerably in number (Papahagi 1925).

A turning point in local history is the period of time between 1949 and 1962, when forced collectivization took place in this region (see Dobeș and Bârlea 2004). At that time, only some of the villages were collectivized, as a form of punishment for the peasant's rebellions against the communist regime and their refusal to renounce Greek-Catholic faith (Kligman and Verdery 2015). In Ieud, the first village to be collectivized in this region, in 1950, this process deeply affected the land use and the proprietary rights and land ownership but not the pastoral system, which remained more or less the same. During collectivization, all arable fields around the village and the forests were seized. The locals remained with little arable fields on steep hills and hay meadows. Nevertheless, the pastoral use of the landscape was maintained in the same way as before the creation of collective farms. The collective farm had its own sheep and cattle with hired shepherds from the community. Before the celebration of the Pentecost, both the flocks of the community and the flocks of the collective farm would go on pendulation outside the village territory to alpine and subalpine pastures of Maramureș and Rodna Mountains (Ivașcu and Rákosy 2017). Botiza and Șurdești villages were not collectivized, thus the land use remained pretty much the same as before 1949. In these two villages, peasants were forced to provide different quotas of all their products (meat, eggs, milk, wool, agricultural products such as cereals, fruits, and so on). The quotas were meant to pay Romania's war debt to the Soviet Union and were established in accordance with their social status (those peasants considered wealthier, because they owned more land, had higher quotas to pay).

Traditional Pastoral System in Maramureș

It should be mentioned that in Romanian scientific literature, both ethnographic and geographic, the term "pendulation" is used to describe short-distance transhumant pastoralism. This activity implies that animals, cattle and sheep, graze during the summer on mountain pastures and then, in autumn, return to the village territory, where they graze in arable fields, meadows, pastures, and in some areas, even forests, depending on the ecological conditions and social organization of each village. During winter, all animals are kept either in barns, or enclosed under open sky, as the local sheep breed, called *țurcana*, is resistant to the cold climate of Maramureș. Altogether, the most important fodder for winter is hay, the locals in this region hold considerable and extremely detailed traditional ecological knowledge about this resource (Ivașcu, Öllerer, and Rákosy 2016; Iuga 2016). Additionally, fodder, like leaves from pollarded broad-leaved trees or spruce branches were quite common in the past, especially for feeding sheep, a frequent practice that was encountered all over Romania (see Hartel, Craioveanu, and Réti 2016).

Long-distance transhumance or simply "transhumant pastoralism" as it is called in scientific literature about Romania (Huband, McCracken, and Mertens 2010; Herseni 1941; Vuia 1964, 1980) is more complex. Similar to distance transhumance, this form of pastoralism entails the movement of large flocks of sheep for summer grazing in the mountains. The difference is that during autumn, shepherds and their flocks would travel to lowland pastures situated on the Danube Riverside, the Danube Delta, but also in the Tisza River plain and lowland Banat (see Vuia 1964; Constantinescu-Mircești 1976; Huband, McCracken, and Mertens 2010; Dragomir 2014). Once they arrived here, after traveling distances of about two to three hundred kilometers, they would spend the whole winter in these regions with milder climate and would reach the mountain pastures in spring.

One of the main differences between transhumance and pendulation is the year-round sheep-grazing (Vuia 1964) on grasslands or arable stubbles. It was a necessity due to the high numbers of animals (around one to two thousand in a single flock according to Constantinescu-Mircești 1976), which made it impossible to procure hay for so many animals for the whole winter. However, once they arrived in the lowlands, transhumant shepherds would also buy fodder from locals if the resources there were insufficient (Vuia 1964). It is also worth mentioning that in the case of transhumance, the flocks of sheep are usually owned by a small number of owners, sometimes it is only one owner. This form of pastoralism has developed mostly in southern Transylvania in the regions of Sibiu, Brașov, and Covasna. As a result, only the wealthier shepherds were practicing transhumance (see Huband, McCracken, and Mertens 2010). Its emergence is linked to the development of the wool industry in the Saxon cities of Sibiu (Hermannstadt) and Brașov (Kronstadt) in the fourteenth century. This is also the time when transhumant shepherds are mentioned in the Danube harbors and the ports of Dobrogea. The high demand of wool in the industry of these cities led to an increase in sheep numbers in the surrounding Romanian villages, beyond the carrying capacity of the grasslands available in this region (Huband, McCracken, and Mertens 2010).

In the Maramureș region, the long-distance transhumance was never practiced by the locals, there is no mention in the historical documents, nor is it a current practice. The most common pastoral system in the region is "the agricultural pastoralism with the sheepfold in the mountains" (Vuia 1964), also called "double cycling pastoral pendulation" (Idu 1999). This practice implies that shepherds will take the sheep and other animals (such as cows and horses) for summer grazing to the nearby mountains. Consequently, herds never leave the region of Maramureș or the territory of the villages during wintertime, as they did in transhumant pastoralism. There are some exceptions, as there are several medieval documents that speak briefly about the presence of shepherds from Maramureș in the Western Beskids Mountains, during summer (Filipașcu 1980), but at the end of summer they would return home. Thus, although there are long distances involved (more than four hundred kilometers), this cannot be included in the long-distance transhumance practice, as winter was spent in the community of origin. Usually, the pendulation, common until very recently, involved small distances, around sixty to one hundred kilometers away, to Rodna or the Maramureș Mountains, although, before the year 1918 shepherds would take their flocks to the Eastern Beskids Mountains.

Pastoral Calendar in Maramureș

The calendar of the pastoral year in Maramureș is divided by local community according to the four seasons: *primăvăratul* (springtime herding), *văratul* (summertime herding or summer grazing), *tomnat* (autumntime herding), and *iernat* (wintertime herding). The English translation of these vernacular names is approximate, because in local speech their meaning is more complex, each of the names deriving from the Romanian name of the seasons (*primăvară*, spring; *vară*, summer; *toamnă*, autumn; *iarnă*, winter).

Although the chapter focuses on drivers of change, we consider that elaborating on the pastoral calendar and the activities for each stage will shed light on the transformations that have occurred lately.

The first quarter of the pastoral year, called *primăvărat*, begins when grass starts to grow, when, as the local beliefs say, "cuckoos start to sing" (end of March) (information from Ieud village, Maramureș, 2016), and lasts until late April, the reference date being the feast of St. George (23 April). Spring grazing is done only with sheep and goats. Cows are kept indoors during all this time and are fed with hay. Grazing happens on a lower scale (Figure 9.1), as each owner grazes its own land (usually former arable fields and hay meadows).

Figure 9.1. Spring grazing in the hay meadows situated in close proximity to the village, Șurdești, 2012. © Anamaria Iuga

It is of great importance to mention that in the villages targeted by our study, the landscape is divided by the community, either by name or by practice, into a number of grazing areas (see Dahlström et al. 2013; Ivașcu and Rákosy 2017). There are three essential borders, marked by several geographic elements (roads, hills, ridges, etc.), demarcating a lower (first) level between 300–650 m in altitude (in Ieud). It is the land situated close to the households and the village, generally used for crops, as it is the best land in the village for agriculture (though lately transformed into hay meadows). Then, a middle (second) level that delimitates the land from the middle part of village land, situated at an altitude of 650–1,000 m (in Ieud), with arable fields (terraced slopes) and hay meadows and secondary forests; and an upper (third) level, 1,000–1,200 m (in Ieud), with permanent semi-natural grasslands used as hay meadows, or currently as pastures and beech and spruce forests. In Ieud, these three divisions bear the name *mejde* (border) (see Ivașcu and Rákosy 2017) and are known as *mejdele de jos* (lower border), *mejdele de mijloc* (middle border), and *mejdele de sus* (upper border). In the other two villages, Botiza and Șurdești, there is no special name for these three different levels, villagers name the landscape with the term *țarină* (land), but use them differently in practice.

In early spring, grazing is done on fields and meadows situated on the first level, mainly the ones from the vicinity of the household, and nearby hilly areas.

After 23 April, grazing will take place on the second border. Animals will spend about four weeks here, attended by their owners. Meanwhile, arrangements are being made for the next step, which means gathering the animals in a large flock, managed, and protected by the shepherds on the high-altitude mountain pastures. The animals will be at a *stâna* (sheepfold) all through the summer, led by a *gazda de stână* or *vătaf* (sheepfold leader), until 8 November, when people celebrate the holiday of *Sânmedru Vechi* (Old Saint Demetrius, calculated according to the Julian calendar). The sheepfold leader will be responsible for the management of the whole sheepfold: coordinating the movement of flocks, animal healthcare, and also milk, cheese, and *urda* (produced from whey, similar to the Italian ricotta) production and redistribution to each animal owner.

Before taking the animals to the alpine pastures, their owners must provide for animal fodder from grazing their own pastures (rent them if necessary) or grazing their own hay meadows. Then, they organize the gathering of the animals, together with the milk measurement custom. This custom takes place at the end of spring, after the first two days of grazing on summer pastures. Over the summer, each owner will receive, periodically, cheese and milk, according to the agreement during milk measurement. It must be highlighted that the production of milk and cheese always takes place at the summer farms.

Milk measurement, which in Șurdești is called *Sâmbra oilor* (meaning "gathering of the sheep"), in Botiza and Ieud, *Ruptul sterpelor* (meaning "separation of the barren sheep"), is an important event of the pastoral year. The three villages are situated at various altitudes, thus, the milk measurement feast takes place at different times of the year, depending on weather conditions: in Șurdești, it is held at the beginning of May, mainly in the first week of the month; in Botiza, it is held in mid-May (around 12–15 May) and, in Ieud, in the second half of May (around 18–27 May). The feast implies that each owner is to milk their animals (sheep and goats, and in Ieud and Botiza also cows) and, according to the amount of milk collected, the amount of cheese that they receive for the whole summer. The milk is measured according to an ancient measurement unit, the *font* (½ liter). In Șurdești, in the summertime, owners receive seventy halves (thirty-five liters) of raw milk for each half liter of milk they are milking at the feast, and one member of their family will turn it into cheese when it is their turn to go for a few days to the mountain farm. In Botiza and Ieud, the owners will receive only the cheese (twelve kilograms of cheese for one liter of sheep milk, or eight kilograms of cheese for one liter of cow milk in Botiza; and in Ieud, they receive ten kilograms of cheese for one liter of sheep milk).

The feast is accompanied by several ritual gestures meant to protect the animals and the shepherds. For example, in Ieud and Botiza a fir tree (*Abies alba*) is brought from the forest with its branches shaped as a cross (called in Ieud "the cross of the fir tree"). It is then placed in the front opening of the corrals (which has been decorated with flowers by young girls only). The sheep will go through this opening when milking begins. In Ieud, the branches of the fir tree are decorated with specific garden flowers (such as the peony, which is also called locally "the flower of the shepherd") and two ritual loaves of bread, which are meant to provide prosperity for the sheepfold. The Lord's Prayer is said before starting to milk. Then, the leader of the shepherds would throw salt above the corralled sheep (the role of the gesture being to protect the udder from injuries and infections). Another important gesture is to thrust an axe in the ground in front of the place where the sheep are milked, to protect the animals from being struck by lightning (in the village of Ieud). To make sure that the sheepfold is protected, the local priest is invited to bless the sheep and shepherds. If the sheepfold is situated too far from the village, holy water is sprinkled by the leader of the shepherds over the sheep before going to graze but after milking, a gesture that would ensure prosperity and protection. In Ieud, before grazing, the owner of the summer farm plays a natural trumpet (a straight tube without valves, originally made of wood, but nowadays made of brass—see Iosif 2016) (Figure 9.2), announcing the

Figure 9.2. Before the sheep leave for grazing, the shepherds' leader plays the natural trumpet, Ieud, 2016. © Anamaria Iuga

end of the milk measurement. The shepherds' leader plays the trumpet also before this custom begins, but during the custom, it is forbidden to play any instrument or even to whistle. After milk measurement, when all owners know how much cheese or milk they get for the summer, a large feast takes place, where main courses are especially cooked: a specific lamb soup prepared with wild thyme (*Thymus* sp.) and a special type of polenta.

After the milk measurement feast, the *vărat* (summertime grazing) begins and the animal flocks go to the mountains, to the alpine pastures. The sheep from Botiza graze on alpine pastures close to Vișeu and Borșa (two cities located 40–52 km northeast; the sheep and cattle from Ieud also spend the summer in Maramureș Mountains and in Rodna Mountains (40–60 km away) and the sheep from Șurdești graze on the mountain pastures of Gutâi, situated around 13 km up north from the village. During the seventeenth century, the noblemen in Ieud owned seven peaks in Rodna Mountains used for summer graze by the whole community. Between World War I and World War II, the community owned two peaks in the Maramureș range (Ștevioara Mică and Ștevioara Mare), but the property rights and grazing rights have changed in the last centuries due to various socioeconomic factors, thus, some of these mountains are not used for grazing anymore.

There is a quite precise grazing calendar followed by sheep owners. In Șurdești, the flocks go to the mountains on 21 May, the feast of St. Constantine and Helen, a feast that has become a landmark of the pastoral calendar. In Botiza, they go higher to the mountains, at the end of May. In Ieud, the system is more complex, due to the division of the village territory in three almost-equal bioeconomic zones, covering an area of 78 km². Thus, animals graze for four weeks on the second level and two more weeks on the third level. Also, flocks go to the mountains at the beginning or mid-June, before Pentecost. Once the mountains are reached, after a trip of twelve or twenty-four hours, summer grazing begins. Sheep are separated here: milking sheep, together with goats, graze the best grasslands; barren sheep, along with lambs and rams graze on other areas. Cows are also brought on these alpine grasslands (in Botiza and Ieud, but not in Șurdești), being watched by a separate herder. Cow's milk is mixed with sheep's and goats' for producing cheese and other products. Oxen and horses could also be brought after ploughing; they may remain in the mountains until the feast of the Beheading of Saint John the Baptist (29 August), when they are brought back to the village, for agricultural works (for hay transportation mostly, but also for crop harvesting). Cows are also brought back to the village for this feast and are kept on the village common pastures (Botiza), or already indoors (Șurdești), grazing only around the household. Individual herders watch every flock (sheep, barren sheep along with lambs and rams, cows, oxen, and horses in the past) during their grazing in the mountain and alpine pastures. Meanwhile, back in the village, farmers harvest hay from the meadows and crops from the fields preparing for the return of the animals.

The end of summer grazing for sheep and goats is marked by the feast of the Elevation of the Holy Cross (14 September), when sheep and goats, together with cattle are brought back to the village territory. Now, the third phase begins, the *tomnat* (autumn herding), with grazing starting from the upper level of the village and moving downwards. In the meantime, this area is already mown, and haystacks are built. Sheep flocks slowly descend through meadows to arable fields, already stubble land when the animals arrive. All the plots in the village territory are communally grazed by the shepherds until 7 November, just before the feast of the Holy Archangels Michael and Gabriel (8 November) or, as mentioned before, according to the "old" Julian calendar, the feast of St. Demetrius. It is the time when shepherds receive their payment from their leader and all the animals return to their owners. However, during this period of time, grazing continues on an individual basis until the fall of the first snow. Each farmer takes their animals to graze within the village territory where they have most of their land. Otherwise, sheep would be taken close to the household, to prepare for the winter (*iernat*) phase. If farmers choose to keep their sheep in the fields during winter, they are to provide hay which is stored in temporary buildings scattered around the landscape, called *colibe* (huts) or *case în câmp* (houses in the field) (Figure 9.3), larger permanent constructions for hay and stables.

During the winter phase, arable fields and hay meadows are manured with the help of livestock. Sheep are enclosed in corrals overnight, and

Figure 9.3. *Casă în câmp* (house in the fields) with *șopru* for hay storage and stables for cows. Sheep are kept under open sky and are moved on the terrain to improve the vegetation, Ieud, 2015. © Cosmin Marius Ivascu

by moving the corrals from one place to another on their properties, the land is manured and prepared for cultivation the next spring. This type of manuring takes place also in spring, when sheep are grazing and are kept overnight outdoors, moving from one field to another. To accomplish this task, locals use movable huts. People with less animals used to pay other villagers who had enough animals to accomplish this task. This is a remnant practice, more intensive in the past, and, in Ieud, as proof of this practice, the terraced slopes used for cultivation, apart from being present all over the lower level of the village, can still be seen at altitudes of one thousand meters.

To increase the hay and crop productivity of these hills, locals fertilized them with the help of sheep kept in corrals during autumn and winter. This is also one of the reasons why this form of pastoralism is also called "agricultural pastoralism with sheepfold in the mountains" (Vuia 1964), agriculture and animal breeding being interlinked and highly dependent on each other.

Nowadays, during the winter phase, most people keep their animals indoors, feeding them with hay, either within their household or in scattered temporary constructions (*colibe*—huts) and barns all over the village territory. Sheep are also kept under the open sky, although they are fenced. Regarding the cattle, these are kept inside barns in the winter, being fed with hay, or second time cut grass, or alfalfa, clover, and also grains. In winter or early spring, the manure they produce is still taken to the fields, mainly on the cultivated plots, to increase productivity.

From the presentation of the pastoral year in the region of Maramureș we can clearly see the interdependence of agricultural activities and pastoral ones and we can better understand how most of the traditional rural households were self-sufficient. Nowadays, although the pastoral practices have remained pretty much the same, the scale at which they are practiced has changed due to several drivers (economic, social, but also political). Along with that, the land-use has also changed. In the following, we present the drivers of change that we encountered in the field and how they have transformed the countryside lifestyle.

Drivers of Change

Agricultural pastoralism with sheepfolds in the mountains was the most widespread practice twenty years ago in Maramureș, but it has been made vulnerable lately by a number of drivers that are urging change. The major change, though, was collectivization (1948–62). In Maramureș, collectivization was imposed in some villages such as Ieud, while in other villages the inhabitants were supposed to provide quotas of their production (Botiza and Șurdești among them). This major change forced the peasants to increase their productivity by intensifying agricultural practices on the land they owned (reduced in size now in Ieud), in order to cover not only for their needs, but also for the state's requirements. For this reason, as resources were limited, many people in Maramureș, especially in collectivized villages, or people who owned small plots of land (in non-collectivized villages), found seasonal jobs outside their region, traveling to the southwestern part of Romania, where they were active in agriculture labor. In the 1980s, people migrated seasonally from Ieud and Botiza to the Banat region (southwestern Romania), but not from Șurdești, where villagers worked at the nearby mining industry in Cavnic town. Migrants returned home either with the money earned, that were used to build a new house, or with products (wheat, corn, rye, etc.), that catered to their household needs.

Due to the fact that people in Maramureș had been used to obtaining necessary cereals or corn from other Romanian regions, the way they worked their land changed after the fall of the Communist regime (December 1989). At the beginning of the 1990s, the land collectivized in Ieud was already requested by the rightful owners. Apparently, there were no drastic changes, as people returned to the lifestyle they had known before Communism; yet, people continued to go to Banat for seasonal work. The generation that had to deal with the restraints of collectivization had gotten old and a new generation of peasants adjusted their work strategies to other rules, migration included. As a result, land-use underwent several transformations: cropland surface was diminished, being transformed into hay meadows, or cultivated with alfalfa or clover. For instance, people had already given up cultivating wheat at the beginning of 1990s, since Maramureș is a hilly region, and the cereals cultivated here were local varieties of wheat, well-adapted to the cold climate and the poor soil composition, but with lower productivity. Over the next two decades (2000–20), the villagers started cultivating cereals exclusively as animal food (rye, oats, triticale, which is a hybrid of wheat and rye), but not on large parcels, as they could always buy more at local markets. In every village there is a local market on a different day of the week, and at the market there are cereal traders from regions such as Satu-Mare or Banat.

Accordingly, the use of parcels situated in different bioeconomic zones changed (see Dahlström et al. 2013) and the dominant transformation concerns the descent of the hay-meadows from the areas situated further away from the village (the third border in Ieud, or the second in Botiza and Șurdești). As a result, these remote hay meadows that used to be grazed in spring and autumn and mown in August are now abandoned or mainly used as pastures. Consequently, as pastures are available that close to the village, there are some summer farms that do not take the animals up to the mountains, instead they remain to graze on the territory of the village. They usually pay the owners of the land they graze on in animal products, mainly cheese, still very much appreciated in the region. In 2010, there was already a summer farm in Șurdești and two summer farms in Botiza that grazed only on the territory of the village, on the remote former hay meadows. Another consequence of this new practice is the abandonment of the alpine pastures situated in the mountains each community owns. In 2010, in Șurdești, villagers mentioned the abandonment of the pastures situated in Gutâi Mountains (around twenty kilometers away from the village, at an altitude of 850–1200 m). In the same year, in Botiza and Ieud, the alpine pastures of Maramureș and Rodnei mountains (thirty-four to sixty kilometers away from the village) were abandoned, because locals could reach them only by walking for one day and one night, or by transporting the sheep in a truck, activity that was deemed too expensive and complicated for the sheep owners, thus, they decided to remain on the village territory.

Another important driver of change is the decrease in the number of animals after the fall of Communism: all over Romania, the number of sheep and goats dropped by 38 percent in twenty-three years, between 1990 and 2013 (Popovici, Bălteanu, and Kucsicsa 2016). In 2014, in Șurdești, there were 498 cattle, and 1,340 sheep and goats (ISUMM 2016); in 2018, there were 401 cattle and 1,372 sheep and goats, according to the data provided by the town hall, and divided into four summer farms. People remember that before 1989 there were more than ten summer farms in the village, and around 3,000 sheep. In 2014 (ISUMM 2016), in Botiza, there were 818 cows and 1,171 sheep and goats, in five summer farms, although people remember that before 1989 there used to be up to nine summer farms, thus, a higher number of animals. The decrease in the number of animals is obvious in Ieud: in 1879 there were 12,000 sheep (Latiș 1993), while in the year 2014, there were only 2,541 sheep and 1,170 cows (ISUMM 2016). This change has two main causes. First, there is the circular migration (see Sandu 2000) to Europe, a repetitive and seasonal migration that has intensified in the last 20 years. Mostly young people leave their birth places to work abroad, leaving behind their children, and also the elderly population to take care of the household, including the animals. This aspect brings us to the second cause, the aging of the local population. Consequently, due to the lack of human resources, villagers started to sell their animals, especially their sheep, which require constant handling by shepherds (information from Șurdești village, 2010). Sheep are sold also because wool is no longer sought for, synthetic fibres being easier to purchase and process.

It must be mentioned that, although the number of animals has been constantly decreasing since 1990, there is another trend emerging within the communities living here. Since 2007, due to the subsidies given by the Romanian APIA (Agency for Payments and Interventions in Agriculture), many local animal breeders have abandoned sheep breeding and replaced them with two to five cows. This trend was noticed also in the neighboring mountain regions of Bucovina and Bistrița-Năsăud. The CAP (Common Agricultural Policy) payments from the European Union have had a considerable role in the change of animal husbandry type. This was motivated by several reasons. The first reason is that cows require fewer operating costs than sheep, and that the subsidies are higher for these animals, thus, being more profitable. Another reason is the lack of shepherds: the professional shepherds in these mountain regions complain about finding seasonal shepherds to help them herd sheep. It is getting harder and harder, as most of the young shepherds prefer to migrate to other EU countries, for seasonal work that provide more income. As shepherding activities also face the lack of workforce, the animals' owners themselves usually look after the sheep, taking turns, and involving only their family members.

In all three villages mentioned, the decrease in the number of animals is linked to workforce migration to the EU and led to the dissolution of centuries-old partition of the landscape for agricultural and pastoral purposes. The first outcome is the abandonment of summer grazing on alpine pastures situated in the high mountains of Gutâi, Maramureș and Rodna. Pendulation to these areas became unprofitable, since many former meadows in the upper level of the village could be used only for this purpose, instead of being abandoned. Nowadays, after the milk measurement, cattle and sheep flocks are moved to the next section of the landscape, where they spend the whole summer, until the feast of The Elevation of the Holy Cross (14 September). After this holiday, flocks start descending to the lower sections of the village, grazing the hay meadows that are already mown, some even a second time. Horses are usually brought for summer grazing in the upper section of the village, when the work they are needed for is done for a while. Yet, there is a significant difference: on the alpine pastures the herd of horses used to be watched by a herder (called *stăvari*), while nowadays they are just brought there and left by themselves on the grasslands, gathering in semi-wild herds that move around freely. The horses' owners come and take them home whenever they have some work to do or take them home for winter, in August.

As mentioned earlier, many villagers completely abandoned sheep breeding and only some farmers are now engaged in this activity. As a result, the collective role of pastoral practice started to diminish, changing it into an individual activity, which reshapes the land use. Some of the animal breeders started buying more land in the first division or on the second level of the village, and completely abandoned moving their animals up to the third level of the village (above one thousand meters) during the summer. To ensure the right amount of land for summer grazing compared to the number of animals, shepherds usually rent from other villagers their former hay meadows for this purpose. In the year 2018, in Ieud, there were at least two mixed cattle and sheep pens—each having more than one thousand sheep—that have grazing animals on this third level. However, there was also a cattle herd summer grazing on the first level (400–600 m), and large sheepfolds on the second one (650–1,000 m).

The fact that peasants specialize in animal breeding, and that the number of families having a few sheep decreased, have brought change to local rituals, namely the milk measurement feast. If there is only one owner of the sheepfold, there is no need to hold the feast, as there is no point in measuring milk and dividing it among the owners. Thus, there are less and less summer farms where this feast is held; it is losing its meaning. Another mutation is the nostalgic approach of the former sheep owners to this feast: as they have no sheep, they do not have any reason to participate in the milk measurement feast; yet, some of them attend local feasts where friends or family members have sheep, just to take part in the spring ritual and to rejoice with their close ones.

An additional change brought to the milk measurement feast is its celebration together with another ritual performed in the past, right before the departure to the alpine pastures. When the livestock and the shepherds passed through the village on their way to the mountains, the priest would perform a special service (called *sfeștanie*), blessing the animals and the herders. The priest is currently invited to the summer farm to perform this service, right after the milk measurement, on the same day.

The changes that have occurred in the past twenty to thirty years are socially driven but have economic and local land-management repercussions. It proves that all human activities are connected and interdependent. Thus, any small change leads to an adjustment in the whole local management system. All alterations are proof of the changing world we are living in, with new sociocultural and economic patterns and values. Maramureș is one of the few places where small-scale agriculture and animal husbandry is still in practice, defining the lives of the locals, although transformations are more intense and frequent than in the nineteenth or the twentieth century. In a sense, the communities in this region, as much as they are bound to and value tradition, are equally eager to change and to embrace all that is new (lifestyle, values, constructions, mechanization, subsidies, and so on). Pastoral practices are but a reflection of the important changes occurring right now.

Conclusion

Among different drivers of change, migration is the most important one, especially migration to Western Europe, an activity with a major impact on the historical herding practices of Maramureș. The result is a rapid change within the structure of the community, which is also reflected by the pastoral practices (organization of movements and specific pastoral customs). The collective role of traditional pastoral practices specific to this region is starting to fade, since decreasing livestock means a small number of locals continue to specialize in this activity. One of the results following all these socioeconomic changes is the increasing number of grasslands (former arable fields) situated now in the vicinity of the village. Thus, mountain hay meadows within the village territory are nowadays used almost exclusively as permanent pastures.

From the pendulation pastoral system, the current pastoral practices are turning more and more into a local agricultural system, where most of the sheep are not moving outside the village territory. Actually, this is the resilient response of the local community to the imminently changing social and natural environment; it is a practice perceived as the only option for using these resources and avoiding the succession of vegetation of these ecosystems, which will eventually turn into forests. However, the old way of managing the land and the old pastoral way of pendulation is inscribed in the landscape, as a place of memory for the local communities. Landscapes are temporal (Ingold 1993) and reflect, by the way they are shaped, the practices our ancestors used to carry them on. Signs of earlier grazing practices are, thus, reflected in the trees, in the biodiversity of the meadows, or in the shape of the hills. Even more, they are still part of the memory of the local population, and they should be valued as such, to remain a vivid component of the local history.

Cosmin Marius Ivașcu is Experienced Researcher (III) at the Advanced Environmental Research Institute (ICAM) and Biology-Chemistry Department, West University of Timișoara. He was a visiting researcher at Kassel University, Kassel University, Department of Landscape and Vegetation Ecology in 2019. His research focuses on traditional ecological knowledge (TEK) and the biocultural heritage of cultural landscapes with remarkable biodiversity and nature conservation.

Anamaria Iuga, PhD, is Head of the Ethnology Studies Department at the National Museum of the Romanian Peasant in Bucharest, Romania. She is also a New Europe College Fellow (2019–20). Her field of research includes the dynamics of material culture and intangible heritage, as well as traditional ecological knowledge.

References

Ardelean, Livia. 2012. *Istoria economică și socială a Maramureșului între 1600 și 1700.* Baia Mare: Ethnologica.

Constantinescu-Mircești, Constantin. 1976. *Păstoritul transhumant și implicațiile lui în Transilvania și Țara Românească în secolele XVIII-XIX.* Bucharest: Editura Academiei.

Crișan, Ion Horațiu. 2007. *Medicina în Dacia: de la începuturi până la cucerirea romană.* Bucharest: Dacica.

Dahlström, Anna, Ana Maria Iuga, and Tommy Lennartsson. 2013. "Managing Biodiversity Rich Hay Meadows in the EU: A Comparison of Swedish and Romanian Grasslands." *Environmental Conservation* 40(2): 194–205. https://doi.org/10.1017/S0376892912000458.

Dobeș, Andrea, and Gheorghe Mihai Bârlea, eds. 2004. *Colectivizarea în Maramureș: contribuții documentare (1949–1962).* Bucharest: Fundația Academia Civică.

Dragomir, Nicolae. 2014. *Oierii mărgineni.* Cluj-Napoca: Argonaut.

Filipașcu, Alexandru. 1940. *Istoria Maramureșului.* Bucharest: Tipografia Ziarul "Universul."

———. 1980. "Vechimea prezenței pastorale în etajul alpin al Carpaților." In *Calendarul Maramureșului,* ed. Ion Bogdan, Mihai Olos and Nicoară Timiș, 57–63. Baia Mare: Asociația Folcloriștilor și Etnografilor.

Hartel, Tibor, Cristina Craioveanu, and Kinga-Olga Réti. 2016. "Tree Hay as Source of Economic Resilience in Traditional Social-Ecological Systems from Transylvania." *Martor* 21: 53–64.

Herseni, Traian. 1941. *Probleme de sociologie pastorală.* Bucharest: Institutul de Științe Sociale al României.

Huband, Sally, David I. McCracken, and Annette Mertens. 2010. "Long and Short-Distance Transhumant Pastoralism in Romania: Past and Present Drivers of Change." *Pastoralism* 1(1): 55–71. https://doi.org/10.3362/2041-7136.2010.004.

Idu, Petru Dan. 1999. *Om și natura în Carpații Maramureșului și ai Bucovinei.* Cluj-Napoca: Napoca Star.

Ingold, Tim. 1993. "The Temporality of the Landscape." *World Archeology* 25(2): 152–74.

Iosif, Corina. 2016. "Trâmbița și rosturile ei în Țara Oașului." *Memoria ethnologica* 58–59: 104–22.

ISUMM. 2016. *Situație privind efectivele de animale din județul Maramureș.* Retrieved 31 October 2019 from https://isumm.ro/wp-content/uploads/2014/07/Anexa-nr.-31-1.pdf.

Iuga, Anamaria. 2016. "Intangible Hay Heritage in Șurdești." *Martor* 21: 67–84.
Ivașcu, Cosmin Marius, Kinga Öllerer, and László Rákosy. 2016. "The Tradition of Hay and Hay-Meadow Management in a Historical Village from Maramureș County, Romania." *Martor* 21: 39–51.
Ivașcu, Cosmin Marius, and László Rákosy. 2017. "Biocultural Adaptations and Traditional Ecological Knowledge in a Historical Village from Maramureș Land, Romania." In *Knowing Our Lands and Resources: Indigenous and Local Knowledge of Biodiversity and Ecosystem Services in Europe and Central Asia*, ed. Marie Roué and Zsolt Molnár, 20–40. Knowledges of Nature 9. Paris: UNESCO.
Johansen, Line, Anna Westin, Sølvi Wehn, Anamaria Iuga, Cosmin Marius Ivașcu, Eveliina Kallioniemi, and Tommy Lennartsson. 2019. "Traditional Semi-Natural Grassland Management with Heterogeneous Mowing Times Enhances Flower Resources for Pollinators in Agricultural Landscapes." *Global Ecology and Conservation* 18(article e00649). https://doi.org/10.1016/j.gecco.2019.e00619.
Kligman, Gail, and Katherine Verdery. 2015. *Țăranii sub asediu: colectivizarea agriculturii în România (1949–1962)*. Iași: Polirom.
Latiș, Vasile. 1993. *Păstoritul în Munții Maramureșului (Spațiu și timp)*. Baia Mare: Marco & Condor S.R.L.
Mihalyi de Apșa, Ioan. 2009. *Diplome maramureșene din secolele XIV și XV*. 4th ed. Cluj-Napoca: Editura Societății culturale Pro Maramureș, "Dragoș Vodă."
Netting, Robert. 1993. *Smallholders, Householders: Farm Families and the Ecology of Intensive, Sustainable Agriculture*. Stanford: Stanford University Press.
Nori, Michele, and Ian Scoones. 2019. "Pastoralism, Uncertainty and Resilience: Global Lessons from the Margins." *Pastoralism* 9(10). https://doi.org/10.1186/s13570-019-0146-8.
Papahagi, Tache. 1925. *Graiul și folklorului Maramureșului*. Bucharest: Cultura Națională.
Popa, Radu. 1997. *Țara Maramureșului în veacul al XIV-lea*. Bucharest: Enciclopedică.
Popovici, Elena-Ana, Dan Bălteanu, and Gheorghe Kucsicsa. 2016. "Utilizarea terenurilor și dezvoltarea actuală a agriculturii." In *România. Natură și Societate*, ed. Dan Bălteanu, Monica Dumitrașcu, Sorin Geacu, Bianca Mitrică, and Mihaela Sima, 329–75. Bucharest: Academiei Române.
Sandu, Dumitru. 2000. "Migrația circulatorie ca strategie de viață." *Sociologie Românească* 2: 5–29.
Vuia, Romulus. 1964. *Tipuri de păstorit la romîni: (sec. XIX– începutul sec. XX)*. Bucharest: Editura Academiei Republicii Populare Române.
———. 1980. *Studii de etnografie și folclor, Volumul II*. Bucharest: Minerva.
Wehn, Sølvi, Anna Westin, Line Johansen, Anamaria Iuga, Cosmin Marius Ivașcu, Eveliina Kallioniemi, and Tommy Lennartsson. 2019. "Data on Flower Resources for Pollinators in Romania Semi-Natural Grasslands Mown at Different Times." *Data in Brief* 25(article 104065). https://doi.org/10.1016/j.dib.2019.104065.

CHAPTER 10

Mountain Pasture in Friuli (Italy)
Past and Present

Špela Ledinek Lozej

Introduction

In mountainous regions, where agricultural activities are constrained by the climatic effects of altitude, edaphic factors, a scarcity of soil, and steep slopes, pastoralism was the most effective and dominant agricultural activity. Large expanses of grassland, which ring the valleys above the tree line, could be made accessible for productive farming activities using the ability of domestic livestock to convert the natural plant cover into nutritious produce. In the Alps—as described also in th Chapter 6 of this volume—a combination of cultivation and herding emerged, which has become known as alpine animal husbandry or the alpine agropastoral system (Italian, *alpicoltura*; German, *Alpwirtschaft, Almwirtschaft*). It is the movement of humans and their livestock between permanent winter settlements in the valley and temporary summer settlements in the alpine and subalpine belt. Two or more spatially segregated spheres of agroproduction are evident—fields and meadows near the village, and (low- or high-altitude) mountain pastures, i.e., alps, with shelters for animals, people, and associated milk processing, where desirable. The summer grazing of animals has many advantages, the most evident is a supplement of (up to one third or more) animal fodder, thus the scarce land in the lower narrow valleys are made available for crop cultivation and haymaking (Kirchengast 2008). Alpine farming is sometimes called "vertical transhumance," and is actually a condensed mountain-adapted variation of short-distance transhumance. In long-distance transhumance, livestock move from the summer pastures in the mountains to their winter pastures in the lowlands, where they stay until summer (Gilck and Poschlod 2019).[1]

Despite the fact that differences between alpine farming, transhumance, and other forms of nomadism and seasonal use of the alpine and pre-alpine belts are difficult to critically assess, archeological discoveries from alpine dairy huts prove the existence of alpine farming in different parts of the Alps during the Bronze age (2200–800 BC); palynological studies have found even earlier (from the early 4500 BC) indicators for high-altitude pasture, but they indicate only pasture, hence it is not possible to distinguish between nomadism, transhumance, and alpine farming (see also Chapter 3). Nevertheless, clear proof in the form of written sources exists from the Middle Ages only, linguistic findings of place names and terms associated with alpine farming indicate beginnings prior to the Middle Ages and even before the time of the Roman Empire (Gilck and Poschlod 2019).

Therefore, it is not surprising that alpine pasture has been dealt with by several disciplines, from agronomy to archeology, history, geography, as well as the more ethnographic approaches of anthropology, ethnology, and folklore studies (German, Austrian, and Swiss, *Volkskunde* and Slovenian, *narodopisje*). Among the anthropological approaches, Robert Burns (1963) identified mixed farming on communally owned alps as a basis for upland subsistence. A chapter was (also) dedicated to mountain husbandry in *The Hidden Frontier*; a classic of alpine anthropology by Cole and Wolf (1999, the first edition was published in 1974). A focal point for such livelihood strategies was given to the relatively closed corporate community as described by Netting in *Balancing on an Alp* (1981). The assumptions discussed in these works were relativized by Viazzo in *Upland Communities* (1989), who was able to demonstrate small scale variations and the importance of other economic activities. These anthropological "views from afar," or, in Viazzo's case, "close up views," are complemented by several collections of "native" ethnological and folklore studies, that have, since the second part of the twentieth century, overcome their patriotic and nationalist past, and given more attention to historical and critical approaches (Baskar 2014, Krauß 2018).

The first specialized studies of the Friuli alps date from the end of the nineteenth and beginning of the twentieth centuries, and were written by the geographers Giovanni Marinelli (1880, 1894) and his son Olinto Marinelli (1902), Musoni (1910a, 1910b, 1914), De Gasperi (1914), Dvorsky (1915), and later also by the agronomist, Marchettano (1908–11). Even in the second half of the twentieth century, geographical and agronomical research of alpine farming took place, with several studies partially dedicated to the local mountain agriculture (Bevilacqua 1960; Bonetti 1960; Valussi 1954), which were even further underpinned by the founding of the University of Udine in 1978.

Pascolini and Tessarin (1985) dedicated a monograph to the peasant alpine herders (*malghese*) and forest workers, which was followed by several works on the alps and alpine farming (Mauro Pascolini 1992, 1997, 2001). That research was more recently complemented by folkloristic interests (Marta Pascolini 2010), by the interests of historians and agronomists about commons (Bassi and Carestiato 2016; Bianchetti 2014; Tagliaferri et al. 1981), by linguistic research (Dapit 1995–2008; Desinan 1982–83), by regional development studies (Bovolenta et al. 2003; Chiopris et al. 2014; Pasut et al. 2006; Pasut, Romanzin, and Bovolenta 2016; Vendrami and Viel 2010), by several local studies (Burelli 1999; Ceconi 2011; Cozzi, Isabella, and Navarra 1998; Danelutto 2003; Depollo 1980; De Zorzi and Mariutto 2015; Furlan 1995; Ledinek Lozej 2016; Madotto 1987; Mauro Pascolini 1994; Pascolini and Tondo 1996; Pasut et al. 2016; Vidrigh 2003), as well as "views from afar," taking several works by Minnich (1989, 1990, 1998) and Heady (1999) into consideration, where alpine pasture is discussed as an important part of the anthropological setting.

The terms *mont, berghe, olbe, Alm, Alp, malga, planina* are used in the southeastern Alps of Friuli Venezia Giulia (FVG) to describe seasonal alps, that is mountain pastures with accompanying agricultural and residential infrastructure—shelters for livestock, people and the accompanying dairying area. The autonomous FVG region actually lies at the crossroads of the Romance, Germanic, and Slavic worlds and borders Veneto Region to the west, Austrian Carinthia to the north and Slovenia to the east. Until recently, it included four provinces—Gorizia, Trieste, Pordenone, and Udine.[2] The chapter focuses on the latter two, i.e., on the hilly and mountainous areas of the Pordenone and Udine provinces, where we find the orographically, morphologically, and geologically diverse Julian and Carnic Prealps, Friuli Dolomites, Julian and Carnic Alps, with Sauris (1400 m) as its highest settlement, and the highest peak at Coglians/Hohe Warte (2780 m). As mentioned, FVG is a meeting point for various language groups: Romance (namely Italian and Friuli), Germanic (either in the form of a compact settlement in the Canale Valley or as language islands in the settlements of Sauris and Timau), and Slovene on the eastern outskirts (in the Canale Valley, Resia, the Torre, Cornappo and Natisone valleys).

Historically, the area has been subjected to constant change and shifting borders between governments, which, together with the natural geographical characteristics of these mountain areas, has affected the demographics, i.e., growth, decline, and (permanent or temporary) migration of the population, property relations, livelihood strategies, and everyday practices (Maniacco 2014). In the Middle Ages, the mountain areas of today's FVG were a border area between the Aquileian Patriarchate and the Austrian provinces, in the Early Modern Period between the Venetian

Republic and the Austrian provinces, in the nineteenth century between the Kingdom of Italy and the Habsburg Monarchy, and in the second half of the twentieth century between Italy, Austria, and Yugoslavia and later Slovenia (Coradazzi and Spinatto 1994). These borders could have either been separated or united, could have been fortified and unpassable dividing lines of areas, or more permeable, enabling cross-border grazing on the basis of pre-established rights. At the end of this abridged sociogeographical outline, it is worth mentioning two disasters that accelerated a number of social processes in the region and, at the same time, brought about a pause to reflect on the future of the area. First, the tragedy at the Vajont dam on 9 October 1963, caused by a landslide in the reservoir lake triggered a tsunami-like wave, which washed away the Longarone settlement lying below the dam. Second, the disastrous earthquakes of 6 May and aftershock on 15 September 1976, with their epicenters in Gemona del Friuli, were the final blows to the remnants of past (alpine) farming livelihoods and practices.

This chapter is based on an overview of the aforementioned expert literature and on fieldwork that took place between April 2016 and September 2017, which included visits to several alps, semistructured interviews with farmers, herders, cheesemakers, and experts, joining several cattle drives, attending different workshops on local development, and several festive events (e.g., the Feast of Transhumance, cheese exhibitions, and fairs).

Historical Overview of Mountain Pastures in Friuli

The past existence of transhumance livestock raising in the Friuli region has been proven by toponyms and archival sources (Desinan 1982–83). Mountain pastures are first mentioned in the eighth century, namely in the donation of the Langobard brothers, Erfo, Ante, and Marco to the monasteries of Salto and Sesto al Reghena from 762, which, among the donated items, also mention a *monte* (meaning "mountain pasture" or "alp" in Friuli language), that could be used for grazing of livestock herds (Faleschini 1970; Mauro Pascolini 2001; Dreossi and Pascolini 2010).

During the Middle Ages, the majority of mountain pastures were in the property of the Patriarchs of Aquileia, and were leased out to vassals and *gastalds*, or even given for use by local inhabitants. For example, in 1275, the patriarch Raimond della Torre made the use of mountain meadows and pastures available to the Carnians in exchange for a tithe on their produce. Some mountain pastures were also the allodial property of nobility, or even the common property of local communities (*usi civici*). The increase of arable land at the expense of pastures in the valley, and as the result of population growth, led to the increase of mountain pastures at the expense of forests (Faleschini 1970; Pascolini and Tessarin 1985; Pascolini 2001).

Only the Venetian Republic (1420–1797) administratively regulated the use of mountain pastures, predominately in order to protect forests (e.g., prohibiting the grazing of sheep and goats in the forests). In some cases during this period, measures were put in place to divide pasture lands that were previously common into shares held by extant households. The owners of the shares formed consortia, composed of members of the "original" families, even though well-off newcomers were permitted to buy shares (Noacco 1959; Faleschini 1970; Mauro Pascolini 2001: 74).

After Napoleon's occupation (1797), French Civil Code was applied under the influence of the new French doctrine. In 1806 the territory was divided into municipalities and commons (*beni comunali*), including mountain pastures, which were given, first only under administrative management, and later as property, to the newly established municipalities. Common property (*beni comuni*) became communal assets (*beni comunali*); commons were transformed into the public property of municipalities (Faleschini 1970; Mauro Pascolini 2001; Carestiato 2014).

The land register of all mountain pastures was elaborated during time of the Austrian Kingdom of Lombardy-Venetia (1814–66). The tax imposed on alpine pastures was higher than that of the valley pastures, thus recognizing the significant economic potential of the alps. Pasture outside the boundaries of the alps (i.e., forest pasture) was abolished. The municipalities were ordered to sell all—once common, and since 1806, municipal—less fertile and uncultivated lands. During this period some lower and/or less fertile alpine pastures were privatized; the higher alps were, for the most part, exempt from being sold off (Faleschini 1970; Mauro Pascolini 2001; Bianchetti 2014).

After the annex of Friuli to the Kingdom of Italy in 1866, the new government showed huge interest in accelerating agriculture, evident from the establishment of itinerant chairs of agriculture for small farmers (*Cattedre Ambulanti di Agricoltura*) in 1869, the Agricultural Chemistry Station of Udine (*Stazione chimico Agraria di Udine*) in 1870, and cooperative dairies (*latteria turnaria*). The first cooperative dairy, in which members assisted the cheesemaker and then took a share of cheese, was founded in 1881 in the extreme northwestern Carnian village of Collina. The government also launched several competitions to reward the best managers of the alps (*malgaro, malghese*) and published expert articles on good practice in alp management in specialist agricultural journals (Tagliaferri et al. 1981; Ceconi 2011; Pasut 2016). And, as mentioned above, the first expert studies by geographers and agronomists (G. Marinelli 1880, 1894; O. Marinelli 1902; Marchettano 1908–11; Musoni 1910a, 1910b, 1914; De Gasperi 1914; Dvorsky 1915) also took place at the end of the nineteenth century.

After World War I, the Canale Valley and Fusine al Lago (until then part of the Habsburg Monarchy's Duchy of Carinthia) were annexed by Italy. In the 1920s, the state of Friuli animal husbandry was in poor condition, due to it being in the vicinity of the Isonzo Front, and the effects of German and Austrian military occupation after the breakthrough of their forces in Kobarid. In the period of postwar renewal, special attention was given to the alps, especially to those on the Montasio plateau in the Western Julian Alps, which were bought from the impoverished municipality of Chiusaforte by the Keepers of the Service Bull Stations of Udine consortium (*Tenutari stazioni taurine di Udine*) (Pasut, Romanzin, and Bovolenta 2016).

In the 1950s and 1960s post-World War II period, alpine animal husbandry was still a viable sector, especially in Carnia. It only started to decline from the 1970s, and in the 1980s in particular, following the earthquake of 1976. If, at the beginning of the twentieth century there were 258 alps in Friuli (De Gasperi 1914) and 132 still active alps in Carnia in 1967 (Faleschini 1970), in 1995 there were only 98 alps across the whole of Friuli (Mauro Pascolini 2001). Among the agricultural reasons that led to such a decline are the intensification of lowland agriculture and, to a lesser extent, the introduction of new animal breeds less adapted to mountain pastures (Bovolenta et al. 2005); among social reasons are the demographic trend of a decline in population and the general processes of urbanization, industrialization, and deagrarization which caused alpine pasturing and dairying to lose its economic and even social *raison d'être* (Pascolini and Tessarin 1985).

In the framework of the "new rural paradigm" and EU Common Agricultural Policy (CAP), which promoted rural development as a multilevel, multiactor, and multifaceted process (Van der Ploeg et al. 2000, Van der Ploeg and Roep 2003), the decline and unsustainability of alpine farming in the mid-1990s, triggered interest in the activity from (regional) government and experts from the field of agronomy and rural development, above all from the Regional Agency for Rural Development (*Agenzia regionale per lo sviluppo rurale*, ERSA). In the last two decades, the ERSA has implemented several European projects,[3] collating information on the state of mountain pastures by using Geographic Information Systems (GIS), organizing them into databases that can be continually updated, and it has also identified concrete and viable development strategies (Pasut 2012, 2013, 2015; Sanna 2013). In addition to collating data and strategic development, the ERSA has also promoted alpine cheeses and other products with a variety of quality labels (certificates, trademarks, and geographical indications). The measures had little substantial impact on alpine dairying and farming production.[4] The one exception however was Montasio cheese that had already been awarded a national Denomination of Typicity in 1955; since 1984 it was under the auspices of the Consortium for the Protection of Montasio Cheese (*Consorzio per la Tutela del formaggio Montasio*) and was finally awarded Protected Designation of Origin at EU level in 1996 (Ledinek Lozej 2016, 2021). A step forward towards resolving the common challenges of alpine husbandry across the wider Alps was made with the establishment of the Society for the Study and Valorization of Alpine Animal Husbandry Systems (*Società per lo Studio e Valorizzazione dei Sistemi Zootecnici Alpini*, SoZooAlp) in 2000.

Several other media have contributed to promoting the alps and alpine pasture to the general public in recent years. Numerous guide books on regional alps have been published (Dreossi and Pascolini 1995, 2010; Chiopris and Pittino 2013; Guida 2016, 2017) and several public and private websites promote visits to the alps.[5] During the 2017 pasture season, the *Messaggero Veneto* regional daily newspaper published a twice weekly supplement called *Journey through the Alps of Friuli-Venezia Giulia* (*Viaggio nelle Malghe del FVG*) written by Nicola Giraldi. This well-known author of travelogues visited the majority of still active dairy alps and shared his impressions with the newspaper's readership. Furthermore, alpine cheese and alpine pasture are (re)presented at many events organized at local and regional level, such as the *Friuli Doc* event, which represents Friuli produce and cuisine, supplemented with enogastronomic and cultural events; Enemonzo's annual Alpine Cheese and Ricotta Market-Fair (*Mostra-Mercato del Formaggio e della ricotta di Malga*); the alpine cheese auction in Sutrio; organized guided tours to the alpine pastures; cattle drive festival; photographic exhibitions and photographic publications on the alps and alpine pastures (Da Pozzo 2004); and television broadcasts on national (*Linea verde*, RAI) and regional television. All the above strengthened the visibility of alpine animal husbandry and the alps to the general public, as well as among breeders, herders, and cheesemakers.

Past and Present Models of Management

Sociohistorical and environmental conditions (location, altitude, morphology, edaphic, and climatic factors) have led to the establishment of differentiated models of use and management in the alps of the FVG region (for the other Italian regions see also Chapters 2, 6, 7, and 12). Seven different models—related to property relations and usage rights—were identified at the beginning of the twentieth century, the time of the greatest expansion of alpine husbandry: the Carnian model, the model of the Claut area, of the Cansiglio area, of the Cavallo area, of the Raccolana Valley, of the Canale Valley, and of Venetian Slovenia (De Gasperi 1914; Mauro Pascolini 2001). The main differences in management regimes were between alps owned by individual private owners, alps in common property, and alps in public ownership, i.e., in the property of local (or other public) authority.

The main feature of the Carnian (as well as the Claut, Cansiglio, and Cavallo models) management regime is that the alps were mostly owned by municipalities. Municipalities leased an alp to the most favorable bidder for three, five, seven, or even ten years at public auction. The tenant herder (*malgaro* or *malghese*) was responsible for the overall management of the alp and, as defined by contract, had obligations for maintaining facilities, fertilizing, and clearing pastures. The contract also defined the date of the cattle drive to and from the alp, the maximum number of livestock, and other eventual duties. The tenant was individually responsible for organizing the work on the alp, such as, hiring a cheesemaker and shepherds, and drawing up individual contracts with livestock owners for the care and grazing of livestock during the alpine pasture season. In the case of dairy cattle, compensation was usually a proportion of milk or cheese. The methods of calculating the share were different, often very complex, based on the calculation of the average on different measuring days (Marchettano 1908–11).

The Raccolana Valley differentiated from the Carnian model, because the alps were divided into plots. Each plot comprised an abode, possibly two units under the same roof, and a cattle shed. A *malgaro* was given a plot of land with an abode; while livestock care, milking, and milk processing were individual, all tenants grazed together (Pasut 2016).

In the valleys of Torre and Natisone in Venetian Slovenia, where less fertile common, and later, municipal land was divided into plots and sold even in the nineteenth century, summer alpine settlements (Slovenian, *planine*) with privately owned huts (Slovenian, *kazon*) were established. Herders grazed and mowed individually, each within their own meadows and pastures; however—after the introduction of cooperative dairies— milk processing was done in common (Musoni 1913; Dvorsky 1915; De Gasperi 1915; O. Marinelli 1915; Mauro Pascolini 1992; Furlan 1995).

A transitional area between the Carnian and Natisone alpine pasture regimes existed in Resia, where the lower mountain pastures (Slovenian, *planine*) were privately owned and managed, whereas high-altitude alps (*malghe*) were the property of the municipality and leased to tenants (Madotto 1987; Rupel 1990).

The management regime in the Canale Valley was more like the regimes in Carinthia (Austria) or in the Eastern Alps (Slovenia). High-alps

Figure 10.1. Herder and cheesemaker Angelo Tessin, Pian Mazzega alp (Malga Pian Mazzega), Piancavallo, 2016. © Špela Ledinek Lozej

were the property of closed local communities (German, *Nachbarschaft*; Slovenian, *srenje, soseke*; Italian *vicinia, consorzio vicinale*). They hired herders and, in the case of dairy cattle, also a cheese maker, who performed the work under the supervision of an elected representative of the community (Slovenian, *olmaister*; German, *Allmaister*; Italian *capomalga*). As elsewhere, the lower mountain pastures and meadows were privately owned and managed (Rupel 1987; Dapit 1997; Ravnik 2015).[6]

In the case of the high-alps, the Carnian model has been preserved to the present day, so that the owner—either the municipality or local community—leases the alp to the most favorable bidder for a period of years. Whereas in the past tenants used to herd two or three cattle from many of the surrounding owners, today they mostly have their own cattle and to a lesser extent the livestock of other breeders. In comparison to the more than 250 dairy alps at the beginning of the twentieth century, in 2012 there were (only) 161 active alps (23 in the Venetian Prealps, 116 in the Carnic Alps, and 22 in the Julian Alps). However, most of these comprised only barren, dry, or nursing animals, and there were only sixty-four dairy alps, that is alpine pastures still accompanied with the milk processing in different dairy products—cheese, whey cheese (*ricotta*), and butter (Pasut 2016). In 2017 there were only forty-nine active dairy alps: eight in the Venetian Prealps, thirty-five in the Carnic Alps, and six in the Julian Alps (Guida 2017).[7] Alongside the general crisis in alpine farming, there are some particular aspects linked to infrastructure (poor road infrastructure, lack of electricity and water supply, problems in telecommunications), human resources (aging work force), production (plenty of unused pastures due to low livestock performance, unused dairying facilities), and sectoral restructuring of the alps for tourism purposes. Based on previous development studies (Pascolini 1997) and strategies (Pasut 2012, 2013, 2015; Sanna 2013), as well as multicriteria analysis (Pasut 2016), the following objectives for the alpine pastures can be identified:

(a) *pasture*: to improve use of available grazing surface; to increase surface of rich pastures; to manage pastures (e.g., rotational or sequential grazing)
(b) *livestock*: to improve livestock suited to mountain pastures
(c) *viability and infrastructure*: to improve external accessibility to the alps, and internal accessibility to pastures (however, the decision to improve viability shall be made by taking into account other territorial objectives, such as sustainability); to guarantee availability of energy and water resources (e.g., rainwater collection tanks in karst areas); to improve worker accommodation; to arrange premises for dairying
(d) *dairying*: to improve milk processing quality; to market alpine dairy products; to facilitate distribution of alpine produce (e.g., in mountain huts or via other channels)
(e) *tourism and other additional activities*: to introduce and balance catering, tourism, and educational activities in the alps

(f) *education, training, and specialized support:* to train staff and to guarantee technical and expert support (on grazing rotation, dairy processing, marketing of dairy products, etc.)
(g) *awareness raising, monitoring, and evaluation*: to disseminate results of expert studies; to raise awareness of the role of Alpine husbandry on the ecosystem and on the meaning of Alpine dairy farming

But all the abovementioned objectives are in the memorylands (Macdonald 2013) of structural nostalgia (Herzfeld 2005) evoking the past—its practices, life, habits, objects, etc.—for present and future purposes (Smith 2006; Harrison 2013); thus interlaced with heritage as a hegemonic idiom (see also Chapter 4). Hence, in the face of considerable heritage discourse (a) *pasture* is no longer just a resource of livestock fodder, but serves predominantly to maintain the cultural landscape and biodiversity; (b) *autochthonous breeds* and (c) restoring of *traditional architecture facets* are encouraged, (d) whereas milk should preferably be transformed into forms conforming to *PDO, Slow Food, the AQUA regional brand*, or other branded produce; and (e) the alps are mostly places for (heritage/cultural/slow etc.) *tourism*, where animals serve as a facet of catering activities. With this, (f) *ethnographers* are forced into the role of experts, that are, among others, also entrusted to (g) raise awareness and explain *past objects/artifacts and practices in completely new constellations* (e.g., exhibits in local museums or displayed on the walls of tourist facilities, cattle drive feasts, etc.). These heritage practices and discourse are supported by several different popular and expert media (from newspapers and events to expert qualification of heritage elements) and operate at various levels—from local (museum collections) to regional (inclusion of alpine cheese into regional territorial brands and local alps into regional tourism offers), and even national (inclusion of alpine dairy products on the quality schemes of the Italian Ministry of Agricultural, Food and Forestry Policies [*Ministero delle Politiche Agricole, Alimentari e Forestali*, MiPAAF]) and supranational level (evident from some echoes of regional media and herders on inclusion of transhumance on the UNESCO Representative List of the Intangible Cultural Heritage of Humanity).

Conclusion

Despite the interest shown in enhancing alpine pasture and dairying by European, Italian, and regional policy through measures under the Rural Development and European Territorial Cooperation programs, endeavors by the ERSA, and considerable heritage discourse, it seems that signifi-

Figure 10.2. Fleons di sotto alp (Malga Fleons di sotto), 2017. © Špela Ledinek Lozej

cant impact on the preservation and enhancement of alpine pasture and dairying is absent. We notice that the number of dairy alps is in decline, that alpine dairy products have difficulties conforming with EU and regional quality schemes, and additionally, that some facilities, although beautifully restored in traditional style and perfectly equipped using EU and regional funds, are not operational. As the alps are part of a wider complex system of alpine agropastoral husbandry established to provide additional fodder and relief from some chores over the summer, to improve alpine pastures requires strategic consideration of the entire alpine husbandry system. The alps—albeit important—have a seasonal character. The sustainability of the system requires integration with broader economic and social frameworks.

In any event, the still operational alps can be divided into four types of subsistence:

(1) enterprise intensification, either in forage or dairy production (e.g., Pian Mazzega alp);
(2) enterprise extensification of agriculture in favor of tourism (catering, accommodation) and occasional educational programs (e.g., didactic farms, the Coot alp);
(3) survival of the past (subsistence) practices (e.g., Fleons alp);

(4) and new forms of cooperation and solidarity-based agriculture (e. g., endeavors of the Consortium of the Valleys and Friulian Dolomites [*Consorzio delle Valli e delle Dolomiti Friulane*] in the Fara and Rest Alps).

It seems that more than the European (for an overview of EU Common Agricultural Policy see also Chapter 1), national, and regional policy measures, the alpine pastures of Friuli are sustained by individuals and their livelihood strategies and passion. Passion for work and life in the alp, as articulated by Ilo Casaro, who has been taking care of Ielma di Sopra alp for years: "If you have a passion for animals, even if it's difficult . . . I was born with cows, have grown up with cows, born and raised with livestock; if you have a passion, you cannot leave them, you have them to the end" (Da Pozzo 2004: 28).

Acknowledgments

The research was funded by the European Social Fund in the framework of TALENTS[3] Fellowship Program, whereby I am grateful to my host institution, the University of Udine and my scientific supervisor, Prof. Roberto Dapit, and co-supervisor, Prof. Donatella Cozzi. However, the chapter was finalized in the framework of the research program *Heritage on the Margins* (P5–0408), funded by the Slovenian Research Agency.

Špela Ledinek Lozej is a Research Fellow of the Institute of Slovenian Ethnology at the Research Centre of the Slovenian Academy of Sciences and Arts (ZRC SAZU). Her research interests encompass heritage studies, food and foodways, mountain pasture and mountain regions, livelihood strategies, and material culture. She has published a scientific monograph on the development of kitchen space in Western Slovenia, contributed to many peer-reviewed journals, and led several European projects. From 2016 to 2017, she was a Research Fellow at the University of Udine. She is currently in charge of the Heritage on the Margins multidisciplinary research program.

Notes

1. This was possible only in the marginal areas of the Alps, especially in the southern part where snow-free winter pastures in the lowlands are accessible (Frei-Stolba 1988).

2. For better readability all toponyms are kept in the official Italian form, although they also exist in Friuli, and in some areas also in German and Slovene.
3. For example: Development Patterns of Agrozootechnical Activities in a Mountain Environment for the Conservation of the Territory and the Enhancement of Local Products (*Modelli di sviluppo delle attività agro-zootecniche in ambiente montano per la conservazione del territorio e la valorizzazione dei prodotti locali*, Interreg IIIa Slovenia–Italy); The Way of the Alps and Lodges (*La via delle malghe e dei rifugi*, Interreg IIIa Italy–Austria); Management and Sustainable Development of Natural Habitats between Italy and Austria: The Way of the Alps (*Gestione e sviluppo sostenibile degli habitat naturali tra Italia e Austria: Via delle Malghe*, Interreg IIIa Italy–Austria); DIVERS—Biodiversity of the Mountain Taste (*Biodiversità dei sapori di montagna*, Interreg IVa Avstrija–Italija); Trans Rural Network (Interreg IVa Avstrija–Italija); MADE—Malga and Alm Desired Experience (Interreg Va Avstrija–Italija).
4. Alpine cheese is actually not included in the ERSA's regional quality Agriculture, Quality, and Environment scheme (*Agricoltura, Qualità e Ambiente*, AQUA), established in 2017 and with seemingly short-lived outcomes. In addition to the European and regional quality scheme there is also a Slow Food presidia formed around Çuç di Mont (Friuli expression for alpine cheese), bringing together three alpine cheese producers (cf. Slow Food FVG 2019).
5. For example: Malghe FVG (2020), Associazione Allevatori Friuli Venezia Giulia (2010), PromoTurismo FVG (2020), and Comune di Sauris (2020).
6. There were also differences between individual villages, e.g., between Uggovizza and Campososso (Dapit 1997), which can be attributed to the morphological distinction between the Carnic and Julian Alps.
7. Recent research of alpine animal husbandry in Friuli distinguishes between the following mountain pasture sections: the Venetian Prealps (subdivided into the Cansiglio and Cavallo areas), the Carnic Alps (subdivided into the Degano Valley, the But Valley, the Chiarsò Valley, the Pontebba Valley, the Pramollo-Cocco, the Mimoias-Cimon, the Crostis-Valsecca, the Rioda-Losa, the Zoncolan-Arvenis, the Col Gentile, the Tinisa, the Varmost-Bivera, the Friuli Dolomites, and the Carnic Prealps zone), and the Julian Alps (subdivided into the Dogna-Miezegnot, the Tarvisio, the Montasio, the Resia, and the Gemonese area) (Pasut 2016). However, there are also other divisions in use, for example ERSA's divison on ten mountain pasture sectors: the Upper Degano Valley, the Upper But Valley, the Chiarsò Valley, the Pontebba Basin, the Julian Alps, the Sauris–Val Pesarina–Ovaro Dorsal, the Upper Tagliamento Valley, the Zoncolan-Arvenis-Dauda Dorsale, the Julian and Carnic Prealps, and the Province of Pordenone (Chiopris and Pittino 2013); and a more recent one in three larger areas: Cansiglio, Piancavallo and Friuli Dolomites; the Carnic and Precarnic Alps; and the Tarvisio and the Gemona area (Chiopris et al. 2017).

References

Associazione Allevatori Friuli Venezia Giulia. 2010. "Malga Montasio." Retrieved 18 July 2020 from http://www.malgamontasio.it/.
Baskar, Bojan. 2014. "Alpe Antropologov." In *Alpske skupnosti: Okolje, prebivalstvo in družbena struktura*, ed. Pier Paolo Viazzo, 435–57. Ljubljana: Studia humanitatis.
Bassi, Ivana, and Nadia Carestiato. 2016. "Common Property Organisations as Actors in Rural Development: A Case Study of a Mountain Area in Italy." *International Journal of the Commons* 10(1): 363–86. https://doi.org/10.18352/ijc.608.
Bevilacqua, Eugenia. 1960. *La Carnia: Saggio di geografia regionale*. Firenze: Olschki.
Bianchetti, Alma. 2014. "I beni comunali in tempo veneto e il Friuli." In *Ville friulane e beni comunali in età veneta*, ed. Alma Bianchetti, vol. 4, 11–62. Udine: Forum.
Bonetti, Eliseo. 1960: *Gli sviluppi dell'insediamento nel bacino del Fella con particolare riguardo all'area linguistica mista*. Trieste: Università degli studi di Trieste.
Bovolenta, Stefano, Virginia Martellani, Carla Fabro, and Piero Susmel. 2003. "L'alpeggio in Friuli Venezia Giulia: Due casi di studio." *Agribusiness Paessaggio & Ambiente* 6(3): 212–22.
Bovolenta, Stefano, Elena Saccà, Michele Corti, and Daniele Villa. 2005. "Effect of Supplement Level on Herbage Intake and Feeding Behaviour of Italian Brown Cows Grazing on Alpine Pasture." *Italian Journal of Animal Science* 4: 197–9. https://doi.org/10.4081/ijas.2005.2s.197.
Burelli, Ottorino, ed. 1999. *Lusevera nell'Alta Val Torre*. Lusevera: Comune di Lusevera.
Burns, Robert K., 1963. "The Circum-Alpine Culture Area: A Preliminary View." *Anthropological Quarterly* 36(3): 130–55. https://doi.org/10.2307/3316628.
Carestiato, Nadia. 2014. "I patrimoni fondiari colettivi in Friuli Venezia Giulia." In *Ville friulane e beni comunali in età veneta*, ed. Alma Bianchetti, vol. 4, 125–192. Udine: Forum.
Ceconi, Tullio. 2011. *Forni Avoltri: 1800–1915; Avvenimenti, risorse locali e mobilità elle persone*. Forni Avoltri: Collana ricerche storiche del Comune di Forni Avoltri.
Chiopris, Giordano, Davide Pasut, Enio Pittino, Maurizio Sanna, and Valentino Volpe. 2014. *The Monitoring of Alpine Farms for the Development of Mountain Agriculture in Friuli Venezia Giulia: Guidelines for Alpine Farm Management* (simultaneously published in German and Italian). Pozzuolo del Friuli: ERSA. Retrieved 18 July 2020 from http://www.ersa.fvg.it/export/sites/ersa/aziende/sperimentazione/Alpicoltura_friulana/Allegati-Alpeggio/libro-Alpeggi.pdf.
Chiopris, Giordano, and Enio Pittino, eds. 2013. *Malga che vai . . . Formaggio che trovi: Le malghe da latte della montagna friulana*. Spilimbergo: ERSA. Retrieved 18 July 2020 from http://www.ersa.fvg.it/export/sites/ersa/aziende/sperimentazione/Alpicoltura_friulana/Allegati-Alpeggio/ERSA_malga_che_vai_formaggio_che_trovi.pdf.
Cole, John. W., and Eric. R. Wolf. 1999. *The Hidden Frontier: Ecology and Ethnicity in an Alpine Valley*. 2nd ed. Berkeley: University of California.

Comune di Sauris. 2020. "Il mondo delle malghe." Retrieved 18 July 2020 from http://www.sauris.org/il-mondo-delle-malghe/.
Coradazzi, Maurizio, and Giovanni Spinatto. 1994. *Antichi termini confinari del Friuli.* Vol. 1–3. Udine: Del Bianco.
Cozzi, Donatella, Domenico Isabella, and Elisabetta Navarra, eds. 1998. *Sauris—Zahre: Una comunità delle Alpi Carniche.* Udine: Forum.
Danelutto, Antonino. 2003. "Le malghe del Montasio." In *Canal del Ferro e Valcanale nel Tempo—Relazioni del convegno "Aspetti storici, economici culturali del Canale del Ferro e della Valcanale,"* 103–16. Padova: CLEUP.
Dapit, Roberto.1995–2008. *Aspetti di cultura resiana nei nomi di luogo.* Vol. 1–3. Padova: CLEUP.
———. 1997. "Mladinski raziskovalni tabor 1997—Kanalska dolina." *Traditiones* 26: 379–83.
Da Pozzo, Ulderica. 2004. *Malghe e malgari.* Udine: Forum.
De Gasperi, Giovanni Battista. 1914. *Studi sulle sedi e abitazioni umane in Italia: Le casere del Friuli.* (Memorie geografiche pubblicate come supplemento alla Rivista geografica italiana al Dott. Giotto Danielli: Supplemento alla Rivista geografica italiana. Sedi e abitazioni umane in Italia 1.) Firenze: M. Ricci.
———. 1915. "Ancora sulla geografia delle casere." *Rivista geografica italiana* 22: 413–15.
Depollo, Vinicio. 1980. "Le malghe dimenticate." In *Lassù sui monti*, ed. Giancarlo Gualandra, 20–29. Udine: Graphik Studio.
Desinan, Cornelio Cesare. 1982–83. *Agricoltura e vita rurale nella toponomastica del Friuli-Venezia Giulia.* Vol. 1–2. Pordenone: Grafiche Editoriali Artistiche Pordenonesi.
De Zorzi, Cristina, and Annamaria Mariutto, eds. 2015. *Monte casone Farra e Fratte.* Andreis: Lis Aganis—Ecomuseo Regionale delle Dolomiti Friulane.
Dreossi, Gian Franco, and Mauro Pascolini. 1995. *Malghe e casere della montagna friulana: Itinerari escursionistici per tutti.* Udine: Co.El.
———. 2010. *Malghe e alpeggi della montagna friulana: Facili escursioni alla scoperta di storia, tradizioni e prodotti tipici.* Udine: Co.El.
Dvorsky, Viktor. 1915. "Sulla geografia delle casere." *Rivista geografica italiana* 22: 298–304.
Faleschini, Giuseppe. 1970. *L'alpeggio in Carnia.* Udine: Regione Autonoma Friuli-Venezia Giulia, Assessorato dell'Agricoltura.
Frei-Stolba, Regula. 1988. "Viehzucht, Alpwirtschaft, Transhumanz: Bemerkungen zu Problem der Wirtschaft in der Schweiz zur römischen Zeit." In *Pastoral Economies in Classical Antiquity*, ed. C. R. Whittaker, 143–51. Cambridge, UK: Cambridge Philological Society.
Furlan, Andrej. 1995. "Svojevrsten način planšarstva v Terski dolini ter analiza dveh primerov stavbne dediščine terskih planšarjev." In *Planšarske stavbe v Vzhodnih Alpah: Stavbna tipologija in varovanje stavbne dediščine*, ed. Tone Cevc, 111–17. Ljubljana: ZRC SAZU.
Gilck, Fridtjof, and Peter Poschlod. 2019. "The Origin of Alpine Farming: A Review of Archaeological, Linguistic and Archaeobotanical Studies in the Alps."

The Holocene. Retrieved 6 July 2020 from https://journals.sagepub.com/doi/10.1177/0959683619854511.
Guida 2016. *Guida malghe Friuli Venezia Giulia.* Udine: ERSA.
Guida 2017. *Guida malghe Friuli Venezia Giulia.* Udine: ERSA.
Harrison, Rodney. 2013. *Heritage: Critical Approaches.* Abingdon: Routledge.
Heady, Patric. 1999. *The Hard People: Rivalry, Sympathy and Social Structure in an Alpine Valley.* London: Taylor & Francis.
Herzfeld, Michael. 2005. *Cultural Intimacy: Social Poetics in the Nation-State.* 2nd ed. New York: Routledge.
Kirchengast, Christopher. 2008. *Über Almen zwischen Agrikultur und Trashkultur.* Alpine Space: Man and Environment Vol. 5. Innsbruck: Innsbruck University Press.
Krauß, Werner. 2018. "Alpine Landscapes in the Anthropocene: Alternative Common Futures." *Landscape Research* 43(8): 1021–31. https://doi.org/10.1080/01426397.2018.1503242.
Ledinek Lozej, Špela. 2016. "Dairying in the Mountain Pastures in the Julian Alps: Heritages, Utopias and Realities." *Studia ethnologica Croatica* 28: 91–111. https://doi.org/10.17234/SEC.27.
———. 2021. "Labelling, Certification and Branding of Cheeses in the Southeastern Alps (Italy, Slovenia): Montasio, Bovec, Tolminc and Mohant Cheese." *Acta Geografica Slovenica* 61(1): 141–56. https://doi.org/10.3986/8746.
Macdonald, Sharon. 2013. *Memorylands: Heritage and Identity in Europe Today.* London: Routledge.
Madotto, Aldo. 1987. *Vivere fra le montagne.* N.p.
Malghe FVG. 2020. "Malga che vai . . . Formaggio che trovi: Le maghe da latte della montagna friulana." Retrieved 18 July 2020 from http://www.malghefvg.it/it.
Maniacco, Tito. 2014. *Storia del Friuli: Le radici della cultura contadina, le rivolte, il dramma dell'emigrazione e la nascita dell'identità di una regione.* 3rd ed. Roma: Newton Compton.
Marchettano, Enrico. 1908–11. "I pascoli alpini della Carnia e del Canale del Ferro." *Bollettino dell'Associazione Agraria Friulana* 52(13–15): 387–400; 52(16–18): 488–96; 53(15–16): 232–36; 53(16–18): 300–5; 53(19–21): 404–10; 54(1–3): 26–33; 54(6–7): 233–38; 54(14–17): 515–19; 54(18–20): 585–91; 55(9–12): 207–26; 55(13–16): 292–319; 55(17–20): 398–418.
Marinelli, Giovanni. 1880. "Le casere in Friuli secondo la loro altezza sul livello del mare." *Bollettino dell'Associazione Agraria Friulana* 20–21: 154–56.
———. 1894. *Guida della Carnia e del Canal del Ferro.* Tolmezzo: Del Bianco.
Marinelli, Olinto. 1902. "Studi orografici nelle alpi Orientali." *Bollettino della Società geografica Italiana* 3(3): 682–716, 757–79, 833–61.
———. 1915. "A proposito di un tipo slavo di casere in Friuli." *Rivista geografica italiana* 22: 502–4.
Minnich, Robert Gary. 1989. "Tradition in the Face of Modernization: Cultural Continuity and 'Deagrarization' in the Village of Ukve." *Slovene Studies* 11(1–2): 97–108.
———. 1990. "At the Interface of the Germanic, Romance and Slavic Worlds—Folk Culture as an Idiom of Collective Self-Images in the South-Eastern Alps." *Studia Ethnologica Croatica* 2(1): 163–79.

———. 1998. *Homesteaders and Citizens: Collective Identity Formation on the Austo-Italian-Slovene Frontier.* Bergen: Norse Publications.
Musoni, Francesco. 1910a. *Studi antropogeografici sulle Prealpi Giulie.* Firenze: M. Ricci.
———. 1910b. *Sulle condizioni agrarie delle Prealpi Giulie.* Udine: Del Bianco.
———. 1913. "Influenza del carsismo sulla vita pastorale del bacino medio del Natisone." *Mondo sotterraneo* 9: 104–5.
———. 1914. *Nuove ricerche di antropogeografia nelle Prealpi del Natisone.* Udine: Del Bianco.
Netting, Robert M. 1981. *Balancing on an Alp: Ecological Change and Continuity in a Swiss Mountain Community.* Cambridge, UK: Cambridge University Press.
Noacco, Eddo. 1959. *Regime giuridico dei boschi e pascoli della Carnia.* Udine: Camera di commercioindustria e agricoltura.
Pascolini, Marta. 2010. "Il palo della Passione nella cultura dell'alpeggio in Carnia." Master's thesis, Università Ca' Foscari Venezia, Facoltà di Lettere e Filosofia.
Pascolini, Mauro. 1992. "L'alpeggio nelle valli del Natisone: La perdita di un originale modello di sfruttamento delle risorse." In *Studi in memoria di Giorgio Valussi*, ed. Vincenzo Orioles, 45–63. Alessandria: Edizioni dell'Orso.
———. 1994. "Evoluzione del sistema insediativo nella Val di Gorto: 'La mont disjamada'; L'abbandono delle malghe." In *In Guart: Anime e contrade della pieve di Gorto*, ed. Manlio Michelutti, 109–26. Udine: Società Filologica Friulana.
———. 1997. "L'alpeggio nella Regione Friuli-Venezia Giulia: Le prospettive di sviluppo dell'attività malghiva, gli aspetti socioeconomici e gli strumenti giuridico-normativi di sostegno nella legislazione regionale, nazionale e comunitaria." Report. Udine: Regione autonoma Friuli-Venezia Giulia and Università degli Studi di Udine.
———. 2001. "L'alpeggio nelle Alpi orientali: Modelli storici e situazione attuale; Una prospettiva geografica." *Erreffe: La ricerca folklorica* 43: 71–81.
Pascolini, Mauro, and Nicoletta Tessarin. 1985. *Lavoro in montagna: Boscaioli e malghesi della regione alpina friulana.* Milano: Franco Angeli.
Pascolini, Mauro, and Giulia Tondo. 1996. "'Fra monti crodosi': Boschi e pascoli del Canal della Torre." In *Tarcint e Valadis de Tôr*, ed. Gianfranco Ellero, 131–40. Udine: Società Filologica Friulana.
Pasut, Davide. 2012. "Studio su alpeggi e pascoli della regione Friuli Venezia Giulia." Finale report. Udine: ERSA.
———. 2013. "Rilievi agronomici e vegetazionali sulle malghe del territorio regionale." Final report. Udine: ERSA.
———. 2015. "Elaborazione di dati agronomici, gestionali e strutturali delle malghe del Friuli Venezia Giulia e realizzazione applicazione GIS." Final report. Udine: ERSA.
———. 2016. *Scenari per l'alpicoltura friulana.* Gorizia: ERSA. Retrieved 19 July 2020 from https://www.sozooalp.it/fileadmin/superuser/altre_pubblicazioni/Scenari_per_l_apicultura.pdf.
Pasut, Davide, Simonetta Dovier, Stefano Bovolenta, and Sonia Venerus. 2006. *Le malghe della dorsale Cansiglio-Cavallo: Un progetto per la valorizzazione dell'attività alpicolturale.* Gorizia: ERSA. Retrieved 18 July 2020 from http://www.ersa.fvg.it/

export/sites/ersa/aziende/sperimentazione/Alpicoltura_friulana/Allegati-Alpeggio/Pasut_malghe_volume_web.pdf.

Pasut, Davide, Alberto Romanzin, and Stefano Bovolenta. 2016. *Malga Montasio: Una storia friulana*. San Michele all'Adige: Edizioni SoZooAlp. Retrieved 19 July 2020 from https://www.sozooalp.it/fileadmin/superuser/altre_pubblicazioni/Malga_Montasio_2016_web.pdf.

PromoTurismo FVG. 2020. "Sauris e il Giro delle Malghe." Retrieved 18 July 2020 from https://www.turismofvg.it/code/87705/Sauris-e-il-Giro-delle-Malghe.

Ravnik, Mojca. 2015. *"Na žegen!": Žegnanje in drugi prazniki z rekruti v Ukvah v Kanalski dolini*. Ljubljana: Založba ZRC, ZRC SAZU.

Rupel, Aldo, ed. 1987. *Tabor Kanalska dolina 86*. Gorica: SLORI; NŠK.

———, ed. 1990. *Tabor Rezija 89*. Gorica: SLORI; NŠK.

Sanna, Maurizio. 2013. *Ricognizione delle strutture e infrastrutture relative alle malghe del territorio della regione Friuli Venezia Giulia*. Udine: ERSA.

Slow Food FVG website. 2019. "Çuç di Mont." Retrieved 18 July 2020 from https://www.slowfoodfvg.it/presidi/cuc-di-mont/.

Smith, Laurajane. 2006. *Uses of Heritage*. London: Routledge.

Tagliaferri, Amelio, Tommaso Fanfani, Giovanni Panjek, and Bruno Polese. 1981. *Elementi per la storia della cooperazione nel Friuli-Venezia Giulia*. Udine: Regione autonoma Friuli Venezia Giulia.

Van der Ploeg, Jan Douwe, Henk Renting, Gianluca Brunori, Karlheinz Knickel, Joe Mannion, Terry Marsden, Kees Roest, Eduardo Sevilla-Guzmán, and Flaminia Ventura. 2000. "Rural Development: From Practices and Policies Towards Theory." *Sociologia Ruralis* 40: 391–408. https://doi.org/10.1111/1467-9523.0015610.1111/1467-9523.00156.

Van der Ploeg, Jan Douwe, and Dirk Roep. 2003. "Multifunctionality and Rural Development: The Actual Situation in Europe." In *Multifunctional Agriculture: A New Paradigm for European Agriculture and Rural Development*, ed. Guido Van Huylenbroeck and Guy Durand, 37–53. Hampshire: Ashgate.

Valussi, Giorgio. 1954. *Evoluzione delle attività economiche nella Val Degano con partico- lare riguardo alla vita pastorale*. Udine: C.C.I.A.A.

Vendrami, Stefano, and Laura Viel, eds. 2010. *Buone pratiche gestionali delle malghe tra Veneto, Friuli Venezia Giulia e Carinzia*. Belluno: Regione Veneto. Retrieved 19 July from http://www.ersa.fvg.it/export/sites/ersa/aziende/sperimentazione/Alpicoltura_friulana/Allegati-Alpeggio/buonepratichegestionalimalgheit.pdf.

Viazzo, Pier Paolo. 1989. *Upland Communities: Environment, Population and Social Structure in the Alps since the Sixteenth Century*. Cambridge, UK: Cambridge University Press.

Vidrigh, Mauro. 2003. "Economia della montagna tra sviluppo e sostenibilità: Il caso della Valcanale – Canal del Ferro." In *Canal del Ferro e Valcanale nel Tempo—Relazioni del convegno "Aspetti storici, economici culturali del Canale del Ferro e della Valcanale*," 133–52. Padova: CLEUP.

CHAPTER 11

From Nomadism to Ranching Economy

Reindeer Transhumance among the Finnish Sámi

Nuccio Mazzullo and Hannah Strauss-Mazzullo

Introduction

In the twenty-first century, reindeer herding continues to be an important livelihood in Finnish Lapland in the Sámi homeland, albeit practitioners cannot entirely rely on herding for their income. The practice of free grazing and the migration of reindeer between summer and winter pastures has endured and adapted to many influences over the last 150 years. Apart from its practical relevance for the local economy, we observe a growing symbolic importance of reindeer herding, and especially, ownership of reindeer in the definition and self-determination of the Sámi nation.

This chapter will describe the routines of Sámi reindeer herding today, followed by a brief overview of the historical and administrative limitations that have restrained old practices, and a discussion of some of the technological innovations that have enabled adaptation. We will then introduce current practices as transhumant practices in a ranching economy as well as the development of tourism for additional income. Having clarified both practices and terminology, as well as the opportunities arising from marketing, we then turn towards Sámi reindeer herding's contemporary cultural and political dimension. Research for this chapter has been conducted by the authors in the communities of Finnish Lapland since 1990 and 2007 respectively. More recently, interviews on touristic activities have been conducted by Nuccio Mazzullo in the context of the Interreg-funded project Arctisen at the Multidimensional Tourism Institute (MTI), University of Lapland (January to April 2019).

Sámi Reindeer Herding through the Seasons

In Finland, reindeer herding society is administratively organized into so-called reindeer herding cooperatives, or *paliskunnat* (pl.), with each *paliskunta* integrating several herding families. The entire Finnish North, a third of the country, about 100,000 km², is currently divided into fifty-four reindeer herding cooperatives of various sizes and reindeer populations. Each district has clearly defined borders, which are marked by high fences, and within which a cooperative's animals can graze without being closely supervised for most of the year.

Within the cooperative's borders, small groups of reindeer gather according to season as well as gender and start foraging together. Especially shortly before and after calving in spring, groups of female reindeer gather in small groups to defend themselves from predators and perhaps aggressive males. While male reindeer lose their antlers after the mating season, females lose theirs only after the calves have become strong enough to be on their own. May is the "month of the calves," or *miessemannu* in Northern Sámi language. It used to be the name for June, but with reindeer starting to calve already in May, the name was moved accordingly (Mazzullo 2012: 220). The location they choose for calving is usually the same. The calves stay with their mothers about six months if they are not slaughtered or prey to predators. The groups of females and their calves stay in the forest until the growing number of mosquitoes become dangerous for the calves. Thus, to protect the young ones, the animals move to higher grounds, or forest clearings and tundra-like areas, and that is the first time that herders are able to intercept with them. For this reason, mosquitoes are often called the "reindeer herder's helper."

During the time when the reindeer come out of the protection of the forest, in early July, the calves then receive their earmarks, and they are in first contact with humans. The earmarking cannot be done earlier because the young ones would be too small to move out of the forest, and not any later, because the warble fly becomes a nuisance in the second half of July, threatening to disperse the herd and making it impossible for herders to gather them again. Therefore, when the reindeer come out of the forest they are driven towards a corral. Herders and their dogs move around on ATVs, sometimes accompanied by a helicopter. The gathered herd is pushed towards (mostly) permanently erected corrals, the funnel fence. To identify the calves, they have to be separated from their mothers and tagged only to be reunited under the eyes of the reindeer herders who then observe together whose reindeer the calf is following and hence whom it belongs to. Back when herders were following the herd throughout the year, this was not necessary, as every calf would be earmarked shortly after they were born. After having been earmarked, all animals are released back into the forest for the rest of the summer. All by themselves, the animals move to the summer grounds where there is less vegetation (tundra-like) and the wind ventilates the grazing area, which makes it more difficult for mosquitoes to fly. During the summer and late summer, the animals eat and get fat before the mating season in autumn.

During early winter migration, when the first snow has fallen in Finnish Lapland, the animals gather in bigger herds and start to follow a herd leader, which is usually a larger male or female reindeer with certain skills (such as physical strength, orientation, or social skills). At the top of the herd, several strong animals take turns trampling the snow to create a path; the rest of the herd, and at some point, the herders will follow them. A line of animals forms on that path, including between one hundred and several thousand reindeer. The herders put a bell (nowadays also a GPS collar) around the neck of the animal that has been chosen as a leader by the herd and which used to be a half-tame animal in the past. To identify the leader requires a trained eye, otherwise the bell is put on the wrong animal's neck. A specific path is suggested by the herders to the lead reindeer through the creation of a track that the animals find appealing to walk on. Today, such suggestion is prepared by snowmobile, leaving compressed snow behind. In earlier days, this would be done by skis.

For the separation and slaughtering in late autumn/early winter, herders move several of these larger herds together. Before Finland's accession to the EU, slaughtering would still take place in field slaughterhouses, in the vicinity of corrals, some consisting of beams on the ground and a makeshift roof. Slaughtering animals in freezing conditions in autumn has proven to be safest as bacteria are unable to spread, especially when handling the entrails of the slaughtered animals. Despite their good hygienic record (Vaarala and Korkeala 1999), the field slaughterhouses had to be replaced with plant slaughterhouses to meet EU requirements. For private use, reindeer herders are still allowed to slaughter reindeer in the field, and this practice continues to be an important social event, particularly in connection with intergenerational sharing of traditional knowledge. Around half of the herd is slaughtered each year, and the other half is released back into nature. The herd moves to the winter pasture deep in the forest by itself, or sometimes with the help of the herder. In the forest, the snow is not exposed to the sun and the snow cover does not melt and freeze so easily. The reindeer continues to graze by digging for edible plants under the snow (changing climatic conditions have caused the snow cover to melt and freeze more frequently, making it harder for the reindeer to dig for food under the snow). In late winter (March/April) Sámi reindeer herders have now started to move around and look for weak animals to provide them with some fodder (hay, artificial feed). Among Finnish reindeer herders, who often keep reindeer as an additional income for their agricultural farms, feeding reindeer is a widely accepted practice. Among Sámi reindeer herders, this is still not considered a viable option, although it is becoming more common. At this point in the year, reindeer survive on their body fat and eat lichen hanging from trees to keep the digestive system going. They are most vulnerable to legally protected predators, such as wolverines, wolves, eagles, and bears.

To summarize this description of the herding cycle, reindeer herder's physical contact with the herd occurs twice a year, during roundups for earmarking in summer and separation/slaughtering in autumn/winter. During other times, animals are mostly observed from a distance, through binoculars, and are approached only if a weak or orphaned reindeer is identified.

While herds can graze freely within a certain area, maximum sizes of herds are set by the Ministry of Agriculture and Forestry and the conditions of the pastures are closely monitored by scientific staff to prevent overgrazing. The "carrying capacity" for a specific pasture is determined for ten years at a time, from which the maximum herd size for each cooperative is then derived. The issue of overgrazed pastures has been discussed since the end of World War II, when the land had been divided into cooperatives and competing land uses started to have a greater impact on the reindeer herding area. The quality of pastures has severely suffered from intensive forestry, as old-growth forests have almost disappeared. During the cold time of the year, and especially towards the end of the long winter, reindeer feed on lichen, either dug from underneath the snow or stripped from trees. The younger the forest, the less lichen grows. In addition to these limitations, which have led to open conflicts between herders and the Finnish state in the past (Strauss and Mazzullo 2014; Mazzullo 2013; Lawrence and Raitio 2006), reindeer often become prey to protected predators. In order to receive compensation for loss to predators, the herder has to provide evidence that a killing has taken place and hope that the animal is still identifiable (i.e., that the ear is still intact). Since reindeer roam the landscape on their own for most of the year, carcasses are not easily found.

Transitions

In the previous section, the aim was to describe reindeer herding as it is practiced today in Finnish Lapland, in particular among the Indigenous Sámi. The changes this practice has undergone are described extensively in other publications (e.g. Bjørklund 2013; Dana and Riseth 2011; Forbes 2006; Mazzullo 2010), however a brief overview is necessary to continue discussion.

The first evidence of a transition from hunting to herding reindeer dates back to the seventeenth century (Leem 1767, cited in Mazzullo 2018) when the Sámi started to keep animals not only for draught or as decoys but as a herd. Over the next centuries, the livelihood developed into a nomadic lifestyle where a community of reindeer herders (*siida*, pl. *siidat*) followed their reindeer herd through the landscape. Animals and people lived and stayed together the whole year on their migratory cycle between inland and coastal areas across the Fennoscandian North. The distance covered throughout the year encompassed several hundred kilometers. This practice was interrupted with the consolidation of state borders at the beginning of the nineteenth century when reindeer and herders were no longer allowed to cross into another national territory. Depending on availability of coastal and inland areas, like in Norway, the seasonal routes and routines remained intact for some communities. In addition, the Lapp Codicil, a bilateral agreement between Sweden and Norway allowed for traditional migration between the inland and the coast on Swedish and Norwegian territory (Koch 2013). In Finland, however, practices changed more dramatically as access to the northern coast was cut off. Pastures available within the newly established borders were used to survive these abrupt changes. The range of movement was thereby reduced significantly, yet, herding families continued to live in the communal organization of *siida* and continued to move around with the animal. In its attempt to treat all citizens equally, the Finnish administration ignored the minority's traditional values and aimed at a reorganization of Indigenous reindeer herding society (Lehtola 2015). In 1932, the Reindeer Herding Act was passed by the Finnish Parliament (amended in 1948 and 1990) which enforced a system of clearly delineated cooperatives that incorporated the existing *siidat* but no longer allowed for the previous degree of flexible land use. The *siida* was made redundant and almost obsolete, and by the end of World War II, completely transformed into herding cooperatives. The traditional communal management of pastures had been flexible within the smaller unit of the *siida*, where decision-making was done according to actual needs and availability of resources. If, for instance, a large herd shrunk in numbers due to overgrazing, difficult winters, the spread of diseases, etc., and pastures were becoming available, then other *siidat* would aim to negotiate with the current land user, who did not need those pastures any longer, and agree on its future use. Often, one *siida* would liaise with another through marriage (exogamy). Accordingly, the use of natural resources was based on needs rather than tenure rights, hence it was dynamic and served the herding community well on these margins of arable land.

Within the newly established, larger herding cooperative, several *siidat* were integrated and headed by a single, democratically appointed herder as determined in the Finnish reindeer herding law, and as a result decision-making became less flexible and more focused on equal opportunity rather than its members' actual needs. In addition, by integrating diverse herding groups together within the same cooperatives, a clash of herding cultures was inevitable, both equally in those cooperatives that included different Sámi groups or Finnish and Sámi reindeer herders. In Finland, unlike in Norway and Sweden, reindeer herding is not an exclusive right to ethnic Sámi, but it can be practiced by all residents of the Finnish Reindeer Herding Area.

With the establishment of nation states, competition for good pastures grew within the area of Finnish Lapland. A greater number of herders was now depending on an increasingly scarce resource within a new, enforced management system. The remedy that reindeer herders came up with was the construction of fences around each *paliskunta* (cooperative) to avoid losing reindeer crossing national borders and to prevent them from moving to other cooperatives' pastures. From there, the reindeer would have to be recovered personally by the owner or otherwise they would have been incorporated into that respective cooperative's stock, slaughtered, and sold as cooperative's common property.

The construction of fences was the vehicle of a major transition in Sámi reindeer herding society. Until the extensive network of wooden structures was erected, herds had to be supervised and guided through the landscape. With the (cooperative's) borders becoming physical obstacles, it was no longer necessary to closely tend the animals. By the time fences were being built, from the 1930s onwards, herding families had settled in winter villages where schools and other facilities had been made available.

With the introduction of snowmobiles in the 1960s, the time that reindeer herders spent on the land was further reduced. Instead of spending several days tracking down the animals, a herder could get an overview where his herd was currently grazing within a single day. The use of helicopters and airplanes as well as GPS collars is a continuation of this development, making the locating of animals even quicker and easier. The use of drones is being considered for this purpose (Länsman and Satokangas 2019).

In much of the literature on innovation in reindeer herding society, the wooden fences do not receive much attention. Probably because they were initially entirely made of wood, and certainly cannot be considered a recent invention. Nevertheless, they allowed a revolution to take place, fundamentally affecting the relation between herders and their animals. The introduction of snowmobiles, which is normally described in revolutionary terms (Pelto 1973), brought less qualitative change but made the sporadic visit to the herd in the forest more efficient, and has thus to be understood as a merely quantitative development. Another of such important changes among reindeer herder societies has been described by Stammler (2009) in the context of mobile phone use among nomadic Nenets. As Nenets continue to move around with their herds far away from urban facilities and infrastructure, the snowmobile with its high dependency on fuel supply and spare parts makes herders extremely vulnerable, while the recharging of mobile phones requires much less resources and has had a significant impact on the organization of people and herds moving in the open tundra. For similar reasons of low capacity and lack of supporting infrastructures, electric snowmobiles are only in limited use in the tourism sector in Finnish Lapland.

Transhumant Practices in a Ranching Economy

On Finnish territory, Indigenous Sámi practiced a nomadic reindeer herding lifestyle up until World War I. With more and more fences being erected along the herding cooperative's borders, from the 1930s onwards, however, it was no longer necessary to stay with the herd constantly. Around the end of World War II, herding practices changed significantly. Until then, two people with dogs would still be sent to circle around the herd, one of them clockwise, the other one counterclockwise. This was done every few days. Nowadays, the herds are still being supervised, however, this is done increasingly remotely, and with the help of most recent technology. Interaction between reindeer and herders has decreased to a minimum and the herder's relationship to the animals no longer resembles the close nomadic connection that was prevalent one hundred years ago. Herders refrain from "becoming-animal" (Palladino 2018: 123, in reference to Deleuze and Guattari 1987), which is constituted by walking and sleeping with the animals, which has been described as a hard, but emotionally rewarding livelihood. At the same time, such a close connection prevails among the older generation of reindeer herders, and it is believed to be impossible to manage a herd without the ability to "think like a reindeer."

Ingold (1988) describes the transition from nomadism to ranching economy meaning that reindeer never stay in a shelter constructed for them, they are always outside on their own and thus, they are actually wild animals, as their keeping resembles hunting (predatory) practices. If they find refuge in an inaccessible place, or spread over vast territory, they cannot be managed by people.

> For the greater part of every year, the animals see little or nothing of man; some, indeed may avoid human contact for years on end. Unlike the pastoralist, who requires regular access to his herd for subsistence, the rancher need not round up his stock more than twice a year: once for branding calves, and once for selecting animals to be sold for slaughter.... By allowing his animals the freedom of the range, he not only minimizes the labor costs incurred in their supervision, but also enables them to make optimal use of available pastures. This last factor is of critical importance in a ranch economy where land constitutes a scarce resource, and where the condition of animals is directly reflected in their market value. (Ingold 1988: 238)

Within reindeer ranching economy, the herds are aided to move between different pastures according to the season, and in this sense, Sámi reindeer herders employ transhumant practices. Their home and families remain settled in one place, while the animals move between the same summer and winter pastures on the same route through the years (see Habeck 2006 on transhumance among the Komi reindeer herders; Luick 2008). In regard to altitude, the movement used to be the opposite of transhumance as practiced in mountainous areas in central and southern European areas: the lower grounds at the coast were sought in summer because they are less frequented by mosquitoes, the higher forest grounds were sought in winter as their snow conditions were more stable and they provided shelter from wind chills. With the closing of national borders, for those Sámi herders who remained on Finnish territory, this practice came to a halt and the seasonal cycle is now a movement between forest and open plain/higher ground pastures with only slight differences in altitude (at least when compared to the differences cattle and sheep overcome in, for instance, the Alps and Pyrenees). Because of its rather recent nomadic past, the terminology to describe current practices in the Finnish context is hardly ever connected to the concept of transhumance. Instead, the term semi-nomadism is often used, while transhumance actually appears to be more appropriate to describe the regular transition between open-plain summer and forest winter pastures.

A clear definition of transhumance is difficult as practices are diverse and in constant adaptation to changing environmental and social circumstances (Costello and Svensson 2018). In his work *The Appropriation of Nature* Ingold discusses various modes of movements of peoples and animals. In his view, the term transhumance could be used "in a much wider sense, to denote any movement of people occasioned by the seasonal exploitation of diverse ecological zones" (Ingold 1987: 182). In many definitions of transhumance, however, limited crop cultivation appears as a means to sustain the livelihood of those living on the margins of arable lands. In a definition delivered by Jones (2005: 358) a transhumant economic system consists of "(1) permanent villages, (2) arable agriculture, and (3) the seasonal movements of livestock." Transhumant communities have permanent settlements and "the village forms the nucleus of transhumant society" (Johnson 1969: 18–19, cited in Jones 2005: 358). They engage in agricultural activities to complement their livelihood.

While interaction with agriculturalists was common, agricultural activities were not practiced among Sámi reindeer herders. Instead, reindeer herding appeared as an additional form of income to those Finnish farmers who were making their living south of the Upper Lapland region. For them, the movement between pastures is only one way of providing feed, and supplementary feeding is common. As pointed out above, among the Sámi, the additional feeding of reindeer is accepted only in emergencies (*hätäruokinto*), not as a regular practice. In the past, Finnish farmers have borrowed reindeer during winter as draught animals, but at the end of the season, those animals on loan would go back to their original herd and migrate together to the coast. Furthermore, reindeer herders never developed the production of cheeses, though they practiced milking and cheese making on a smaller scale for own consumption considering the little milk reindeer give. A group of intermediaries, called *verdde*, existed, helping to trade products other than dairy between Sámi herders and Finnish farmers. These were mostly farmers settling close to herders and keeping close relations with them, far away from farming villages. In addition, many Sámi reindeer herders used to work as seasonal forest workers with the effect that there was less conflict between forestry and herding during that time. Fundamentally, by being employed, herders had more control over the felling of trees as they were directly involved in the process.

Breeding in reindeer herding has not focused on the development of individual traits. Traditionally, Sámi reindeer herders have put emphasis on the composition of the herd rather than the reproduction of specific traits across the group of animals. A herder put effort to maintain a diverse herd including also "non-productive" animals. These have "particular roles which contribute to the productivity of the herd as a whole" (Tyler et al. 2007: 197). Hence, profit is derived not from the output of a single animal. But just as a diverse landscape with a variety of pastures fulfills different functions in the seasonal adaptation to weather, so do breeding strategies gain their strength from the diverse composition of the herd. The Sámi term *čappa éallu* describes "a beautiful herd" in terms of its composition (Oskal 2000). For instance, in a herd that normally features few males, strong, old female reindeers are important to break the ice to access plants underneath or trample the snow in order to make a path (Tyler et al. 2007). This would generally not be understood as an "efficient form of breeding," but it is most appropriate in maintaining and sustaining herding practices in the long term.

How does reindeer herding as it is practiced by the Sámi in Finnish Lapland fit into the discussion of the heritage of transhumance? While it fits the definition of making use of land that cannot be harvested with agricultural techniques (the growing of crops is hardly possible in the southern parts of the reindeer herding region and it is impossible in the northern parts where only berries and mushrooms can be collected from the forest, as well as occasionally riverside grass as fodder), it is far from being a marginal practice. In fact, it is the only viable livelihood within an extensive region. The emergence of transhumant practice in Finland, however, is due to administrative changes, aided by the installation of fences around newly demarcated territories. Thus, the tradition is only about one hundred years old. Previously, herding communities followed a nomadic livelihood with flexible routes and grazing opportunities, covering a larger area and crossing with other herding communities' routes.

As it is practiced today, reindeer herding is highly dependent on the market economy for living on the land. During times of nomadic livelihood, slaughtering was done for the herders' own consumption, or to trade meat and hides for tools. With the growing importance of nation states, these increasingly pushed for a modernization and rationalization of reindeer husbandry "to promote sustainable, but maximal, economic yields" (Beach 1997: 123 in reference to the Swedish case), a process that is not yet considered complete in current discourse. In the Finnish context, still in the 1970s, most of the meat was consumed in the area of production, whereas since the turn of the millennium, most is consumed in the Helsinki region (Saarni et al. 2007). After a period of declining profits from meat in the mid-1990s a reappropriation of centralized meat-producing facilities has taken place in the reindeer herding area. The business became more profitable for small enterprises run by Sámi herders by training their own butchers at the Sámi education center in Inari, *SAKK*, and by avoiding the use of middlemen to sell the product. In the attempt to use all parts of the animal, meat, hides, leather, and also antlers are being turned into profitable products.

Functional items produced from natural material in Sámi handicraft are called *duodji*. The concept describes the process of production, which is taught in and through personal relationships, as well as the different products (Guttorm 2017). The production of handicrafts continues to be an important cultural marker and economic sector, with growing business opportunity in tourist markets. Guttorm (2017: 166) argues that a process of "autonomization" has occurred in the relationship between the item and the crafter, but in a certain respect, it will always resemble the art of crafting as it was done when items had practical use. For instance, the selection of suitable wood for the making of a reindeer milking bowl requires the *duojár* (the maker of *duodji*) to search the woods for a birch tree with a burl (deformed growth) just like in the old days.

Marketing Reindeer Transhumance in the Tourism Sector

Duodji items as authentic products are rather expensive due to their traditional production methods, and hence mainly affordable to well-off tourists. A small wooden cup made traditionally from birch burls may easily cost EUR 200. For decades, cheap souvenirs, produced elsewhere on industrial scale, have entered the market and since then been contested. Smaller, more affordable items are also produced in Finland, and the Sámi are pondering ways to regain control over the production and sale of souvenirs in Sámi reindeer herding area by, for instance, producing authentic items of lower quality than *duodji* items. Rather than buying a pen that reads "I was in Lapland," made in China, visitors could buy a locally produced matchbox made from birch wood. The sale of (untreated) antlers directly to tourists is another opportunity.

It has to be stated, however, that most herding routines do not allow for (slow) tourism experience. A growing number of tame reindeer are frequently used in tourism encounters, to stand for pictures in city centers or take tourists for a short sledge ride (but so are husky dogs, which are foreign to Scandinavia). An increasing number of reindeer "farms" in the vicinity of tourist resorts invite visitors, who can experience animals on leashes, or within fenced areas in closer vicinity. The selling of such experience is a lucky opportunity in the sense that weak reindeer which have to be pampered and made fit to be send back to the forest can recover and bring additional income as they are visited by tourists in zoo-like conditions. In some cases, the older generation of herders run the reindeer "farms." Small enterprises are organized in the association of reindeer farms that coordinates tourist activities and that watches over animal rights to prevent exhaustion.

These zoo-like conditions are seizing the opportunity to let tourists meet reindeer, but it is not the transhumant practice that allows for such an encounter, as it involves too many risks for all participants, humans as much as animals. Any person unfamiliar with the environmental conditions of migrating or rounding up the animals in sub-zero temperatures, or in muddy conditions without any facilities in convenient proximity, would be at risk themselves, and require too much attention from the herders. Moreover, the uncoordinated movement of people and vehicles within the herding area can easily lead to the dispersal of a herd, or the exhaustion of young reindeer who might die if pushed along a route too quickly. During migration and roundups, a group of herders and their family members work as a team that relies on decades of common experience in the context of centuries-old traditions. It is mostly implicit knowledge that makes a person know when, where, and how to move. Growing up in reindeer herding society lets a herder observe animal needs with a trained eye (Ingold 2000). For these reasons, tourists are welcome at established points of encounters and as consumers of product, but they are considered a disturbance to work routines outside those points. While a very limited number of visitors (explorers, adventurers, scientists) has always managed to find entrance to Sámi herding society by working hard to learn the skills required for such activity, the average, regular visitor today expects comfort and care which reindeer herding cannot offer along the transhumant path. On the contrary, a visit to the real world of reindeer herding may be rather disillusioning, considering the use of modern technology, nontraditional clothing, and the potential killing of animals. This side of the story is underrepresented in (social) media accounts of tourism experiences, which tends to romanticize and gear everything towards a Santa Claus winter wonderland, with Sámi in traditional clothes in silent and peaceful snowy landscape on a sled pulled by reindeer (or huskies).

The numbers of tourists are still rising. An approximately 1.5 million tourists visit Finnish Lapland each year, which is significantly more than the 180,000 inhabitants of Finnish Lapland, and a fourth of Finland's total population (Veijola and Strauss-Mazzullo 2018). Airports are being extended everywhere across Lapland, allowing more tourists to visit remote places regularly. Additional capacities are being installed, and an increasingly greater diversity of activities is offered, serving different ideas, needs, and budgets. With more and more tourists entering and thus appropriating the landscape, conflicts arise. The lakes commonly reserved for local elderly have seen hobby fishers from abroad, and consequently the younger generation of the local population has started to fish, too, in order to prevent the lake from being harvested mainly by foreigners, leaving little fish behind for those elderly who used to fish there. Similar disruption is being experienced from husky safaris crossing reindeer herding area. While routes for snowmobiles are fairly established and indicated, there are no such routes that dictate to the owner of husky sledges where to go. If the dogs get too close to reindeer herds, they leave considerable mess behind, as a herd might disperse completely in the attempt to escape.

It cannot be denied that the income from tourism has been gaining momentum in reindeer herding society among the Sámi of Finnish Upper Lapland. For example, over the past five years, the town of Inari has changed significantly. Especially winter is not as it used to be. Previously, the town would be silent and calm in winter, mostly inhabited by locals. More recently, it has been bustling with activity in the coldest time of the year, and only two months of the year have been less frequented. Taking tourists to the forest in winter has a good side. Because of the snow cover, reindeer pastures do not suffer from the impact of vehicles. In addition, silent tourism is being experimented with: snowmobiles equipped with electric motors allow for a less disturbing trip on the land. Due to short battery life however, it is only a niche so far.

During the spring and summer season of 2020, the COVID-19 pandemic has prevented foreign tourists from entering Finland, and the additional income that many herding families tend to rely on has fallen away, with resuming domestic tourism hardly filling the gap. Regarding the future, it is already clear that mass tourism as Finnish Lapland knew it will not resume in the same way. This means that more fish will be available to local fishermen, but also clearly needed as rising unemployment drives people to rely on subsistence measures.

Reindeer Herding's Symbolic Functions

The contemporary image of Sámi livelihood centers around reindeer herding, however, previously subsistence practices were more diverse, and they continue to be vital for the local economy. Fishing, hunting, and trapping, as well as gathering are at least part of every family's (formerly *siida*) economy, and depending on the availability of resources or the season of the year, one or the other activity can be pursued. Within a family, activities are divided by gender, age, and skill. For instance, among the Sámi there is the common agreement that the lakes near a village are reserved for the elderly, while the younger generation is encouraged to seek places less easily accessible. Independent of their actual importance regarding monetary gain, the Sámi livelihood consisting of reindeer herding, fishing, hunting, and gathering techniques entails nutritious sources, local activity, social encounter, and symbolic practice at the same time.

Thus, on the one hand, Sámi livelihood has been "normalized" towards reindeer herding. On the other hand, most Sámi today live far away from where reindeer roam, as sixty percent of approximately ten thousand ethnic Sámi are living outside their traditional homelands. For them, reindeer herding is no longer a daily routine. Nevertheless, reindeer herding remains an important cultural marker also for the "city Sámi," who often still keep reindeer earmarks even though these have little practical relevance. The connection to reindeer herding livelihood is emblematic in conversations about identity and ethnic belonging, and it has become a powerful attribute. Until more recently, reindeer herding was highly important to sustain Sámi language, because it was the only place where the language was spoken regularly. Currently, concerted efforts are being made to preserve the language, and material and schooling in Sámi language is now available all over Finland. Hence, if herding disappeared, the language would persist without the livelihood.

Among currently active Sámi reindeer herding families, the tendency is to divide the herd among the children, which means that the herds are becoming smaller and smaller (in terms of their legal ownership), with as little as ten or fifteen animals belonging to a person today. For all those who own such a small herd, income from other activities is vital. Tourism often appears as the most opportunistic activity to avoid moving to other places with more employment opportunities, mainly cities or industrial as well as mining towns. Knowing the land and bringing the cultural characteristics of Sámi livelihood to the market is considered a crucial resource, but it also has been claimed that this needs to be protected. Harsh and open criticism is voiced especially towards the use of Indigenous dresses by non-Sámi and the sale of cheap copies of Sámi dresses and handicraft. "Ethical tourism" business also presents an opportunity for those coming back from the city who did not grow up in a reindeer herding community.

Conclusion

Transhumant migration of reindeer herds between summer and winter pastures among the Finnish Sámi is the result of legal restrictions of a former nomadic livelihood where communities of *siida* lived with and followed herds through the land. Considering the emergence of transhumance in the first decades of the twentieth century, we argued, following Ingold (1988) that the practice of reindeer herding resembles a ranching economy and is very much alive and a viable economic sector among Indigenous Sámi in Finnish Lapland today.

In particular, technological changes (e.g., fences, snowmobiles, and GPS collars) have determined the conditions for a transition of reindeer herding from nomadic livelihood to ranching economy requiring contact between herder and the relatively wild herd of animals only during certain times of the year. As such, restricted by natural and bureaucratic limitations, the livelihood has nevertheless resisted attempts to rationalize established practices to an even greater extend. Part of this resistance is the refusal to provide reindeer with supplementary feed on a regular basis, which would allow herders to have more control over the herds' movements, bringing some independence from climatic variabilities and competing land uses. Rather than submitting to this model practiced regularly by Finnish herders, Sámi herders have fought for the preservation of winter pastures where forestry has had highly destructive effects (Mazzullo and Miggelbrink 2011). While they have not always succeeded with their legal claims and continue to live with the threat of forest clear-cuts on state-owned land that hosts important pastures, they have succeeded in inscribing transhumant reindeer practice into public memory far beyond the domestic context.

In addition, contemporary reindeer herding as it is practiced in Finland has found a promising counterpart in the tourism sector. It is, however, a balancing act between allowing visitors to encounter Sámi reindeer herding society without destroying the natural resources and traditional practices of the very same. In the attempt to protect while not completely avoiding the influx of tourists, it seems viable to continue to work with tourists along certain, restricted paths. The establishment of reindeer farms is an example of allowing visitors to experience animals and landscape without interfering with the transhumant migration cycle, that is, with the rounding up, selecting, slaughtering, and migrating animals. If these limitations can be observed, the traditional livelihood will continue to provide income and consolidate the cultural standing of the Nordic countries' Sámi Indigenous minority.

Acknowledgments

This chapter has received the financial support of the Academy of Finland's project, WIRE (Fluid Realities of the Wild), decision number 342462.

Nuccio Mazzullo is an anthropologist, who has been working as a senior researcher for the Anthropology Research Team at the Arctic Centre, University of Lapland, Finland since 2007. He received his PhD at the University of Manchester in 2005. Since 1990 he has conducted extensive fieldwork for different research projects, focusing on indigeneity, space and territoriality, and oral history and narratives in Finnish Lapland working mainly with Sámi people. His topical interests include human-environment relationships, reindeer herding, landscape and perception, environmental politics and Indigenous rights, learning skills and handicraft, Sámi narratives and identity, and anthropology of circumpolar peoples.

Hannah Strauss-Mazzullo, PhD, is a senior researcher and lecturer at the University of Lapland, Finland. She has previously published on energy policy and management and the role of scientific advising in public decision making. In her current work, she focuses on everyday activities from a sociological perspective. Hannah studies how people make-do on the margins from Lapland to Sicily, and this includes animal husbandry, tourism, and general urban/rural lifestyle.

References

Beach, Hugh. 1997. "Negotiating Nature in Swedish Lapland: Ecology and Economics of Saami Reindeer Management." In *Contested Arctic. Indigenous Peoples, Industrial States, and the Circumpolar Environment*, ed. Eric Alden Smith and Joan McCarter, 122–49. Seattle: University of Washington Press.

Bjørklund, Ivar. 2013. "Domestication, Reindeer Husbandry and the Development of Sámi Pastoralism." *Acta Borealia* 30 (2): 174–89. https://doi.org/10.1080/08003831.2013.847676.

Costello, Eugene, and Eva Svensson. 2018. "Transhumant Pastoralism in Historic Landscapes. Beginning a European Perspective." In *Historical Archaelogies of Transhumance across Europe*, ed. Eugene Costello and Eva Svensson, 1–13. Oxon: Routledge.

Dana, Leo Paul, and Jan Åge Riseth. 2011. "Reindeer Herders in Finland: Pulled to Community-Based Entrepreneurship and Pushed to Individualistic Firms." *The Polar Journal* 1(June): 108–23. https://doi.org/10.1080/2154896x.2011.568795.

Forbes, Bruce C. 2006. "The Challenges of Modernity for Reindeer Management in Northernmost Europe." In *Reindeer Management in Northernmost Europe. Linking Practical and Scientific Knowledge in Social-Ecological Systems*, ed. Bruce C Forbes, Manfred Bölter, Ludger Müller-Wille, Janne Hukkinen, Felix Müller, Nicolas Gunslay, and Yulian Konstantinov, 11–25. Ecological Studies 184. Berlin: Springer.

Guttorm, Gunvor. 2017. "The Power of Natural Materials and Environments in Contemporary Duodji." In *Sami Art and Aesthetics: Contemporary Perspectives*, ed. Svein Aamold, Elin Haugdal, and Ulla Angkjær Jørgensen, 163–77. Aarhus: Aarhus University Press.

Habeck, Joachim Otto. 2006. "Experience, Movement and Mobility: Komi Reindeer Herders' Perception of the Environment." *Nomadic Peoples* 10 (2): 123–41. https://doi.org/10.3167/np.2006.100208.

Ingold, Tim. 1987. *The Appropriation of Nature: Essays on Human Ecology and Social Relations*. Iowa City: University of Iowa Press.

———. 1988. *Hunters, Pastoralists and Ranchers. Reindeer Economies and Their Transformations*. Cambridge, UK: Cambridge University Press.

———. 2000. *The Perception of the Environment. Essays on Livelihood, Dwelling and Skill*. London: Routledge.

Johnson, Douglas L. 1969. *The Nature of Nomadism: A Comparative Study of Pastoral Migrations in Southwestern Asia and Northern Africa*. Chicago: University of Chicago Press.
Jones, Schuyler. 2005. "Transhumance Re-Examined." *Journal of the Royal Anthropological Institute* 11(2): 357–59.
Koch, Peter. 2013. "Sámi-State Relations and Its Impact on Reindeer Herding across the Norwegian-Swedish Border." In *Nomadic and Indigenous Spaces. Productions and Cognitions*, ed. Judith Miggelbrink, J. Otto Habeck, Peter Koch, and Nuccio Mazzullo, 113–36. London: Routledge.
Länsman, Kaija, and Gabriela Satokangas. 2019. "Poronhoitajien Uusi Renki Työskentelee Ilmassa—Drooni Vähentää Kuluja Ja Työtaakkaa." YLE News, 22 May. https://yle.fi/uutiset/3-10796250.
Lawrence, Rebecca, and Kaisa Raitio. 2006. "Forestry Conflicts in Finnish Sápmi: Local, National and Global Links." *Indigenous Affairs* 4: 36–43.
Lehtola, Veli Pekka. 2015. "Sámi Histories, Colonialism, and Finland." *Arctic Anthropology* 52(2): 22–36.
Luick, Rainer. 2008. "Transhumance in Germany." *Report to the European Forum on Nature Conservation and Pastoralism*. Lampeter, UK: EFNCP. Retrieved 11 January 2021 from http://www.efncp.org/hnv-showcases/south-west-germany/swabian-alb/facts-and-figures/.
Mazzullo, Nuccio. 2010. "More than Meat on the Hoof? Social Significance of Reindeer among Finnish Saami in a Rationalized Pastoralist Economy." In *Good to Eat, Good to Live with: Nomads and Animals in Northern Eurasia and Africa*, ed. Florian Stammler and Hiroki Takakura, 101–19. Sendai: Center for Northeast Asian Studies, Tohoku University.
———. 2012. "The Sense of Time in the North: A Sami Perspective." *Polar Record* 48(03): 214–22.
———. 2013. "The Nellim Forest Conflict in Finnish Lapland: Between State Forest 'Mapping' and Local Forest 'Living.'" In *Nomadic and Indigenous Spaces. Productions and Cognitions*, edited by Judith Miggelbrink, J. Otto Habeck, Peter Koch, and Nuccio Mazzullo, 91–112. London: Routledge.
———. 2018. "'A Dog Will Come and Knock at the Door, but Remember to Treat Him as a Human': The Legend of the Dog in Sámi Tradition." In *Dogs in the North. Stories of Cooperation and Co-Domestication*, ed. Robert J. Losey, Robert P. Wishart, and J. Peter L. Loovers, 251–66. London: Routledge.
Mazzullo, Nuccio, and Judith Miggelbrink. 2011. "Winterweide Und Holzlieferant. Interessenkonflikte Bei Der Waldnutzung in Nordfinnland." *Geographische Rundschau* 07–08: 36–42.
Oskal, Nils. 2000. "On Nature and Reindeer Luck." *Rangifer* 20: 175–80.
Palladino, Paolo. 2018. "Transhumance Revisited: On Mobility and Process Between Ethnography and History." *Journal of Historical Sociology* 31(2): 119–33. https://doi.org/10.1111/johs.12161.
Pelto, Pertti J. 1973. *The Snowmobile Revolution: Technology and Social Change in the Arctic*. Menlo Park: Cummings Publishing.
Saarni, Kaija, Jari Setälä, Leena Aikio, Jorma Kemppainen, and Asmo Honkanen. 2007. "The Market of Reindeer Meat in Finland, Scarce Resource—High-Valued Products." *Rangifer* 27(3). https://doi.org/10.7557/2.27.3.273.

Stammler, Florian. 2009. "Mobile Phone Revolution in the Tundra? Technological Change among Russian Reindeer Nomads." *Folklore* 41: 47–78.

Strauss, Hannah, and Nuccio Mazzullo. 2014. "Narratives, Bureaucracies and Indigenous Legal Orders: Resource Governance in Finnish Lapland." In *Polar Geopolitics? Knowledges, Resources and Legal Regimes*, ed. Richard C. Powell and Klaus Dodds, 295–312. Cheltenham: Edward Elgar.

Tyler, Nicholas J. C., Johan Mathis Turi, Monica A. Sundset, Kirsti Strøm Bull, Mikkel Nils Sara, Erik Reinert, Nils Oskal, et al. 2007. "Saami Reindeer Pastoralism under Climate Change: Applying a Generalized Framework for Vulnerability Studies to a Sub-Arctic Social–Ecological System." *Global Environmental Change* 17(2): 191–206. https://doi.org/10.1016/j.gloenvcha.2006.06.001.

Vaarala, Auli M., and Hannu J Korkeala. 1999. "Microbiological Contamination of Reindeer Carcasses in Different Reindeer Slaughterhouses." *Journal of Food Production* 62(2): 152–55.

Veijola, Soile, and Hannah Strauss-Mazzullo. 2018. "Tourism at the Crossroads of Contesting Paradigms of Arctic Development." In *The Global Arctic Handbook*, ed. Matthias Finger and Lassi Heininen, 63–81. Cham: Springer.

CHAPTER 12

Wandering Shepherds
New and Old Transhumances in Sardinia and Sicily

Sebastiano Mannia

Preliminary Observations

The agricultural policies promoted in the second half of the twentieth century had the main purpose of modernizing and rationalizing the primary sector and intensifying the quantity of production to encourage economic growth, leading over time to the development and transformation of agropastoral farms but also leading to land exploitation, to an uncontrolled production of various agri-food products, and the dependence of shepherds and farmers on the dynamics of global markets. In this period, therefore, this sector has become more specialized and its general structure has changed: the number of companies has gradually fallen, the agricultural area has halved, and the countryside was gradually abandoned, causing various sociocultural and environmental problems. Such a development model has fallen into decline over the last decades, showing all its limits, and it has prompted the reorganization of the agricultural system in different areas through new forms of business planning and production, the so-called New Peasantries model (Barberis 2009; Van der Ploeg 2006, 2009, 2018; Milone and Ventura 2009).

Within this general framework, specific cases such as Sardinia and Sicily show peculiar characteristics and they need to be properly contextualized. In Sardinia, since the end of the nineteenth century and especially after World War II, the number of sheep has progressively increased, due to favorable economic conditions linked to the sale of milk by the shepherds and to the increased sales of Pecorino Romano in the markets by cooperatives and dairy industries, which led to the specialization of the island sheep sector in a highly competitive monoculture that is made up today of about twelve thousand companies, mostly aimed at the production and sale of dairy products (Angioni 1989; Farinella 2018; Mannia 2014). In Sicily, on the other hand, pastoral farming faces various economic and structural problems, caused by erratic historical processes resulting in a favorable cycle at the beginning of the last century, a block from the World War I until after the World War II, and a recovery in the 50s and 60s, despite delays accumulated over time (Astuto 2011; Mannia 2013; Rochefort 2005).

The different dynamics that affected the Sardinian and Sicilian pastoral systems have contributed to the creation of well-defined economic structures and to the remodeling of practices and principles of traditional pastoralism. Several cultural traits have disappeared, others still persist, others have adapted. Among the latter figures also transhumance, and while in Sardinia nowadays pastoral movements, if present, constitute forms of mobility over short distances aimed at the exploitation of pasture resources, in Sicily, on the other hand, transhumance—short-distance here as well—still represents a necessary practice for many breeders, due to the high fragmentation of pastoral property and a poor rationalization of the sector.

Therefore, pastoral mobility is still a common practice in the abovementioned contexts and in the last decades there has been a growing interest in this practice (Mannia 2010, 2018)[1] and more generally for pastoral cultures, which has led to the promotion of a varied and articulated regulatory framework aimed at safeguarding and enhancing environmental and cultural heritages. Locally, there is also a new attention to the multifunctionality of agricultural companies and to the development and enhancement of forms of sustainable tourism.

However, limited with respect to the current production direction, these new forms of use of rural spaces are significant. They are innovative methods of use and enhancement of pastoral areas created by breeders who have a strong connection to the territory, but also by entrepreneurs who have decided to focus on forms of ecotourism or by shepherds who have rethought rural realities and have returned to the countryside with new ideas related to typical productions, farm camping, and agritourism.[2] This change in heritage perspective opens up new horizons for transhumance as a cultural and touristic heritage, particularly in inland areas.[3] If, on the one hand, pastoral systems are involved in the dynamics of contemporary markets and policies, on the other, the need to develop new models of pastoralism and new ways of enhancing its practices emerges more and more clearly, always focusing on pastoral communities as the main resource for local development and for the protection of environmental and cultural landscapes.

Pastoralism in Sardinia and Sicily

Ethnographic research carried out in Sardinia and Sicily[4] has shown that the two forms of pastoralism, as they are structured today, are the outcome of the political-economic and sociocultural changes that have occurred since World War II. What emerges with particular evidence from the analysis of the two contexts is the clear contrast between pastoral farms employing traditional systems and modern enterprises. And even so, while in Sardinia the livestock sector—in all its components—is a main contender on the national and European scene due to its high level of competitiveness, the Sicilian livestock sector shows several problems due to an incomplete rationalization of the sector.[5] In Sardinia, in the late twentieth century, there was an almost total disappearance of cereal farming and the definitive affirmation of pastoral activity. This process, spread to varying degrees on the island, is reflected in the shepherds who settle down, acquiring large land properties in areas once used for cereal farming and for the wintering of transhumant flocks. Pastoral development on the island is essentially due to the interaction between political intervention, with the allocation of contributions and subsidies, and the change of mentality of pastors—with obvious differences between the various segments of Sardinian society—in relation to the modernization of the sector. The mechanization of the countryside, the progressive disappearance of long transhumances and the consequent sedentarization (see the next section), the construction of modern infrastructures, and technological innovation are the main factors that have contributed to the epochal change in the island pastoralism in the last century (Angioni 1989; Mannia 2014; Murru Corriga 1990).

At the end of the nineteenth century, the arrival of dairy industries from the Lazio region was another significant factor that influenced, both positively and negatively, the growth and progress of the sector. Dairy industries, as already mentioned, have stimulated the expansion of sheep flocks (Idda, Furesi, and Pulina 2010: 64–72), causing a profound identity transformation of the shepherds and a remodeling of traditional roles: for breeders, in fact, direct sales of milk proved to be more profitable, and this has brought important economic and cultural changes that led the shepherds themselves to pass from cheese producers into milk sellers (Mannia 2014: 135–63).

Sicily, on the other hand, has followed a rather different path: cereal farming and horticulture have expanded and been rationalized, elevating the island to one of the most important agricultural realities in the Mediterranean; pastoralism, on the other hand, does not seem to have benefited from, or only marginally, the advantages offered by the innovation processes which, following World War II, were introduced for the revitalization of local economies. Many pastors, still today, do not benefit from the annual contributions allocated by the European Union, refuse any subsidies for the modernization of companies, and make very limited personal investments to rationalize their activities. For these reasons, Sicilian pastoralism presents several problems, especially in inland areas, and in certain contexts, for example in the Madonie Mountains, pastoral activity is gradually disappearing. The companies, mostly based on purely traditional business models, present a series of structural deficiencies and limitations—in some cases now historical (Astuto 2011; Rochefort 2005; Uccello 1980)—relating to the organization and business management, which emerge clearly in relation to the modernization process which has taken place in the last sixty years, as in many other Mediterranean countries.

The lack of functional zootechnical structures persists, as does the fragmentation of land and property, the modest number of sheep per farm, the lack of technological innovation, and the lack of technical training, all due to a lack of economic-zootechnical and sociocultural innovation but also to a low anthropological attention from the institutions. In many cases the company structures are incorporated in the old farms, where the shepherd takes care of milking, the transformation of the milk, and the cheese conservation. Both electricity and water supply are often a problem, the premises are often inadequate, and the equipment used to process cheese sometimes does not comply with the hygiene-sanitary directives imposed by the European Union. From a commercial point of view, moreover, the products are uneven and variable both in terms of quality and quantity. The rental of pastures is widespread, and the lack of owned land together with the constraints imposed on grazing in different mountain areas lead to frequent pastoral mobility. For these reasons, Sicilian pastoralism is predominantly wild and transhumant (Giacomarra 2006; Mannia 2013).

Some data better clarify, in a comparative perspective, the current reality of the two types of pastoralism. Since the last census in 2019, 3,067,522 sheep have been bred in Sardinia, 813,525 in Sicily (the two regions occupy the top two positions in Italy for sheep flocks). In Sardinia, out of 15,001 herds, the dairy ones are 11,126, the breeding ones 86, and the mixed ones 3,431 (the remaining 346 are for self-consumption production); in Sicily, out of 8,759 farms, the dairy ones are 1,097, the breeding ones 5,565, and the mixed ones 1,531 (the remaining 554 are for self-consumption production).[6] The data highlight, on the one hand, the historical path taken by the Sardinian shepherds in the specialization of dairy farming, and on the other, the diversified orientation of Sicilian farmers aiming at the production of milk and meat, with a greater, however recent, inclination towards the latter.[7]

From the cases analyzed in Sicily, it emerged that the coexistence of the various production orientations can be related to a type of inherited traditional management rather than related to the result of a rational choice aimed at creating added value. In Sardinia most of the milk production is reserved to the production of the three PDO (*Pecorino Romano, Pecorino Sardo*, and *Fiore Sardo*), destined for international markets and in particular for the American market. In Sicily, only a marginal share of milk is reserved for the production of the three PDO (*Pecorino Siciliano, Piacentinu Ennese*, and *Vastedda della Valle del Belice* [D'Amico 2011]). Most shepherds prefer to make their own cheese and sell it on informal markets. Milk is also processed in local industries, even though it is not very profitable for the shepherds. Basically, while in Sardinia pastoralism is aimed at the production and sale of milk from shepherds to dairy industries and cooperatives, in Sicily traditional forms of pastoralism still aim at the sale of cheeses, ricotta, and meat in local markets (Mannia 2013, 2014, 2016).

Transhumance in Sardinia and Sicily

This brief overview is useful to understand how the transhumance phenomenon is collocated in the two islands. In Sardinia and Sicily, pastoralism has historically been structured around two fundamental constants: the lack of land for grazing animals—due to the fragmentation and dispersion of property derived from environmental, historical, and sociocultural factors—and the climatic variables that still influence the quantity and quality of available resources. To cope with these inevitable constraints, shepherds have developed specific cultural and economic strategies, and among these, transhumance is certainly the most significant outcome (see Angioni 1989; Caltagirone 1986; Giacomarra 2006; Mannia 2013, 2014; Meloni 1984; Murru Corriga 1990, 1998; Ortu 1988). In the Sardinian case, pastoral mobility has become a necessary practice, and, due to the progressive and over-sized increase in the livestock population, a process that began at the end of the nineteenth century. The search for pasture lands became urgent, especially between the 1950s and 1970s, when the number of sheep increased further following a new development of the markets and political interventions which, starting after the World War II, although with alternating results, were directed in favor of the sector. The resulting relationship between animal capital and the carrying capacity of the land used for grazing has thus been severely disrupted, on one hand pushing many shepherds to emigrate—many of whom, for example, have occupied the land abandoned by the Tuscan sharecroppers (see Meloni 2004; Solinas 1989–90)—and on the other to find an appropriate solution to the problem of wintering livestock—since municipal and/or private pastures were no longer sufficient to sustain the number of animals. These are the reasons why until the 1970s, before the process of sedentarization of shepherds began, in the autumn-winter months, it was possible to see thousands of sheep crossing the rural roads of the island, from the mountains to the plains and coastal areas.

Therefore, for centuries transhumance has been a practice characterizing the agropastoral economy of Sardinia and Sicily, with similar elements among the numerous communities, in particular with regards to the rules and functions underlying the social structure. To clarify, long transhumances—or long-distance displacements—have disappeared, while short-range transhumances persist, i.e., forms of mobility of animals from one pasture to another usually within the same territory with journeys of a few kilometers, helping to define local pastoral practices, and the rural landscapes and sociocultural references of the communities involved (Bergeron 1967; Giacomarra 2006; Le Lannou 1992; Meloni 1984, 1988). For greater clarity, we will discuss first the Sardinian transhumance and then the Sicilian one.

Long-range transhumance in Sardinia was an inverse transhumance and involved the movement of animals from the mountains and high hill areas to the plains and coasts where climatic conditions were more favorable. The Sardinian transhumance has mainly affected the counties surrounding the mountain areas of the center of the island; in fact, the shepherds moved from the communities of Gennargentu, Supramonte, and Montalbo. The routes were variable: towards Iglesiente the route was about 100–150 km; for Oristanese the route was about 80–100 km; for Sarrabus the route was 100–120 km (Caltagirone 1986: 44). More widely, the Campidani, the Nurra, the plateaus of Bonorva and Macomer, the plain of Chilivani near Ozieri, the valley of Coghinas, the coastal areas that extend from Olbia to the Gulf of Orosei, the Baronies were the main arrival places of transhumant shepherds of the mountain areas of central Sardinia.

In turvera, a turvare, in tràmuda, tramutanne are the terms and ways in which transhumance was designated depending on the place. *A nos ponnere in caminu*, or "to walk," was the common reference for those who annually moved with the flock. For the people of Fonni the transhumance was *s'isverrare* (to winter) or *s'istrangiare* (to go among foreigners), while the return to the community was *sa muda* (the renewal) (Murru Corriga 1990: 29). Antoon Cornelis Mientjes pointed out that "the term transhumance does not exist in the dialect of Fonni, although currently shepherds understand its meaning. The expression *in viaggiu* was used to indicate the seasonal transfer of shepherds and flocks to distant places" (Mientjes 2008: 200).

The departures—and this seems to be the most significant fact—were planned on the basis of defined economic-productive times, in particular on the births of lambs. Between the end of June and the beginning of July the sheep were mated, and the births were planned for when the animals would reach the winter pastures. They always tried to leave before the sheep gave birth in order not to have problems during the journey, even if the fatigue of the journey often hastened the births.

Moreover, the periods of departure for transhumance varied annually according to climatic and environmental conditions. In some cases, transhumance was already taking place in October, although the months in which the animals were to be moved were November and December. The duration of the transhumance depended on the distances that had to be covered and the unexpected events that could occur on the way. From Austis it took two or three days to cover the approximately one hundred kilometers that separate the mountain village from the plains of Campidano (Meloni 1984). The shepherds of Fonni who traveled towards Solarussa took two or three days of walking, crossing the countryside of different communities in addition to the towns of Neoneli, Busachi, and Fordongianus (Mientjes 2008: 202). A shepherd from Orgosolo reported:

> Transhumance was in November. To go to Isalle [wintering area] we started from the Orgolese plateau, descended into the countryside of Mamoiada and slept there. The following day we went down to Oliena where we spent the night, and on the third night we arrived in Isalle. It took three days because the road was long. The overnight stops, for just a few hours, were already planned because they were places we knew well from years of *tràmuda*, and we also had to avoid thefts and trespassing. We often knew someone, the so-called *sos cumpanzos de posata*, that would host us and help us with the flock. (N. P., Orgosolo, interview of 28 December 2015)[8]

The return from the wintering places was scheduled for May—usually between the 15th and 20th of the month—and the date changed in relation to the geographical position of the mountain pastures, the expiry of the lease of the land where they overwintered, and especially the conditions of access to municipal pastures. The time of return was also planned on the basis of economic variables and depended on climatic and environmental factors, the needs of the flock, and the cereal cycle. In many communities, in fact, the pastoral year was intersected with the agricultural one and the return from transhumance coincided with the work of reaping, threshing, and wheat harvesting. After the harvest, the land was opened for grazing *s'istula*, or stubble.

Despite the fact that the return journey included the duties and uncertainties of the outward journey—to which was added, among other things, the milking of the sheep and the delivery or processing of milk—

Figure 12.1. Transumanza in Sardegna, 1959. © János Reismann

the return to the village was characterized by a relaxed and festive climate. In Fonni the day of return was *sa die primargia*, while in Desulo it was *sa die prima*, that is, the first day, to define a specific temporal span typical of the beginning of a new cycle. In several communities it was common to donate the milk of the first milking to all the families when they returned from the pastures. In Orgosolo, I am told, "we came the 20th of May. Before going up to the municipal lands, people milked inside a ruined church at the entrance of the village and while shepherds presented their milk (in Sardinian, *presentavan su latte*) or rather offered the milk to neighbors, friends, and relatives, women brought *sos macarrones lados* which were consumed in a joint lunch" (B. S., Orgosolo, interview of 4 January 2016). On this particular day, the village was reunited, and a large part of the male population reintegrated into the sociocultural and economic community.

In the second half of the last century, one of the most important transformations of the Sardinian zootechnical sector was completed: the sedentarization. In the lands purchased by the shepherds—many of which were first reserved for the wintering of animals—the systematic rationalization of the sector began. Mechanization and, therefore, the cultivation of grasslands, the construction of infrastructure and technological innovation are some factors that have favored the modernization of the island's pastoralism and the subsequent disappearance of long transhumances. They are gradually abandoned in the 1940s and 1950s, although the process was completed in the 1970s and 1980s together with the frequent use of trucks to transport animals. There are sporadic episodes of transhumant shepherds even in the following decades (see Mientjes 2008: 205), but today there are very few who resort to pastoral mobility and in most cases, vans are used to greatly reduce the hours of travel. Thus, one informant: "From the mid-Seventies people put a stop to transhumance, someone continued still in the 80s, but really a few people, and by the way they moved them by the truck. Since the 80s, there were a very few people who did transhumance by walking. Me too I did my last transhumance with a truck, that was the last year I took them in the Nurra" (B. R. Orune, interview of 20 December 2015).

The lands that became property—and not just the flat lands purchased in the Campidano, in the Nurra, etc.—have allowed us to express the capacities and productive potential of pastors and companies.

Basically, the Sardinian pastoral economy has changed from a transhumant to a sedentary model. This is an important change that is not just economic but mainly cultural. From the words of those who have transhumed for many years emerges a sense of revenge against an emigration that annually pushed the shepherds to abandon the community, the family and, more generally, their own microcosm of reference. A shepherd from Orune told me that "today everything is changed, what people have bought: there are very large rooms that allow you to winter; then there is the animal feed. The time in which you knew when you would leave but not when you would come back it's over" (C. B. 1954, interview of 20 December 2015).

Despite the fact that today in Sardinia there are more than three million sheep, the movement of thousands of sheep from the mountains to the pastures is a practice of the past that only remains in the collective memory of the different pastoral communities. Instead, what is still possible to observe are the short displacements, which follow the seasonal climatic trends but above all the availability of pastures and fodder crops. In addition to owned and rented land, municipal lands take on particular importance: to every shepherd who requests it—through the payment of a sum of money that varies in relation to the number of sheep to be fed in the pastures—a pasture quota is assigned. So, for example, an Ogliastrian shepherd told me:

In May I leave from Arzana, where I have a farm with the barns, the sheep milking machine, the vehicles, and I move to the municipal lands, where the sheep can graze. I do it every year in May and October. It is a 12 km-long journey and it takes a couple of hours, always on foot. Other shepherds do the transhumance as well, in October-November (the winter pastures are called *s'accordiu*)—depending on lambing periods. Some go as far as Tortoli, Cea, Quirra, Villassalto but they move the animals on trucks. Then they go up to the mountain, in May-June, some bring their sheep to the Gennargentu mountains. The land is either owned or municipal. Among the shepherds who have pastures far from Arzana, some sleep in the countryside because they have the sheep pen and maybe return every three days. Others travel every day because they have assistant shepherds who live on the farm. Transhumances to Trexenta, Campidano, Baronie no longer exist; before there were no barns or pasture, today you can have forage delivered on trucks. (V. L., Arzana, interview of 9 June 2020)

A shepherd from Lula told me: "Unfortunately, I don't have a single, large land, and I have to move the sheep from one pasture to another, which are my property. To get to the farthest one, however, it doesn't take more than a couple of hours. I usually move them in October-November, the first months of the year and in summer after the forage harvest, although I am trying to invest to avoid moving so frequently" (D. M., Lula, interview of 3 January 2016).

Even in Sicily the long transhumance, *mutari l'armali*, was an inverse transhumance, with the movement of the animals from the mountains to the coastal areas (each change of grass was called *muta*). Depending on the altitude, the grazing areas were divided into plain (*marina*), hill (*minzalina* or *mezzalina*), or mountain (*muntagna*) pastures and the flocks were moved in winter, autumn, and spring, and in summer. The departures and the stops of the shepherds were planned in advance depending on the availability of pastures and shelters, as well as in relation to the temperatures: "Autumn was the least reliable period: the organization completely depended on the intuition and the experience of the shepherds and their predictive ability" (Giacomarra 2006: 47). Therefore, in September-October pastures were grazed in the hill areas, in November-December the flocks were led to the coast, in March they returned to the hill pastures and in late spring they were moved to the mountain areas until August. During the summer, after the wheat harvest, many shepherds led their animals to the fields for stubble grazing. The traditional day of shifting of the winter flocks was 25 November, the anniversary of Saint Catherine of Alexandria. As the shepherds' proverb goes: *"Ppi Santa Caterina, vacchi e pecuri a la marina"* (On Saint Catherine's day, sheep and cows leave for the *marina*).

The diverse conformation of the island has led to the identification of specific grazing areas and to the establishment of an extremely articulated road network, the *trazzere*. From Agrigento, for example, for winter transhumance the shepherds moved in October to the coastal areas of Porto Empedocle, Sciacca, Licata, and to the flat pastures of Catania and Gela. Here they stayed for a few months, usually until December-January, after which they returned to the hill pastures. Conversely, the internal mountainous areas were reached in late spring and summer or by going up the western side to the mountains of Cammarata and Lercara Friddi (some shepherds went as far as the Madonie) or following the eastern internal side towards the Peloritani, the Nebrodi, Etna. Depending on the weather, these shifts could begin in May. Even the shepherds from the Nebrodi—due to the orographic characteristics of the area—moved on long routes, in some cases reaching the flat areas near Catania and Siracusa: from May-June to October-November the animals remained in the mountain pastures of the Nebrodi, while in the remaining months they were moved towards the plains and coasts. In the Madonie, in addition to the mountain routes, there were essentially three itineraries of movement: the first, to the north, led to pastures that approach the sea; the latter departed towards the medium and low hills of Nisseno and Agrigento; the last, mostly chosen by the shepherds of Sclafani and Caltavuturo, descended towards the valleys on the west. These routes extend about ten to twenty kilometers north and sixty to seventy kilometers south (Giacomarra 2006: 44–45). Here is what a shepherd told me: "Usually my father was in the mountains from June to September, then he returned to the pastures near Polizzi for a short time and from September to December he moved in the Nucitedda district, about 12 km from the town. From January to March he brought the flock downstream, and then he returned from April to May to the lands of Polizzi in Sciumiranni. My father had many animals, sometimes he stopped for a whole week" (G. G., Polizzi Generosa, interview of 12 July 2009).

The life of Sicilian shepherds was characterized by a permanent mobility which inevitably influenced family and social relationships. In addition, the constant movements required the presence of many people to control and take care of the animals. The largest flocks were divided into groups of about two hundred sheep, the so-called *guardia*, entrusted to a guardian, *picuraru*, helped by an assistant, *cumpagnu*, or by an apprentice, *garzuni*. For the transhumance, some shepherds used horses, but most moved on foot using mules and donkeys for the tools they needed for milking and making cheese and for arranging shelters. Often the shepherds formed associations of medium or large companies, to optimize the workflow on the one hand, and also to save on the rental of pastures, on the salaries of the shepherd assistants, and on the transfer on the other (Giacomarra 2006). Long transhumances gradually disappeared in the years following World War II and the shepherds who still own or rent land far from their farms mostly use trucks to move their flocks. Conversely, short-distance journeys often influence the decisions of shepherds much more than in Sardinia. Seasonal cycles and environmental variables, ownership and/ or land rental regimes, the presence or lack of infrastructures are elements that influence individual farms, and the shepherd must possess a set of knowledge and skills that allow him to guess and consequently plan the movements of livestock and the periods of grazing in certain areas, especially in relation to the cycles of vegetation and crop rotations.

Transhumance is also widely practiced by shepherds who have their own land, and this underlines how this practice is intimately linked not only to the search for pastures but also to climatic variables. A shepherd from Polizzi Generosa reported:

> Sheep need transhumance to produce. If you *muti i piecuri* (move the sheep), they always make milk. If you leave them in one place, it's different. Today I move less because the animals are fewer, but the transhumance is still there. Finally, the park has influenced the pastoralism with the constraint of the land. The mountain pastures are bound and the forest ranger no longer rents, so we have to move, this is why in the summer I bring the sheep in Contrada Gagliardo, which is higher. Therefore: from October to December, I bring them near Polizzi, from January to March in Scillato, in the valley pastures, from April to June I come back to the village. (V. G., Polizzi Generosa, interview of 10 April 2013)

Even in the province of Enna transhumance is still practiced, both by the landowners and by the shepherds who rent the land. One of them reports that he practices transhumance in May and September, moving in May by car and in September on foot, towards Leonforte. These are short-haul trips. Long-range walking transhumance, on the other hand, has disappeared, and some shepherds say: "both because of the laws and because there are few animals. Before the land was not all owned, most rented land and then moved. Now, thanks to the public funding and to the devaluation of the land anyone can buy it and then will not have to move" (A. B., Villarosa, interview of 14 June 2018).

The main objective of the shepherds, in the past as well as today, has always been to optimize milk and meat production. In this perspective, the continuous movement of the flocks and the perennial search for pastures are justified, which is why the movement of animals is not exclusive to the hilly and mountainous pastoral systems. Field research carried out in Palermo and Trapani also confirms the use of transhumance in these areas, which is still practiced mainly by shepherds who carry out their activities on a traditional basis. A shepherd of Ventimiglia di Sicilia told me: "In the summer, but already at the beginning of May, we take the sheep to the mountains because in Ventimiglia, if it doesn't rain, everything is dry. The pastures are not far away. They stay there until June and then we bring them back here. From October to December, they are moved again to the mountain pastures and, in January, we bring them back here. We often move them and it is mainly because of the lack of pasture" (S. C., Ventimiglia di Sicilia, interview of 30 April 2013). Similarly, a shepherd from Marsala told me: "Depending on the availability of grass, we move the animals. Usually, when it is necessary to move, the sheep are divided into groups from September to March, while the whole flock is moved in the spring. The movements do not take longer than a couple of hours; they range from 2–3 km to the neighboring pastures, to 11–12 km to the pastures I own in a nearby town. When the lands are not enough, I rent them" (S. C., Marsala, interview of 21 November 2016).

The most rationalized farms, at least the ones I analyzed, do not practice transhumance: the flock is grazed on lands close to one another or in rented pastures close to the farm. While today many shepherds have invested in the modernization of their farms, in general the percentage of businesses that have modern equipment for the cultivation of lands, the production of fodder, and premises to housing the animals is still low. Transhumance, in fact, continues to play a fundamental role in the organization of Sicilian pastoralism.

The New Paths of Transhumance

As seen in many other forms of pastoralism in the Euro-Mediterranean area, in Sardinia and Sicily transhumance is still common in local breeding systems and in recent times—especially following the candidacy process that led the movement of flocks to be inscribed by UNESCO among the Intangible Cultural Heritages of Humanity—it is increasingly becoming a touristic attraction. In Sardinia, the first real property interests[9] date back to the early 2000s, when the traditional transhumance routes attracted the interest of the LAGs (Local Action Groups) Barbagie and Mandrolisai, Mare Monti and Ogliastra, who, with the collaboration of political institutions, local authorities, and scholars, started the *Tramudas* project, with the aim of making transhumance known, to promote the communities involved through the rediscovery of places (the routes wind through archaeological emergencies, places of worship, natural oases, landscape peculiarities, museums), to promote rural identity and create an opportunity for tourism and cultural development through the recovery of ancient pastoral routes and more extensively of the culture of the pastoral world. The programmers have also prepared a guide to let visitors know the routes and steps to follow. A further initiative, launched in 2006 by the Region of Sardinia, is *Camineras de tramuda* (The paths of transhumance) which encourages the collaboration of different partners in the implementation of an integrated design of horseback trails along the shepherds' paths. A more recent project named *Sardegna Sentieri*, includes the section, "Places of transhumance," which proposes four "places of interest": one in the geographical area of Baronie and Montalbo, one in Barbagia di Seulo, and two in Sarcidano. The routes include visits to the *pinnettos* (the traditional shepherds' sheepfolds) restored and equipped for tourists, to farms where the agropastoral activity is carried on, as well as picnic areas and places of particular naturalistic, cultural, and historical-religious interest.

While the initiatives described above have had a rather marginal practical development and have resulted in the identification of the transhumance routes, in the preparation of guides, brochures, and advertising materials and in the installation of signs along the routes, in some areas transhumant practices are being becoming touristic attractions by promoting walks on foot, on horseback or with donkeys on the paths traveled by the flocks. Equestrian tourism is one of the main driving forces of the tourism industry and represents an alternative to the classic and inflated routes, with the aim of developing the economy of the areas concerned and of safeguarding and enhancing the rural, environmental, and cultural heritage of the island. In this regard, the project promoted by the Sardaigne en Liberté ecotourism agency is particularly interesting. Its main purpose is to promote eco-responsible, ethical, and fair tourism through the enhancement of rural landscapes by reducing the impact of tourist activity on the environment, supporting the consumption and sale of local products, and promoting the role of communities and their culture.

Among other activities, the travel agency also offers trekking with donkeys, an eight-day tour that traces the roads of transhumance in Ogliastra. The director of the agency told me:

> The idea of transhumance arose because foreigners expressed interest in this practice and in the pastoral world. Transhumance is a phenomenon that has always allowed sustainable activities, the fight against fires, the protection of the environment. Enhancing it today means enhancing the territory and local communities, improving the environment, promoting local products, helping the shepherds develop a brand of transhumance products with an added value. It is a political matter that should be made heard in Sardinia. In addition to the tours we offer, we have been working on a project for al-

Figure 12.2. Transumanza, 2020. © Anna Piroddi

most three years and in 2019 we organized the first transhumance in which more than 100 people participated. In 2020, because of Covid, we could not repeat it and we created a live broadcast on Facebook: more than 12 thousand people followed the event. We hope to organize it again next year, in different periods and in different areas of Sardinia. (J.-L. M., Paris, interview of 12 June 2020)

The shepherd leading the transhumances promoted by Sardaigne en Liberté is a thirty-five-year-old young shepherd, who says:

What I have been doing in October and May, I have been doing for 13 years every year. In 2019 there were a hundred tourists who followed the flock and me and the guide answered their curiosities. After a couple of hours, at 12:30 p.m. we arrived at the farm, my relatives waited for them with drinks and sweets, then we had lunch with sheep meat and roasted goat and cheese. Now many are taking an interest in transhumance, especially after the boom in live streaming on Facebook. As soon as possible, once Covid is over, we will do it again. We want to organize a kind of transhumance village festival with the mayor of Arzana, while with a friend of mine we were thinking of doing a transhumance in May with my animals and one in June, a little longer, with his. (V. L., Arzana, interview of 9 June 2020)

In Sicily, pastoral mobility is bound to strict norms that regulate it, aimed at containing epizootic diseases. To date, the entire island is subject to a plan for the eradication of brucellosis that imposes the control of animal movement. For reasons of health emergency, at the end of May 2018, the event "La transumanza da Gioiosa Marea a Longi" (Transhumance from Gioiosa Marea to Longi), scheduled from 1 to 3 June and organized by some cultural associations, was cancelled. The event is part of a large project involving several coastal and inland locations of the Nebrodi affected by the routes of the flocks, it started in 2016 and it is called "From the shores to the mountains: for the rediscovery of ancient traditions." It included an excursion on foot and on horseback, following the transhumance of animals, on the *tratturi* (tractors) of nine villages from Gioiosa Marea to Longi. The event was created with the intention of expanding the tourist offer of Sicily and in particular of the territory of Messina.

A similar event, "La Transumanza—Un'Esperienza da Pastore nei Sicani," was proposed by some associations in Santo Stefano Quisquina on 21 May 2018. The description of the event reads:

> That's right: the great attraction of transhumance, a unique opportunity to stay in close contact with the shepherds and their flocks, between stories and lots of fun until you reach the farm where we will taste cheese and ricotta prepared on the premises as it once was and where we will experience the festivities of transhumance as real protagonists.
>
> It will be a journey not only for the body, but also for the mind: during the slow journey we will learn about the widespread organization that, over the centuries, has characterized the Civilization of Transhumance. A journey during which we will follow shepherds and herdsmen from the pastures of the mountains to the valley and that will give us the opportunity to rediscover our spirituality through contact with nature and slow movement. (Sicilia on Press 2017)

While the events mentioned above are mostly incidental and aimed at enhancing transhumance through excursions and tastings of typical products, the town of Geraci Siculo in the province of Palermo has been presenting more articulated projects for several years. Here, the herds of cattle and the flocks of sheep are moved to the mountains in the summer months and to the pastures downstream in the winter: from 24 May to 20 November the shepherds *sgavitànu 'a muntagna*, i.e., they access the closed public pastures, *u gavitatu*, from 6 March to 24 May. In 2009 the local publishing house Edizioni Arianna, supported by the municipality, the regional province of Palermo, the Madonie Park Authority, and Geraci associations, decided to transform transhumance into a moment of cultural rediscovery and tourism promotion and created "The festival of the transhumance of the shepherds of Geraci Siculo, *Si sgavìta a muntagna*," which reached its eleventh year on 18–19 May 2019. The event aims to support the safeguarding and enhancement of pastoral mobility and more extensively of the intangible cultural heritage of Geraci Siculo. Among the planned activities are guided tours of some farms where you can attend the gathering of the animals and tasting of dairy products, and participate in conferences and nature excursions.

> We wanted to give greater visibility to something that already existed and that everyone already knew, yet somehow it often went unnoticed, as if it only concerned the shepherds. So, we wanted to make people focus on the cultural value of such a historical practice, choosing a different theme every year to inspire artists, visitors, and writers. The festival was held from 2009 until 2019, this year we had to cancel because of Covid. Shepherds, producers and local associations are involved: the whole Geraci community is deeply invested. (A. A., Geraci Siculo, interview of 16 June 2020)

Basically, the historical-economic and sociocultural dynamics that affected Sardinian and Sicilian forms of pastoralism in the last century have influenced and transformed pastoral mobility practices. Transhumance, in the two contexts, is a phenomenon with a double function: on the one hand it continues to represent a necessary and fundamental practice within the annual pastoral cycle; on the other, even if it is just the beginning, it is becoming an asset to be valued and used in the development of an alternative tourism which, especially in inland areas, helps the protection of rural areas, the creation of alternative incomes for local operators and in particular for the shepherds, and the expansion of the offer of agri-food products linked to the pastoral tradition. In this regard, within two forms of pastoralism whose objectives are based on the creation of income through the production and sale of milk, meat and derivatives through the dynamics of contemporary capitalistic markets and in which the new peasant model occupies a marginal position precisely because of the structural characteristics on which the two pastoral systems are based, transhumance becomes more and more an icon that collects and re-functionalizes a cultural heritage, through an effective process aimed at recovering, safeguarding, and endowing the constitutive features with a new sense of the past to adapt them to new needs. Equipped with symbols that have the power to attract and convey specifically researched messages, transhumance thus becomes an important resource that catalyses symbolic practices, common practices, and capitalization processes.

Sebastiano Mannia is Assistant Professor at the University of Palermo, where he teaches cultural anthropology, history of popular traditions, and intangible heritage. His publications include: *Il pastoralismo in Sicilia* (2013); *In tràmuta. Antropologia del pastoralismo in Sardegna* (2014); and *Questue e figure vicariali in area euromediterranea* (2015).

Notes

1. The discussion is not limited to Sardinia and Sicily only. See in this regard: Ballacchino and Bindi 2017; Bindi 2019a, 2019b, 2020; Fossati and Nori 2017; Nori 2016, 2018; Verona 2016. More generally, transhumance has historically characterized and influenced farming practices and social structures in the Mediterranean area (and beyond) and on these aspects a large literature has been consolidated which I would like to highlight, among others: Aime, Allovio, and Viazzo 2001; Arbos 1922; Braudel 1976; Brisebarre 2007; Davis 1980; Fabre, Molénat, and Duclos 2002; Petrocelli 1999. For further bibliographical references see Mannia 2014.
2. On multifunctionality and rural development see: Balestrieri, Cicalò, and Ganciu 2018; Idda and Pulina 2006; Meloni 2006; Meloni and Farinella 2013; Meloni and Pulina 2020; Tola 2010.
3. On the development of inland areas I refer, among others, to: Carrosio 2019; Cois 2020; De Rossi 2018; Marchetti, Panunzi, and Pazzagli 2017; Meloni 2015.
4. The research on pastoralism in Sardinia started in 2005 and is still ongoing. Interviews, company surveys, and informal conversations were conducted in numerous communities in the four historical provinces. In Sicily, the research, started in 2011 and not yet completed, has involved several towns near Palermo, Agrigento, Enna, and Trapani. The information in this chapter is the result of a multilocation search (see, among others, Falzon 2001; Marcus 1995).
5. It is necessary to clarify that in the Sicilian case, unlike the Sardinian one, numerous companies coexist, and they are characterized by a traditional pastoral management system with few modern and rationalized enterprises. The company typologies, of course, are not homogeneous but rather diversified in terms of production, land property, technology, etc. For these reasons, from the analysis of the data collected in the course of my ethnographic research, an attempt was made to trace representative models of the two pastoral systems, given that the dynamics and problems are common to both individual contexts. For a comprehensive framework of the two forms of pastoralism, see Mannia 2013, 2014.
6. Data provided by the BDN of the Zootechnical Registry established by the Ministry of Health at the CSN of the "G. Caporale" Institute of Teramo, retrieved 3 June 2020 from http://www.izs.it/IZS/.
7. On 30 April 2015 the composition was instead: 1,072 dairy farms, 3,585 breeding farms, and 3,582 mixed farms (National Zootechnical Registry of the Teramo Zooprophylactic Institute of Teramo, http://www.izs.it/IZS/, retrieved 5 August 2015).
8. To ensure their privacy, I will report only the initials of the interviewed shepherds.
9. In the last decades, the discussion on the safeguarding and enhancement of cultural and landscape heritage has broadened new perspectives for the involvement of local institutions and communities in the management of human and land resources. In particular, the Local Action Groups, but also local

institutions and associations, have been involved in various types of projects with the aim of promoting a new development of rural territories, pastoral practices and identities, and transhumance routes.
10. I attended and documented the second annual Festival of the transhumance of the shepherds of Geraci on 22–23 May 2010.

References

Aime, Marco, Stefano Allovio, and Pier Paolo Viazzo. 2001. *Sapersi muovere. Pastori transumanti di Roaschia*. Roma: Meltemi.
Angioni, Giulio. 1989. *I pascoli erranti. Antropologia del pastore in Sardegna*. Napoli: Liguori.
Arbos, Philippe. 1922. *La vie pastorale dans les Alpes françaises*. Paris: A. Colin.
Astuto, Giuseppe. 2011. "La pastorizia e il problema dei pascoli in Sicilia tra età moderna e contemporanea." In *La pastorizia mediterranea. Storia e diritto (secoli XI-XX)*, ed. Antonello Mattone and Pinuccia F. Simbula, 78–93. Roma: Carocci.
Balestrieri, Mara, Enrico Cicalò, and Amedeo Ganciu, eds. 2018. *Paesaggi rurali*. Milano: Franco Angeli.
Ballacchino, Katia, and Letizia Bindi, eds. 2017. *Cammini di uomini, cammini di animali. Transumanze, pastoralismi e patrimoni bioculturali*. Campobasso: Il Bene Comune.
Barberis, Corrado, ed. 2009. *La rivincita delle campagne*. Roma: Donzelli.
Bergeron, Robert. 1967. "Problèmes de la vie pastorale en Sardaigne. Premier article." *Revue de géographie de Lyon* XLII(4): 311–28.
Bindi, Letizia. 2019a. "'Bones' and Pathways. Transhumant Tracks, Inner Areas and Cultural Heritage." *Il capitale culturale* 19: 109–28.
———. 2019b. "Walking Knowledge, Transhumant Practices. Intangible Cultural Heritage as a Multi-Situated and Multi-Disciplinary Fieldwork." In *Between Folk Culture and Global Culture in Contemporary Europe*, ed. Anna Brzozowska-Krajka, 28–37. Sharjah: IOV/Bahrein.
———. 2020. "Take a Walk on the Shepherd Side: Transhumant Narratives and Representations." In *A Literary Anthropology of Migration and Belonging*, ed. Cicilie Fagerlid and Michelle A. Tisdel, 19–46. London: Palgrave Macmillan.
Braudel, Fernand. 1976. *Civiltà e imperi del Mediterraneo nell'età di Filippo II*. Torino: Einaudi.
Brisebarre, Anne-Marie. 2007. *Bergers et transhumances*. Romagnat: De Borée.
Caltagirone, Benedetto. 1986. "Lo studio della transumanza come dispositivo di analisi del mondo pastorale." *Études Corses* XIV(27): 27–44.
Carrosio, Giovanni. 2019. *I margini al centro. L'Italia delle aree interne tra fragilità e innovazione*. Roma: Donzelli.
Cois, Ester, ed. 2020. *Aree rurali in transizione oltre la crisi economica*. Torino: Rosenberg & Sellier.
D'Amico, Mario. 2011. *Le produzioni agro-alimentari tipiche in Sicilia*. Acireale-Roma: Bonanno.

Davis, John. 1980. *Antropologia delle società mediterranee. Un'analisi comparata*. Torino: Rosenberg & Sellier.
De Rossi, Antonio, ed. 2018. *Riabitare l'Italia. Le aree interne tra abbandoni e riconquiste*. Roma: Donzelli.
Fabre, Patrick, Gilbert Molénat, and Jean-Claude Duclos. 2002. *Transhumance. Relique du passé ou pratique d'avenir?* Turquant: Editions Cheminements.
Falzon, Mark-Anthony, ed. 2001. *Multi-Sited Ethnography. Theory, Praxis and Locality in Contemporary Research*. Farnham: Ashgate.
Farinella, Domenica. 2018. "La pastorizia sarda di fronte al mercato globale. Ristrutturazione della filiera lattiero-casearia e strategie di ancoraggio al locale." *Meridiana* 93: 113–34.
Fossati, Laura, and Michele Nori. 2017. "Pastori in movimento. L'evoluzione di una pratica fra cambio generazionale e manodopera straniera." In *Per forza o per scelta. L'immigrazione straniera nelle Alpi e negli Appennini*, ed. Andrea Membretti, Ingrid Kofler, and Pier Paolo Viazzo, 149–59. Roma: Aracne.
Giacomarra, Mario. 2006. *I pastori delle Madonie*. Palermo: Fondazione Ignazio Buttitta.
Idda, Lorenzo, Roberto Furesi, and Pietro Pulina. 2010. *Economia dell'allevamento ovino da latte*. Milano: Franco Angeli.
Idda, Lorenzo, and Pietro Pulina, eds. 2006. *Paesaggio e sviluppo rurale in Sardegna*. Milano: Franco Angeli.
Le Lannou, Maurice. 1992. *Pastori e contadini di Sardegna*. Cagliari: Edizioni Della Torre.
Mannia, Sebastiano. 2010. "In turvèra. La transumanza in Sardegna tra storia e prospettive future." *Archivio Antropologico Mediterraneo* XII/XIII(12–1): 97–107.
———. 2013. *Il pastoralismo in Sicilia. Uno sguardo antropologico*. Palermo: Officina di Studi Medievali.
———. 2014. *In tràmuta. Antropologia del pastoralismo in Sardegna*. Nuoro: Il Maestrale.
———. 2016. "Sardegna e Sicilia: pastoralismi a confronto tra intervento politico, dinamiche di mercato e variazioni culturali." In *Etnografie del contemporaneo*, ed. Rosario Perricone, 31–46. Palermo: Edizioni Museo Pasqualino.
———. 2018. "Traveling Shepherds. Transhumance in Sardinia and Sicily Between Historical Dynamics, Today's Practices and Future Prospects." *Etnografie del contemporaneo* 1: 59–68.
Marchetti, Marco, Stefano Panunzi, and Rossano Pazzagli, eds. 2017. *Aree interne. Per una rinascita dei territori rurali e montani*. Soveria Mannelli: Rubbettino.
Marcus, E. George. 1995. "Ethnography in/of the World System: The Emergence of Multi-Sited Ethnography." *Annual Review of Anthropology* 24: 95–117.
Meloni, Benedetto. 1984. *Famiglie di pastori. Continuità e mutamento in una comunità della Sardegna Centrale 1950–1970*. Torino: Rosenberg & Sellier.
———. 1988. "Forme di mobilità ed economia locale in Centro Sardegna." *Mélanges de l'Ecole française de Rome* C(2): 839–55.
———. 2004. *Migrazione di Sardi nei poderi mezzadrili della Toscana*. Buonconvento: Museo della Mezzadria Senese.
———. 2006. *Lo sviluppo rurale*. Cagliari: Cuec.

———, ed. 2015. *Aree interni e progetti d'area*. Torino: Rosenberg & Sellier.
Meloni, Benedetto, and Domenica Farinella, eds. 2013. *Sviluppo rurale alla prova*. Torino: Rosenberg & Sellier.
Meloni, Benedetto, and Pietro Pulina, eds. 2020. *Turismo sostenibile e sistemi rurali locali. Multifunzionalità, reti d'impresa e percorsi*. Torino: Rosenberg & Sellier.
Mientjes, Antoon Cornelis. 2008. *Paesaggi pastorali*. Cagliari: Cuec.
Milone, Pierluigi, and Flaminia Ventura. 2009. *I contadini del Terzo Millennio*. Perugia: AMP.
Murru Corriga, Giannetta. 1990. *Dalla montagna ai Campidani. Famiglia e mutamento in una comunità di pastori*. Cagliari: Edes.
———. 1998. "Il nomadismo dei pastori sardi." *Annali della Facoltà di Lettere e Filosofia dell'Università di Cagliari* LIII: 301–30.
Nori, Michele. 2016. "Shifting Transhumances: Migrations Patterns in Mediterranean Pastoralism." *CIHEAM. Watch Letter 36—Crise et résilience en la Méditerranée*. Retrieved 15 May 2020 from www.iamb.it/share/integra_"les_lib/"les/WL36.pdf.
———. 2018. "TRA_MED—Mediterranean Transhumances Immigrants Shepherds in Mediterranean Pastoralism." In *Water as Hazard and Water as Heritage*, ed. Maria Boştenaru and Alex Dill, 93–95. Karlsruhe: Karlsruher Institut fur Technologie.
Ortu, Gian Giacomo. 1988. "La transumanza nella storia della Sardegna." *Mélanges de l'Ecole française de Rome* C(2): 821–38.
Petrocelli, Edilio, ed. 1999. *La civiltà della transumanza. Storia, cultura e valorizzazione dei tratturi e del mondo pastorale in Abruzzo, Molise, Puglia, Campania e Basilicata*. Isernia: Cosmo Iannone.
Rochefort, Renée. 2005. *Sicilia anni Cinquanta. Lavoro, cultura, società*. Palermo: Sellerio.
Sicilia on Press. 2017. "La Transumanza – Un Esperienza da Pastore nei Sicani." 19 May. Retrieved 3 June 2020 from https://www.siciliaonpress.com/2017/05/19/la-transumanza-un-esperienza-da-pastore-nei-sicani/.
Solinas, Pier Giorgio, ed. 1989–90. *Pastori sardi in provincia di Siena*, 3 volumes. Siena: Laboratorio etno-antropologico, Dipartimento di Filosofia e Scienze Sociali.
Tola, Alessio, ed. 2010. *Strategie, metodi e strumenti per lo sviluppo dei territori rurali*, Milano: Franco Angeli.
Uccello, Antonino. 1980. *Bovari, pecorai, curatuli*. Palermo: ACMP.
Van der Ploeg, Jan Douwe. 2006. *Oltre la modernizzazione. Processi di sviluppo rurale in Europa*. Soveria Mannelli: Rubbettino.
———. 2009. *I nuovi contadini*. Roma: Donzelli.
———. 2018. *I contadini e l'arte dell'agricoltura*. Torino: Rosenberg & Sellier.
Verona, Marzia. 2016. *Storie di pascolo vagante*. Roma-Bari: Laterza.

CHAPTER 13

The Coexistence of Transhumance Shepherding Practices and Tourism on Bjelašnica Mountain in Bosnia and Herzegovina

Manca Filak and Žiga Gorišek

Introduction: Written and Visual Research

This chapter analyzes the coexistence of transhumance shepherding practices and tourism in Lukomir, the highest village (1472 m above sea level) in the Federation of Bosnia and Herzegovina, located on the southern slopes of the Bjelašnica mountain massif (see Figure 13.1). In order to understand the village of Lukomir, we must consider it as an integral part of the Bjelašnica mountain, therefore in the text we usually refer to both, the Bjelašnica mountain in general and the village Lukomir in particular.

From a long-term ethnographic fieldwork conducted from April 2014 to May 2017, both a written anthropological analysis *Transhumance at the Crossroads of Changes: Transhumance and Tourism as Strategies of Survival in Lukomir on the Bjelašnica Mountain (BIH)* (Gorišek 2017), and an ethnographic film *Lukomir, my home* emerged (Filak and Gorišek 2018). The film is a visual ethnography of the daily lives of an elderly couple, Ismet and Tidža Čomor who live in the village. They are the main protagonists in the film as well as our hosts in the village.

The dissertation presents social, historical, and geographical contexts describing how daily life in Lukomir has changed due to many different factors. The film conveys the couple's connection to the land and the animals as well as the general changes in their social world, tracing the various spatial and material dimensions of their annual migration from

Figure 13.1. Lukomir village, Bjelašnica mountain, 2014. © Žiga Gorišek

the Bjelašnica ridge to the villages near Sarajevo and their relationship to a lifestyle that is slowly disappearing.

To a certain extent, these written and visual ethnographies complement each other. Nevertheless, they must be experienced (read or watched) separately. For this reason, we advocate multimodal representations, as they allow multiple identifications and multilayered understandings (see also Pink 2011; Collins, Durington, and Gill 2017). The writing itself can be multimodal, as it includes field research diaries, dialog transcriptions, interviews, evocative descriptions, and photographs to add qualitatively richer information (Lunaček Brumen 2018: 97; see also Turk Niskač 2011).[1] The Slovenian visual anthropologist Naško Križnar emphasizes the peculiar paradox of where to publish the findings of our visual research (based for instance on filming), we have to translate this visual information into words (Križnar 2002: 91; see also Biella 1993). The combination of different methods (filming, participant observation, semi-structured and unstructured interviews, fieldwork diary notes), therefore serves as a basis for discussion of the anthropological understanding that can be gained through audio-visual material in comparison to written research (see also Filak 2019).

Each research topic leads us to expand our methodologies in different ways in order to explore how to *look at* a particular topic. The mutually constitutive examination of one's own research topic and the use of different visual media can reveal connections, sensorial dimensions, and worldviews that might otherwise not be recognizable. In the following sections we will analyze the anthropological insights we have gained through the use of written and visual ethnography in order to understand *when* and *how* we create new meanings and new knowledge through participant observation and visual methods. More than just data from which we can read/observe cultural meanings, we consider both the written text and the final film as complementary processes from which new meanings and knowledge can emerge.

Transhumance on Bjelašnica Mountain

> It is the same for me as for those who get a job somewhere and go to work every day. You cannot leave your job. It is the same for me, I have to herd sheep every day, I feed myself from them, buy my daily bread, buy flour and similar. I have to wake up every day and do the same thing as you—look at my watch so that I am not late for work. I have no schedule, but every morning and every evening I have to herd sheep so that I can sell them later to make money. That is my life. (Interview with an older shepherd from Lukomir, 19 July 2014)

The Federation of Bosnia and Herzegovina is located in the central part of the Balkan Peninsula. With its varied relief, the Bjelašnica mountain massif is part of the Dinarid mountain system. Due to the mixing of Mediterranean and continental air masses on the Bjelašnica mountain, there is a lot of rain, constant wind and snow, which can remain on the northern slopes and certain sinkholes until the beginning of June (Sarajlić 1983: 6). The mountains of the Dinarides were already inhabited in the Late Neolithic and Bronze Age when sheep breeding was one of the most important economic activities (Čović 1990: 73; Marković 2003: 13). Livestock breeding, seasonal migration, and transhumant shepherding[2] allowed the communities in this area to gradually settle in the Bjelašnica region, which would otherwise be much more difficult in these harsh conditions.

Shepherding communities on Bjelašnica mountain have practiced vertical transhumance at different altitudes and built seasonal or temporary settlements with sheds standing on the edges of the grazing areas and smaller parcels that were used for farming. In these sheds, shepherds lived, processed milk, and stored milk dairies during summer. Most of the sheds enabled only the necessities of survival. The locals call them *stanovi*, *katuni*, or *mahale* (Chabbouh Akšamija 2009: 159–80). Until 2010, there were nine shepherd settlements or "permanently"[3] inhabited villages on Bjelašnica mountain and ten seasonal settlements where shepherds lived during the summer season, as in the past most shepherds moved their herds to lower-lying villages during the winter. The reverse process took place on Bjelašnica mountain in spring. Because of the lack of food in winter, the shepherds were forced to seek food for their herds outside their usual place of residence. In this sense the shepherds on Bjelašnica mountain could have more animals than their parental territory allowed them

(Perović, Čopić, and Milišić 1990: 604; see also Bartosiewicz and Greenfield 1999). The transhumant practices on Bjelašnica mountain enabled contacts and exchanges between the shepherds who settled "permanently" on the mountain and those who settled "permanently" at the foot of Bjelašnica mountain. In search of pasture, both groups of shepherds spent part of the year away from their "permanent" residence, which in fact contributed to a mixing of people[4] and customs.[5]

It is important to recognize that traditional sheep breeding as we know it today has changed over time. In the last fifty years, transhumant shepherding on Bjelašnica mountain has experienced a sharp decline or transformation and adaptation to new forms of animal husbandry for various economic, political, and social reasons. Similar processes can be observed in other parts of the world (see Bartosiewicz and Greenfield 1999: 9). Therefore, we see transhumant shepherding on Bjelašnica mountain as a survival strategy that is constantly changing and shifting its form over time.

The Case of Lukomir

From the abovementioned seasonal settlements on the steep slopes, nucleated mountain villages gradually emerged.[6] *Dolnji Lukomir* (Lower Lukomir) and *Gornji Lukomir* (Upper Lukomir) are examples of transformations of seasonal settlements on Bjelašnica mountain between which the shepherds migrate. Most of the houses in Upper Lukomir, now known only as Lukomir, were built of rocks, the longer side being sunk into the steep ground. The space dug out of the ground was for the animals, while the space above ground was for the shepherds and their families (Chabbouh Akšamija 2009: 159–80). Up to ten people could live in such a dwelling. New houses and barns next to them began to appear in the 1970s and it was common to use former houses as barns. In 1985, forty-three permanent households were still active in (Upper) Lukomir, while Lower Lukomir was already abandoned (Općina Konjic 2017).

The villagers of Lukomir have experienced various waves of migration between the different villages, valleys, and settlements on Bjelašnica mountain. During the period of gradual settlement between 1952 and 1974, a school with compulsory first four years of primary education and a mosque were established in Upper Lukomir. As there was less food for the animals during winter, the shepherds still had to look for better pastures on other parts of the mountain. Therefore, the "permanence" of the place of residence we mention is always in some sense temporary. We have to consider the movements between different settlements and the associated transhumance shepherding as constantly changing practices over time.

Slowly, in the second half of the twentieth century, many of the seasonal settlements like Lower Lukomir on Bjelašnica mountain were abandoned. Consequently, shepherding communities were "permanently" settled in villages and towns, especially in Sarajevo and its surroundings. This was mainly due to the industrialization of Yugoslavia and various measures taken by the communist party, such as the collectivization of agriculture, taxes on animals and land larger than ten hectares, and the ban on nomadic grazing in the forest (Halpern 1975: 86–90, 163). During the same period, the inhabitants of various villages on Bjelašnica mountain, such as Lukomir, began to work in the new factories in Sarajevo and received social security, which they had not known as shepherds (Ljiljana Beljkašić Hadžidedić, pers. comm., 3 June 2014). The villages on Bjelašnica mountain, where their parents had normally stayed, became places where they returned during holidays or when help was needed on the mountain.[7] When most of the young people left, Bjelašnica's population began to age and consequently schools were closed, which until then had had a great impact on the literacy of the rural population.

The biggest change on Bjelašnica mountain came with the disintegration of Yugoslavia and the Yugoslav War between 1992 and 1995. Lukomir is one of the few villages on Bjelašnica mountain (as well as Čuhovići and seasonal settlements like Gradina above Umoljani) that was not burned down during the war, although the frontline between the Bosnian Army of the Republic of Bosnia and Herzegovina, and the Serbian Army (Army of Republika Srpska) traversed the area. Despite the long-term tendency of depopulation from Lukomir to the urban settlements around Sarajevo, many returned to Lukomir during the Yugoslav War, and stayed until the end of the war, as it was known to be safer there than around Sarajevo.

In the postwar period many people in Bosnia and Herzegovina were left without work, pensions or other means of earning a living. During this period, many people from Bjelašnica mountain, who worked in Sarajevo or the surrounding area before the war, decided to return to shepherding and transhumant shepherding practices, which served as a main source of income for those who were not able to earn a living in the cities. Nevertheless, the process of emigration continued in the second half of the 1990s, mainly to the growing town of Hadžići, as well as Iliđa, Tarčin, Pazarić, and similar towns at the foot of Bjelašnica and Igman. Many of the villagers from Lukomir and other villages from Bjelašnica mountain had built their houses in a settlement above Hadžići, where Orthodox residents had lived before the war. They also bought pastoral land on which they built barns. Despite the official change of their "permanent" (in this case winter) residence in a relatively urban area, they continued the tradition of transhumant shepherding. And despite the migration of the population and the increase in tourism in recent decades, transhumant shepherding has remained one of the most important economic strategies in the villages of Bjelašnica as well as an important social aspect throughout the year.

Due to the aging of the population and the poor transport connections in winter, the inhabitants of Lukomir decided not to spend the winter of 2010 in the village. Since then, the village has been inhabited only in summer, with around twenty-two households still active. The seasonal migration of families and their flocks of sheep characterizes the life of the villagers, which may be divided roughly into two seasons: summer on Lukomir (see Figure 13.1) and winter in the lower settlements near Sarajevo.

The summer season consists of bringing sheep to the mountain pastures, drying hay, and doing various jobs that provide the inhabitants of Lukomir with their livelihood all year round. In winter they continue their grazing, mainly in the area called Bare near Hadžići, where they have to deal with problems concerning the grazing land (see Figure 13.2). In Bare, for example, investors from Dubai are building a so-called Ourika Resort, a luxury settlement with fifty-eight plots and up to 996 m² of land (Ourika 2017), which will use a lot of shepherds' grazing land. There are also many locals from the area who disapprove of grazing, as the land is mostly private and already divided among the population.

Figure 13.2. Ismet while shepherding in Bare, with snowy Bjelašnica in the background, Hadžići, 2014. © Manca Filak

Tourism in Lukomir, a New and Innovative Survival Strategy

> The season for agrotourism here runs from May to autumn. In winter it lasts for only two months. I came here mainly to make money, to survive. Later everything else came. I came to Umuljani [on Bjelašnica mountain] because I could earn more money here than in Sarajevo. If I had a better salary in Sarajevo, I would build myself a weekend cabin here. (Interview with a caterer, Umuljani, 25 September 2014)

Most of the villages on Bjelašnica mountain were burned down during the Yugoslav War. After the war, many donations came from abroad, which enabled various NGOs and small entrepreneurs to begin a gradual reconstruction of the houses. Most of them got running water, indoor toilets and the like for the first time. Stones and wood were replaced by newer materials such as bricks, concrete, and sheet metal.[8]

Lukomir, Gradina, and Čuhoviči were among the few settlements that were not burned down during the war. Therefore, Lukomir has preserved some of its traditional architecture and appearance, which is the main attraction for the increasing number of visitors. In brochures for domestic and foreign tourists, Lukomir is presented as a picturesque village above the Rakitnica Canyon, one of the most authentic and untouched villages in Bosnia and Herzegovina (see for example Crevar 2018). In 2009 the village was protected as a monument of cultural importance, but there are nevertheless many new buildings and reconstructions, sheet metal roofs, and new catering facilities. The whole area is popular among hikers as well as skiers in the winter season, who can stay in some of the huts and eat or drink in Lukomir's catering facilities.

The first forms of organized tourism on Bjelašnica mountain appeared in the twentieth century with the development of mountaineering and the 1984 Winter Olympics in Sarajevo. Before that period, the area saw mainly regional tourism. The first regular visitors to Bjelašnica mountain and Lukomir were mountaineers from the countries of former Yugoslavia, who stayed in some of the mountain huts or with the locals in their barns. Besides the introduction of electricity, a very important contribution to the modernisation of the villages was the development of Olympic infrastructure.

We can only speak of larger and more organized forms of tourism in Lukomir after the end of the war in 1995, when many international and nongovernmental organizations came to the city of Sarajevo to help repair the war damage. Due to the large number of foreigners living and working in the city, the need for organized and safe trips to the countryside arose. One of the first agencies in Lukomir was Green Visions, established in 2000 in cooperation between people from BiH, Holland, and the US to offer their guests safe travel to areas where there were no mines or other dangers. Since its foundation Green Visions has been promoting so-called responsible tourism and stands for the protection of the natural environment and cooperation with the local population. The groups they take to Lukomir are small, usually comprising less than twenty people. The highlight of their trip is an overnight stay with locals in their house (Interview with one of the founders of Green Visions, 20 August 2014).

There are many different and complex views on the development of tourism on Bjelašnica mountain. On one hand, we see some of the government plans and strategies that, with the help of European funds and NGOs such as Green Visions, follow the guidelines of sustainable development for tourism, especially on the southern side of Bjelašnica mountain. The growing number of tourists in recent years has led to an increase in so-called heritage tourism, which is based on the desire of tourists for genuine contact with the villagers, a taste of homemade food, and insight into local traditional stories, etc. (Dinero 2002: 69–73; Brandth and Haugen 2011: 41). Tourist agencies and locals are following these global heritage trends promoting ecotourism, rural tourism, heritage tourism, or even slow tourism, as noted by Ledinek Lozej in this volume (Chapter 10). In rural areas, these trends are often emerging because of the needs of urban consumers who want to spend their time outside urban areas (see also Kozorog 2012). This is one of the main reasons why ecological and agro forms of tourism have emerged in the context of rural development in Lukomir and the village of Umuljani in the last decade. These uses of rural spaces connect two aspects: on one hand locals from Lukomir are returning to their land with new or rethought ideas about how to make a living related to sheep farming or gastronomic establishments, and on the other hand small (urban) businesses and entrepreneurs are developing ecotourism and other agro forms of tourism in mountain areas. In this perspective, we can see transhumance as a cultural and touristic heritage, as Mannia notes (Chapter 12).

Tourism with connotations such as alternative, responsible, green, sustainable, conscious, etc., is moving away from the normal practices of mass tourism (see Weber 1997; Skočir 2011). This is not so on the northern side of the mountain and in the valleys around Sarajevo, where standard practices of mass tourism focus on the development of ski slopes and hotels by (mostly) Arab states. For example, in the Babin dol ski resort new hotel complexes are being built and skiing capacity is being expanded. Investors from various Arab states like Kuwait, Saudi Arabia, or Turkey are investing in infrastructure and want to bring more people to Bjelašnica mountain to cover the large investments that are being made.

There are several cases where the locals in Lukomir have started to use tourism as their survival or economic strategy, especially in the last five years. The most interesting case is that of the Lijetna Bašta (Summer Garden) catering facility, which was built by a local couple.

Since they were shepherds themselves in the past, they understand the needs of local villagers and often work with them. In the catering facility, which they run together with their children, they sell various items, including wool products made by the local people. They also buy kaymak, cheese, and meat from the villagers. In recent years, Lijetna Bašta has become a local meeting place, where villagers socialize daily and on special occasions. They cooperate with the Green Visions tourist agency, which brings guests to Lukomir two or three times a week. In 2017 they moved from a simple wooden building in the center to a new one at the main entrance of the village, which is visited by most of the tourists who come to Lukomir (see Figure 13.2.). Here they offer meals to guests, as well as toilets with running water, a real rarity in Lukomir. Two rooms of the restaurant have been furnished in bed-and-breakfast style, and as the number of tourists continues to grow, they plan to set up shared beds in the attic. In 2019, many other tourist facilities similar to Lijetna Bašta were built in the village, which is obvious even at the main entrance of the village, from where visitors are directed to many of the "ethno" houses. Competition for tourists is increasing in the village, as many more locals try to eke out a living from tourism during the summer months.

Figure 13.3. New location of Lijetna Bašta at the main entrance of Lukomir, 2017. © Žiga Gorišek

Mostly older villagers decide to set up tourist activities, usually with help from their children. Tourism allows them to earn extra money and presents opportunities for them to stay on the mountain. At the same time, they can later help their children who live in Sarajevo or in the surrounding area. In recent years, in addition to the catering industry, wooden spoons and wool products with traditional patterns have been produced in Lukomir and in other villages on Bjelašnica mountain. The locals themselves sell wool products in Lukomir, but the informal market causes many disputes among the women, which can be unpleasant for tourists who are often annoyed by the pushiness of the locals. Otherwise, the locals in the village are very hospitable and willing to accept people as guests in their homes. In return for coffee, food, and a bed, guests usually give some money or buy wool products.

The Effects of Tourism on Life in the Village

By promoting Lukomir as one of the most authentic, traditional, isolated, and remote "ethno" villages in Europe, tourism agencies as well as various bloggers, articles, and similar media content (see for example Viator 2020; Meet Bosnia 2020; Green Visions 2020; Funky Tours 2020) have created a myth of the "Bay of Peace" (a literal translation of the name Lukomir is *luka miru*). Due to the growing number of visitors, the purpose of many agricultural buildings and plots of land has changed and adapted to the new requirements and desires of tourism. This has often been accompanied by the process of creating the heritage and identity of the place, which includes local hospitality, cuisine, wool products, music, singing, and storytelling, etc. (see also West, Igoe, and Brockington 2006; Brandth and Haugen 2011; Grasseni 2013). Interestingly, the abovementioned model of ecological and agrotourism only became successful when part of the local population took the initiative to switch from shepherding to catering for tourism. At the same time, due to the high unemployment rate and the low level of social welfare and retirement in the country, an important change had taken place. The land on Bjelašnica mountain, especially in Lukomir, had become valuable for many, enabling villagers or their descendants to return to the village and earn additional income from tourism. Therefore, for many families and individuals, tourism enables modest survival under otherwise rather harsh conditions, which are prevalent throughout the country.

Lukomir and several other villages like Umoljani on Bjelašnica became important starting points or destinations for many local and foreign visitors. The development of tourism in this part of BiH has caused many changes that will have a long-term effect on the local population. As one of the most visited places in the region, Lukomir faces a great challenge. The phenomenon of tourism has created many new opportunities to earn extra money and new survival strategies. Infrastructure has been expanded and the social conditions of the local people have improved. At the same time, however, there are many negative consequences such as environmental pollution, increased use of natural resources, rising prices of land and real estate, and financial disputes. All of these aspects can impact not only the daily life of the village but also the shepherding practices. Namely, their preservation, modification, and gradual abandonment.

Reactions to the growing number of tourists are of course diverse and complex. The locals usually like to receive guests and are not against tourism per se. A problem for most of them is the conflict between the main summer tourist season and work that needs to be done on the mountain, such as haymaking for winter. Today there are fewer shepherds in the village, but more of them have a larger flock (more than one hundred sheep on average).[9] Because of the larger number of sheep, as well as the larger number of tourists, there is less space in the village. Therefore, bigger sheep breeders have their flocks on the outskirts or in front of the village. Among them are some who do not like tourists and oppose the development of tourism with various techniques to deter visitors, such as the accumulation of animal entrails and garbage in the places where most tourists pass by. While most villagers support tourism, some do not really know what would be best for the village. Most of the older inhabitants shrug their shoulders when asked such questions and say that it is always better for the village to be alive and full of life than empty.

Working with animals is extremely hard work, most of the families in Lukomir do not have time to take a holiday, let alone go to the sea. Sometimes they go to the *Boračko jezero* (lake) or to the Baščaršija market in Sarajevo. They do not experience the kind of tourism that is enjoyed by the tourists who visit Lukomir. In this sense, many locals would change their way of life for something different at the first opportunity.

Tourism in Lukomir through the Camera's Lens

> On the way to Lukomir we met a group of shepherds who were bringing their sheep down to the village. I had to film the scene because it was so picturesque. The villagers were angry with me. They asked me why I was filming and what I was going to do with the footage. I apologized and explained that I only wanted to film the sheep on their way to the village. That was the truth. I took a long shot so that the shepherds were not recognizable in the video. When they saw that I spoke their language, they were reas-

sured. They asked me where I was sleeping. When I told them at Ismet's, they started laughing and said that he was the most important "glumac" (actor) in the village. Since he often stands in front of the camera, the villagers often make jokes like that. (Notes from the fieldwork diary, Lukomir, 12 July 2014)

The use of a camera has helped us to better understand the possibility of coexistence between shepherding practices and tourism in Lukomir. The topic we initially wanted to explore with the camera was the influence of tourism on the way of life in the village. But while we continuously participated in the daily life routines of our protagonists, we slowly started to see the camera as an obstacle. We began to compare ourselves with the other tourists who came to the village and carelessly used their cameras without asking for permission or thinking about their position with regard to the inhabitants. We felt restrained because we felt like them, taking images *of* and *from* the people without their permission.

Similarly, we began to realize that the focus on tourism as an obstacle or potential threat reinforced the binary relationship of *hosts and guests* (Smith 1977) that is often mapped onto tourists and locals. Consequently, after some uncomfortable situations, we decided not to use the camera in the initial phase of fieldwork. We were also too concerned in some ways for the process of visual ethnography to affect our friendship, so we turned our camera instead to the landscape surrounding us and looked at Lukomir from a more photographic angle. Later, we began to follow the natural flow of events regardless of whether tourists were present or not, and instead adapted ourselves to the relationship we had with our main protagonists, Ismet and Tidža. The locals also seemed to be afraid of foreigners because of many previous experiences (including those on film). Because Lukomir is often portrayed in different ways by various tourist and commercial organizations, the villagers are aware of the power images can have (see Koevorden 2010).

Later on, the protagonists themselves began to insist and point out things that should be recorded and documented, such as pie making, pulling wool thread, *mevlud* (the biggest festival of the year), etc. These are elements of their everyday lives that are considered to be traditional and are perceived as such by them as well as outsiders. Therefore, the creative and relational use of the camera helped us to involve our protagonists in the process of meaning-making, and thus to bring about the processual aspects of social relations instead of just documenting things "out there" (Favero 2013: 70). With the help of the camera, we were able to understand more about what is important to the villagers, for example *mevlud* as a form of social display or elements of daily life that they perceive as traditional.

Another good experience was connected with the use of the small handheld camera with built-in projector, with which the film material was shown on the wall inside their small house. In this way family members could see the things we recorded and they were excited to see themselves and the landscape in the film; also, through their comments and enthusiasm when watching the footage, we could see what was important to them (the nature, animals, other villagers, etc.). Each visual representation is consumed (not just created) in different social contexts that evoke certain feelings of similarity, distance, recognition, or empathy (Banks and Ruby 2011: 9; Vávrová 2014: 3; see also MacDougall 1992: 25, 32). Thus, viewing visual ethnographies is not only about *looking at* but also about positioning yourself in a particular time and space through the sensory experiences and perceptions of other people. This enables one to better understand, relate to others, and create new meanings about the topic (MacDougall 2005: 4, 58; Vávrová 2014: 25).

When protagonists become their own audience, they become phenomenologically bound to their own representation in a way that is not possible for those who are not part of their community (Banks 1998: 124 in Grossman 2010: 186). When we showed the video material, or in this case the final film to our protagonists, they did not appreciate the aesthetics or the narrativity of the film so much, but reacted to details that were more significant to them, for example which period of the year it was according to the greenness of the grass in the footage, which sheep are still alive, etc. Overall, they appreciated the final film as a form of personal inheritance for their descendants, namely their children and grandchildren. The final film was a result of our cooperation, as Tidža and Ismet proposed that we film the abovementioned tasks or areas which were important to them. Therefore, we saw the potential of a filmmaking approach to depict and explore transhumance and tourism, and to create a common understanding of their everyday lives (see also Barabantseva and Lawrence 2015: 23).

Peter Ian Crawford suggests that the strategy or aesthetics of a particular film fits a particular culture by basing his argument on the connection between the narrative quality of a film, the culture portrayed, and the audience (1992; see also Postma 2006; Henry and Vávrová 2016). In our case, the video material follows various spatial and material dimensions of seasonal migration in general, and shepherding in particular. It follows the intrinsic flow of the annual cycle of the shepherds and depicts village life. The film material also shows the slow pace of our protagonists' everyday lives in space and time. It contrasts stillness and movement, work and leisure, mountains and city, summer and winter, waiting and working.

The "how" to film and "how" to show a certain phenomenon (i.e., cinematographic strategy) comprises a combination of one's own views with the protagonist's worldview (Piault 2006: 372). The camera has encouraged us to examine more closely what Lukomir and shepherding means to our protagonists, through which elements they identify themselves, how they see the future of Lukomir, how they anticipate ascending the mountain the following spring, and how they enjoy the clear air, routines, and exchanges with neighbors and tourists, and visiting family members. In this way, the process of visual ethnographic research (including the fear of the influence of the camera) and the re-viewing of the video material led us to look beyond our original theme: the impact of tourism. Visual ethnography made it easier to shift the focus of our interest to the tactile aspects of their lifestyle. These included the gentle and close relationship with the animals, the routines of daily sheep care and the hard work that seems to be embodied in their movements. By adapting our cinematographic strategy (Piault 2006; Postma 2006)—refocusing on everyday life and shepherding—we realized that tourism is in fact only one of the changes taking place in Lukomir. Although tourists are often present in the area, moving in and out of their lives, as well as from the footage, we do not believe that this is the only important aspect of change.

By using a camera in Lukomir, we have understood what it feels like to take images of and from the people in the village, the impact of tourism, and the interpretation of a specific traditional, local culture by tourists. We could also feel the intrinsic rhythm and flow of the people in Lukomir by observing different elements of their daily lives. By letting them show us what we should film (wool threading, *mevlud* festivities, flowers, sheep, etc.), we came to better understand what is important to them and what they consider important for their way of living. Furthermore, we were inspired to reflect on whether there are differences in the methodological and analytical procedures of obtaining anthropological understanding. A camera offers visual particularities of a certain time and space that are concrete, visible, and audible. Finally, it also captures what is happening in the background, which can provide an excess of information that can be useful for research (see also De Bromhead 2014: 234).

Conclusion

There are numerous factors that change the cultural landscape of Bjelašnica mountain, which throughout history has been shaped mainly by transhumance shepherding practices. The biggest obstacles to the maintenance of these practices are not only tourism and its infrastructure, but also the aging population and lack of pastoral land near Hadžići (the suburbs of Sarajevo), where most of the villagers from Lukomir migrate to in winter.

The pasturelands are shrinking due to the construction of many luxurious settlements built by investors from Turkey, Saudi Arabia, United Arab Emirates, and Kuwait, as well as local disputes over land. The state subsidy system is active, but not strong enough to help all farmers to maintain and expand their shepherding activities. As a result, the village community in Lukomir has an important decision to make for their future—whether or not to maintain a lifestyle linked to sheep breeding and the seasonal transhumance practices of herding.

Like written text and visual material, we see tourism (despite its different impacts) and transhumance practices in Lukomir as complementary to each other rather than conflicting. The research with the camera and the subsequent viewing of the film material helped us to better understand this coexistence. Field notes are useful in this sense, but the writing always takes place after the experience, while every film recording is made at the moment of shooting (Devereaux 1995: 72; see also Barabantseva and Lawrence 2015: 9). With the help of both, the camera and usual fieldwork methods, we have understood the traditional transhumance practices as a persistent survival strategy in this area, where—due to many different factors in the country, such as the lack of political unity, inconsistent funding, political conflicts, poor welfare state—there is often no other or better option.

Even though shepherding and its distribution area has changed throughout history, the people of this area have always kept it as part of their way of life, which usually involves the help of extended family members. Despite changes in the country's policies, changes in financing, the type of subsidies, varying interests, the financial and political crises, there are still people who insist on this seasonal way of life. That is why we do not see transhumance exclusively as an economic strategy, but also as a livelihood strategy, with an emphasis on its cultural, social component, as the villagers of Lukomir migrate to the Bjelašnica mountain in the summer even when they do not have sheep. And when the villagers move to the valley in autumn, they mostly move together into the same area.

We also see the seasonal shepherding practices as an important attempt to be self-sufficient. During the past ten years, these practices have been complemented by new tourist offers and facilities, which are increasing in scope and number. Although there are many different local reactions to tourism (rejection as well as acceptance) and although tourism is a relatively new aspect of everyday reality in the village (changed infrastructure, new facilities, etc.), it does not refute or hinder the transhumance practices on Bjelašnica mountain. The question that remains open is therefore related more to the future of tourism in Lukomir. If fewer tourists come to the mountain due to the COVID-19 pandemic, there is a good chance that people will decide to increase their herds and thus the practices of transhumant shepherding. In the opposite case, with a possible expansion of tourism, these shepherds are in the line of fire, as they are still one of the groups most at risk due to their age, lack of insurance, and retirement planning, etc.

We can therefore see that the practices of transhumance shepherding change over time (and in space) and should not be understood as rigid. These changes are a consequence of various factors such as tourism, valuations, emigration of young people, privatization of public spaces, diseases, urbanization, etc., but the practices of transhumance shepherding will continue in this area, as history has proven.

Acknowledgments

This research was conducted during the master's studies at the Department of Ethnology and Cultural Anthropology (University of Ljubljana, Faculty of Arts) under the academic supervisor Prof. Bojan Baskar. However, the chapter was completed within the Young Researcher Program at the Research Centre of the Slovenian Academy of Sciences and Arts, Institute of Slovene Ethnology.

Manca Filak is a Young Researcher at the Research Center of the Slovenian Academy of Sciences and Arts at the Institute of Slovenian Ethnology. She is the director of award-winning ethnographic films and documentaries that have been screened at various festivals in Europe and abroad. She is co-organizer of the festival Days of Ethnographic Film and the Summer School of Visual Ethnography in Ljubljana.

Žiga Gorišek is an ethnologist and cultural anthropologist, photographer and award-winning ethnographic and documentary filmmaker. His passion for mountains and high mountain cultures led him to devote himself to the study and documentation of transhuman pastoralism.

Notes

1. Žiga's thesis is an example of this, as he included photographs and fieldwork diary excerpts in the text.
2. In the broadest sense, we understand transhumance and transhumant shepherding to mean seasonal migration of people and their livestock between different vertically (at different altitudes) or horizontally separated grazing areas. Both movements are adapted to the season in which the shepherds

look for suitable pastures. Therefore, transhumance is an important factor in the Alpine world and in Southern Europe, where it is mainly associated with sheep breeding (Burns 1963: 140). Transhumance on European ground is interesting mainly because of its diversity, as there are many different organizational structures and strategies that have survived throughout history to this day (Chang 1993: 699).
3. We do not use the word "permanent" in a literal sense. For the people of Bjelašnica mountain, "permanence" is more related to a sense of belonging or identity. "Permanent" settlement is a term also used to refer to the place or land where part of the family lives, farms, or gathers hay for the winter.
4. Bjelašnica mountain was a meeting place for shepherds from the northern, Bosnian side of the mountain and shepherds from the southern side, who came from Herzegovina. According to our interlocutors, many people came to Bjelašnica from the area of Ljubuški, Nevesinje, Podvelež, Konjic, and other places in Herzegovina. Interestingly, there are only two surnames in Lukomir, Čomor and Masleša.
5. Some shepherds who were also herding sheep for other people did not practice shepherding as their main occupation.
6. The nucleated mountain villages characteristic of Bjelašnica mountain can be divided into two types, the southern one, where the houses were mostly built of rocks, and the northern one, where the houses were made of wood.
7. In addition to transhumant shepherding practices, Claudia Chang speaks of social transhumance, where people, like shepherds with their animals, move from urban settlements to rural settlements every year, usually to places where they have previously practiced transhumant shepherding (1993: 690–91). In this way, these communities maintain a community identity that is shaped by life in different places.
8. On the southern, Herzegovinian side of the Bjelašnica mountain, the houses were mostly covered with wooden shingles or *šindle* (Sarajlić 1983: 46). In the second half of the twentieth century, these wooden shingles were often replaced by sheet metal from old barrels.
9. Parallel to the livestock subsidies, that is, the increase in the size of the flock, the number of families renting pastures is growing. Most of the younger sheep breeders see the subsidies as something positive, especially in combination with renting pastures from other villagers and the possibility of grazing on Bjelašnica mountain. By comparison, in the 1990s, before the Yugoslav War, people had an average of fifty sheep per family.

References

Banks, Marcus, and Jay Ruby. 2011. "Made to Be Seen: Historical Perspectives on Visual Anthropology." In *Made to Be Seen: Perspectives on the History of Visual Anthropology*, ed. Marcus Banks and Jay Ruby, 1–18. Chicago: University of Chicago Press.

Barabantseva, Elena, and Andrew Lawrence. 2015. "Encountering Vulnerabilities through 'Filmmaking for Fieldwork.'" *Millennium: Journal of International Studies* 43(3): 911–30.

Bartosiewicz, László, and Haskel J. Greenfield, eds. 1999. *Transhumant Pastoralism in Southern Europe: Recent Perspectives from Archaeology, History and Ethnology.* Budapest: Archaeolingua alapítvány.

Biella, Peter. 1993. "Beyond Ethnographic Film: Hypermedia and Scholarship." In *Anthropological Film and Video in the 1990s*, ed. Jack R. Rollwagen, 131–76. Brockport: The Institute, Inc.

Brandth, Berit, and Marit S. Haugen. 2011. "Farm Diversification into Tourism—Implications for Social Identity?" *Journal of Rural Studies* 27: 35–44.

Burns, K. Robert. 1963. "The Circum—Alpine Culture Area: A Preliminary View." *Anthropological Quarterly* 36(3): 130–55.

Chabbouh Akšamija, Lemja. 2009. *Autentičnost ruralnog graditeljstva kao preduvjet aktivne zaštite: Pilot projekat bjelašničkog sela Ledići.* PhD diss., University of Sarajevo.

Chang, Claudia. 1993. "Pastoral Transhumance in the Southern Balkans as Social Ideology: Ethnoarchaeological Research in Northern Greece." *American Anthropologist* 95(3): 687–703.

Collins, Samuel, Matthew Durington, and Harjant Gill. 2017. "Multimodality: An Invitation." *Multimodal Anthropologies* 119(1): 142–53.

Čović, Borivoj. 1990. "O strukturi stočarstva na sjeverozapadnom Balkanu u bronzanom i željeznom dobu." In *Godišnjak knjiga 28: Centar za balkanološka ispitivanja*, ed. Alojz Benac, 65–73. Sarajevo: Akademija nauka i umjetnosti Bosne i Hercegovine.

Crawford, Peter Ian. 1992. "Grass, the Visual Narrativity of Pastoral Nomadism." In *Ethnographic Film Aesthetics and Narrative Traditions: Proceedings from NAFA 2*, ed. Peter Ian Crawford, Jan Ketil Simonsen, and the Nordic Anthropological Film Association, 121–40. Aarhus: Intervention Press in association with the Nordic Anthropological Film Association.

Crevar, Alex. 2018. "High in the Mountains, this European Village Stands Frozen in Time: Lukomir is Home to 17 Families and Medieval Traditions." Retrieved 26 November 2018 from https://www.nationalgeographic.com/travel/destinations/europe/bosniaherzegovina/things-to-do-high-mountain-village-lukomir/.

De Bromhead, Toni. 2014. *A Film-Maker's Odyssey: Adventures in Film and Anthropology.* Aarhus: Intervention Press.

Devereaux, Leslie. 1995. "Experience, Re-presentation, and Film." In *Fields of Vision: Essays in Film Studies, Visual Anthropology and Photography*, ed. Leslie Devereaux and Roger Hillmam, 56–76. Berkeley: University of California Press.

Dinero, C. Steven. 2002. "Image is Everything: The Development of the Negev Bedouin as a Tourist Attraction." *Nomadic Peoples* 6(1): 69–94.

Favero, Paolo. 2013. "Picturing Life-Worlds in the City: Notes for a Slow, Aimless and Playful Visual Ethnography." In *Archivio Antropologico Mediterraneo—AAM* 15(2): 69–85.

Filak, Manca. 2019. "A Long-Term Visual Ethnography in a Bosnian Village: Tracking Epistemological and Methodological Issues." *EthnoAnthropoZoom/ЕтноАнтропоЗум* 16(16): 251–303.
Filak, Manca and Žiga Gorišek, dir. 2018. *Lukomir, My Home*. Ethnocinema production (Slovenia). Color, 61mins.
Funky Tours. 2020. "Lukomir Nomad Village Hiking—Full Day Tour from Sarajevo." Retrieved 10 June 2020 from https://sarajevofunkytours.com/tour/lukomir-hiking-tour/.
Gorišek, Žiga. 2017. "Transhumantno pašništvo na razpotju sprememb: transhumanca in turizem kot strategiji preživetja v vasi Lukomir na Bjelašnici (BIH)." Master's thesis, University of Ljubljana.
Grasseni, Cristina. 2013. "Re-inventing Food: Alpine Cheese in the Age of Global Heritage." Retrieved 14 June 2013 from http://aof.revues.org/6819.
Green Visions. 2020. "Lukomir Highland Village Tour." Retrieved 10 June 2020 from https://greenvisions.ba/en/activity/240100/lukomir-highland-village-tour.
Grossman, Alyssa R. 2010. "Choreographies of Memory: Everyday Sites and Practices of Remembrance Work in Post-Socialist, EU Accession-Era Bucharest." PhD diss., University of Manchester.
Halpern, M. Joel. 1975. "Some Perspectives on Balkan Migration Patterns (with Particular Reference to Yugoslavia)." In *Migration and Urbanization: Model and Adaptive Strategies*, ed. Brian M. Du Toit and Helen I. Safa, 77–115. Paris: Mouton Publishers.
Henry, Rosita, and Daniela Vávrová. 2016. "An Extraordinary Wedding: Some Reflections on the Ethics and Aesthetics of Authorial Strategies in Ethnographic Filmmaking." *Anthrovision* 4(1): 1–19.
Koevorden, Niels van, dir. 2010. *Winterslaap in Lukomir*. Fermfilm (Holland). Color, 30mins.
Kozorog, Miha. 2012. "Proti 'urbanosti': Ohlapno strukturne misli o ohlapno definirani temi." In *Antropološki načini življenja v mestih*, ed. Jaka Repič and Jože Hudales, 59–74. Ljubljana: Zupaničeva knjižnica.
Križnar, Naško. 2002. "Etnologija in vizualna antropologija." *Traditiones* 31(2): 85–92.
Lunaček Brumen, Sarah. 2018. "V iskanju skupne antropologije." *Bulletin of the Slovene Ethnological society* 58(3–4): 91–106.
MacDougall, David. 1992. "Whose Story Is It?" In *Ethnographic Film Aesthetics and Narrative Traditions: Proceedings from NAFA 2*, ed. Peter Ian Crawford, Jan Ketil Simonsen, and the Nordic Anthropological Film Association, 25–42. Aarhus: Intervention Press in association with the Nordic Anthropological Film Association.
———. 2005. *The Corporeal Image: Film, Ethnography, and the Senses*. Princeton: Princeton University Press.
Marković, Mirko. 2003. *Stočarska kretanja na Dinarskim planinama*. Zagreb: Naklada Jesenski i Turk.
Meet Bosnia. 2020. "Lukomir Village Tour from Sarajevo." Retrieved 6 June 2020 from https://meetbosnia.com/tour/lukomir-village-tour-from-sarajevo/.

Općina Konjic. 2017. "Kulturni krajolik—selo Lukomir." Retrieved 22 September 2017 from http://www.konjic.ba/index.php/o-opcini/nacionalni-spomenici/item/278-selo-lukomir.
Ourika. 2017. "Ourika Hills—3D Modelling and Animation by BNpro Sarajevo." Retrieved 16 July 2017 from https://www.youtube.com/watch?v=enYFn7xK9nY.
Perović, M., Č. Čopić, and M. Milišić. 1990. "Organizacija ovčarskih kretanja iz sela sa područja planine Bjelašnice." In *Zbornik Biotehniške fakultete Univerze Edvarda Kardelja v Ljubljani*, ed. M. Perović, 603–12. Ljubljana: VTOZD za agronomijo Biotehniške fakultete Univerze Edvarda Kardelja.
Piault, Colette. 2006. "The Construction and Specificity of an Ethnographic Film Project: Researching and Filming." In *Reflecting Visual Ethnography: Using the Camera in Anthropological Research*, ed. Metje Postma and Peter Ian Crawford, 358–75. Leiden: Routledge.
Pink, Sarah. 2011. "Multimodality, Multisensoriality, and Ethnographic Knowing: Social Semiotics and the Phenomenology of Perception." *Qualitative Research* 11(3): 261–76.
Postma, Metje. 2006. "The Construction to Narrative: What's Left of Ethnography?" In *Reflecting Visual Ethnography: Using the Camera in Anthropological Research*, ed. Metje Postma and Peter Ian Crawford, 319–57. Leiden: Routledge.
Sarajlić, Vesna, ed. 1983. *Bjelašnica: Priroda i ljudi*. Sarajevo: Svjetlost.
Skočir, Melita. 2011. "Turizem kot dejavnik ohranjanja planinskega pašništva v Zgornjem Posočju." PhD diss., University of Primorska.
Smith, Valene L., ed. 1977. *Hosts and Guests: The Anthropology of Tourism*. Philadelphia: University of Pennsylvania Press.
Turk Niskač, Barbara. 2011. "Some Thoughts on Ethnographic Fieldwork and Photography." *Studia Ethnologica Croatica* 23(1): 125–48.
Vávrová, Daniela. 2014. *Skin Has Eyes and Ears: Audio-Visual Ethnography in a Sepik Society*. PhD diss., James Cook University.
Viator. 2020. "Lukomir Village Hiking Day Trip: Outdoor Activities in Travnik." Retrieved 6 June 2020 from https://www.lonelyplanet.com/bosnia-and-hercegovina/central-and-northern-bosnia-and-hercegovina/travnik/activities/lukomir-village-hiking-day-trip/a/pa-act/v-71408P7/358726.
Weber, Irena. 1997. *Kultura potepanja*. Ljubljana: Mladinska knjiga.
West, Paige, James Igoe, and Dan Brockington. 2006. "Parks and Peoples: The Social Impact of Protected Areas." *Annual Review of Anthropology* 35: 251–77.

Afterword
Desire for Transhumance

Cyril Isnart

In the social sciences, there is nothing quite like transhumance. It is unique in that it comprises so much: mobility, human-animals relations, rurality, traditions, mountains, shepherds, cheese making, local economy, authenticity, nostalgia, tourism, international comparisons, conflict over land, *longue durée*, territoriality, know-how, and, obviously today heritage-making. Since it allows us to investigate a large number of human practices and human-nonhuman relations, transhumance represents a classic object of study for a broad range of disciplines, such as history, anthropology, geography, or biology to name but a few. In many regional contexts, a significant number of studies and investigations have been carried out over the years, some considered reference works in their respective fields.

We must now consider a triple revision of the literature. The first aspect to address is related to the Anthropocene and climate change dynamics, which notably transform the very material conditions of shepherds and their animals. The second aspect, the ontological turn, especially the field studying human-animal relations, puts transhumance, and more generally pastoralism, into full view, and invites us to reconsider the widespread contemporary opinion that transhumance is just a disappearing rural practice in a more and more urban world. Thirdly, approximately a decade ago, transhumance was entered onto the long list of activities classified as cultural heritage in Europe and the Mediterranean region. In 2019, Greece, Italy, and Austria completed the inscription of transhumance onto the UNESCO Representative List of Intangible Cultural Heritage of Humanity under the title of "Transhumance, the Seasonal Droving of Livestock along Migratory Routes in the Mediterranean and in the Alps."

It is true to say that the heritage domain is still in expansion and has incorporated no end of social, spiritual, biological, and (other) cultural items. It is thus no surprise to see transhumance competing for official international heritage status. Transhumance is a phenomenon that can readily encourage, challenge, and generally play a role in tourism development, cultural investigations, territory mapping, and symbolic representations of identity. In turn, the very heritagization process of transhumance offers a vivid and fascinating point of view on heritage processes, mixing social, environmental, and biological dimensions that rarely merge so intimately in the field of critical heritage studies.

One could say that many types of cultural heritage, such as a religious site, a musical performance, or an industrial building, could also open the field of heritage to various other domains and stand at the intersection of several thematical studies. In that respect, transhumance is not a unique case and can be analyzed as an ordinary heritage object. Nevertheless, something uniquely fascinating does set transhumance as heritage apart from the rest. The fact that transhumance entered the artistic domain very early on is a sign of this peculiar re-enchantment of the world we are seeing today (Isnart and Testa 2020). Countless writers, musicians, painters, and documentary film makers use ethnography of transhumance as a means and pretext of expression and turn transhumance into the object to be documented, shot, described, and preserved.

To recall one artistic production among many images and writings involving transhumance, the work of Jean Giono (1895–1970) took a prominent place when he described and poeticized peasant life in rural Southern France. The novel *Le Grand Troupeau* (1933) depicts two parallel views of, on the one hand, the retreat of injured and exhausted World War I soldiers leaving the front line, and, on the other, numerous sheep and cattle left in poor health due to the low number of shepherds looking after them during an instance of transhumance. Linked to the antiwar convictions of the author, who suffered as a soldier in the trenches, the evocation of transhumance exemplifies the pain and distress of men impacted by the war. Another novel, *Le Serpent d'Étoiles* (1931) includes an immersion into the daily life of a shepherd's family, a play enacting the creation of the world performed by the family and a somehow mysterious and magical contemplation of a spectacular transhumance passage in the bottom of a valley. Giono's writings on transhumance merge an ethnographic perspective of the material conditions of peasants in Southern France, with antiwar and pre-ecological claims through the poeticization of the presence of livestock and the imaginary of local people. Today, the legacy of Giono's interpretation of transhumance is a key element in a wide range of heritagization processes. Different agents and institutions borrow from it to create festivals, develop activities for school children, or carry out photographic and ethnographic surveys. Such heritage actors continue the artistic transformation of transhumance into new and original ways and contribute to the re-enchantment of this agricultural activity today.

In sum, what we seek to find answers to is not so much why transhumance has survived so-called modernity, but better how the fascination for transhumance is vivified, experienced, and transmitted, and what the arguments for a continued or even strengthened attachment to transhumance are today. There are many questions to ask in order to better grasp why transhumance has been selected as cultural heritage in our societies: Why do people and institutions choose to defend transhumance in a mainly urban Europe? How do they materially and ideologically respond to the complicated process of heritage bureaucracy, while promoting a declining rural activity? What motivations move them and what kind of opportunities can they seize through transhumance? What are their arguments, their strategies, and their representations of transhumance? These questions change the direction of the analysis from an ideological rationale opposing "tradition" and "modernity" and lead us towards a more emotional and representational function of transhumance. The new orientations could drive us to a more nuanced and profound understanding of its characteristics.

In many of the cases described in this book, we find a type of strong desire, a sort of profound attachment to transhumance beneath the movement of heritage-making. There is also a sincere determination to overcome the multiple difficulties, dangers and risks of heritage bureaucracy and constraints. The theme of desire is not usually linked to heritage-making in its formal and organizational dimensions. Power, domination, conflicts and diplomacy are the major and more common themes in the vast literature on cultural heritage. However, a recent trend of literature has begun to examine and question emotions, attachment techniques, and the role of individuality in heritage activities, especially when looking at small-scale local projects (Smith 2006; Fabre 2013; Tornatore 2014). Clearly, desire is not the definitive concept to use to think about transhumance, but it takes its place in the picture and deserves to be addressed.

The following paragraphs underline some of the preliminary thoughts perhaps necessary for drawing such a pluralistic picture of transhumance as heritage. These considerations are a partial synthesis of the many ideas put forward by the authors of this book. They also represent an attempt to comprehend transhumance and its appeal in our societies, by addressing the following: the way we, as social scientists, perceive this phenomenon; the combination of politics and policies that transhumance involves; the narration of the past it carries; and lastly, the question of desire. Each of these themes leads to questions that could help to better grasp the tangible and imaginary presence of transhumance in our world.

Reconfiguring the Perception of Transhumance

This concerns the *way* we look at transhumance. The authors of this book provide us with new perspectives on transhumance and describe and analyze how people, institutions, NGOs, and communities participate in the heritage transformation of transhumance. One can read in these chapters about cultural heritage, art, politics, role-playing games, ethics, human-nonhuman relations, tourism, health of cattle, EU as policy maker, conservation practices of nature and culture, and education. Today, transhumance is no more and not only a simple matter of geography, history, or economics. We cannot ignore the many social, scientific, and cultural domains in which transhumance is embedded and with which transhumance exists as a human activity. If transhumance is not only a rural or farm folk phenomenon, then it is possible to think about it and examine it out of the classical frameworks of environmental studies or peasant studies.

Changing our focus and tools implies, on the one hand, changing the way we see actors and institutions of transhumance and the way they modify the spectrum of our interest; on the other hand, a complete social approach to transhumance involves complementing the most classic devices that previous scholars used before, and maybe inventing others. Once more, this sophistication of the methodology is not only necessary in the case of transhumance. Performances of traditional music, for instance, are certainly studied by musicologists and ethnomusicologists, who work on musical structures with musicological tools to understand melody, harmony, rhythm, or instruments. However, anthropologists or geographers can also question music making and social influence on music with their own concepts and apparatus, like power, identities, kinship, policies, or even culture.

So, as transhumance indeed enters other fields, it could be interesting to add new instruments to the toolbox, but also to compare this particular case with other emblematic objects of various disciplines. Thus, the question is: how could we implement a more composite approach to transhumance?

Politics of Transhumance

The second point of investigation that emerged when speaking about transhumance deals with frictions and conflicts, *politics*, so to speak. Many of the chapters have demonstrated that transhumance is a human and landscape regulator that can be included in sustainable development plans and celebrated as a virtuous and ecological practice. Nonetheless, this activity is not easy to carry out. Moving with cattle from one space to another comes with difficulties, strategies, oppositions and conflicts, with landowners, state or local authorities, and with other users of the landscape like tourists or natural park administrators.

The same or similar areas of conflict can be found in any project of heritage-making. Such difficulties come from the fact that many stakeholders are involved in the management of the item selected as heritage and this often prevents a smooth process. The causes of rifts are numerous: some people are owners of the item and others do not want the item to survive; those who need it to survive complain if some practical restrictions are imposed on them; or simply, some people love the item and don't want other people using it. One could list a broad and contrasting range of relations between people and the item, as well as between people themselves. These different modalities of attachment to or dislike of the item provoke tensions and conflicts. The stakes could be power, property, legitimacy, money, social status, honor, or economy but all encompass competition and conflict as a substantial and significant dimension of social life.

In sum, a second avenue for transhumance studies could be how tensions and conflicts shape its practices and its social world. In fact, transhumance is not a consensual object, and perhaps, the essential conflictual nature of transhumance deserves to be addressed as a central topic of the field. Why are contestations and conflicts so crucial to address in transhumance and what is the role of agonistic dimensions in transhumance as a social practice?

Nostalgia, Conflicts, and Transhumance

As the readings on cultural heritage underline, heritage-making always relates to a certain version of the past. A museum, an archive, or a monument never carries all the narratives of the past, but the version of the past that some people and some institutions in particular want to maintain, in society and in the community. Usually, cultural heritage is a national one, in the hands of public cultural administrations. With the UNESCO ICH convention of 2003, with cultural diversity policies, with regionalization processes of Europe, and with the empowerment of minorities, we are facing an increasing transformation of heritage actors and modalities of involvement in heritage matters. Therefore, the question now becomes not only what does the state want to transmit through cultural heritage but who is telling the story of the past? What is the rhetoric and what are the narratives used? What groups are fighting for the past? And, why do actors use a certain past and a certain cultural element?

The plurality of narrations of the past transmit various and often contradictory idealized images of the past that shape the present of the social agents. The fact that transhumance's past is today used for tourism, economy, heritage, and rural development purposes, and the fact that conflict always features in the transhumance landscape, both evoke the concept of *structural nostalgia* of Michael Herzfeld. In Herzfeld's book *Cultural Intimacy* (1997), we encounter shepherds on the island of Crete whose conflicts are regulated by the invocation of rules coming from a partly imaginary past in which the community existed peacefully. In short, the pastoral past legally frames the present relationships.

This is merely an intellectual and theoretical intuition, but the presence of the past and the permanent dimension of conflict in transhumance could be analyzed through the lens of structural nostalgia. This would allow us to appreciate why and in which ways people and communities living off transhumance continuously recall the past to experience and talk about their activity. And then comes a third general question: to what extent could specialists on transhumance apply Herzfeld's structural nostalgia concept in their analysis?

Sharing Desire for Transhumance

The last point is not a small question; it deals with scholars' engagement and involvement, and directly relates to desire. In my own fieldwork as an anthropologist, I do not make numerous formal interviews; I prefer to cooperate for certain tasks, chat during informal occasions, share moments with the people I work, and let them state, defend, and explain *in situ* what their worldviews are. This methodology, definitively not a new one in the history of ethnography, acknowledges people's ability to state and argue their point of view, as well as their individual or community coherence or inconsistencies. In the field of cultural heritage, the aim of my methodology is to enter the heritage-making process and to drop myself somewhere in the assemblage of people who work and act for heritage. This is not always easy, and sometimes actually impossible, but that is part of the fieldwork challenge of entering the heritage system (Isnart 2020). This methodology makes me sometimes a collaborator, sometimes an expert, sometimes an activist. In any case, I am thus able to understand and experience the fight for the cultural good, at the side of the people. Above all, this situates me as a person among the others, a voice among the others, engaged in conflict, alliances, and negotiations. I share at least the interest, or at most the will, to see the element celebrated. In fact, people often ask me to demonstrate my engagement with their cause.

In my role as investigator with this methodology, I am neither an external expert appointed by UNESCO, nor a national administrator of culture, nor a cynical scientist searching for weaknesses or paradoxes. This particular ethnographic position avoids the blind implication of the "heritage believer" or the judgment from the top of the "heritage atheist," as well as the difficult neutral position of the "heritage agonistic," as Christoph Brumann defined the three potential approaches in heritage studies (Brumann 2014). Alternatively, I try to build anthropological knowledge of a heritage experience according to the people around me, in order to understand what caring for the heritage means from their point of view. Today, and until a further reflexive reorientation of this ethnographic methodology emerges, I will argue that engagement in heritage activity—or collaborative ethnographic presence—remains the most adequate way to critically think about heritage-making. Such an engagement implies sharing the desire that animates the people we study, and here for this book, the desire for transhumance.

In sum, if there is nothing quite like transhumance, then heritage actors, shepherds, artists and scientists should consider, reflexively, why and how their paths will cross on the fields of transhumance.

Cyril Isnart is an anthropologist and CNRS Research Fellow at the Institute for Mediterranean, European, and Comparative Ethnology at Aix Marseille Université/CNRS. He investigates heritage practices within traditional music and religious contexts in southern Europe and the Mediterranean. As a research associate at MUCEM (Museum of European and Mediterranean Civilizations, Marseille), he is in charge of an international seminar on the materialities of minorities on the Mediterranean shores (*Singuliers. Les objets des minorités en Europe et en Méditerranée*, 2021–22). He is cofounder of the international network *Respatrimoni*, he coordinated the program *Merap-Med* on religious memory and heritage assertion in the Mediterranean (2013–15), and he served as board member of SIEF (2017–21). He has published thirty articles and forty book chapters and coedited, among others, *Les vocabulaires locaux du "patrimoine"* (2014), *Fabrique du tourisme et experiences patrimoniales au Maghreb* (2018), and *The Religious Heritage Complex: Legacy, Conservation, and Christianity* (2020).

References

Brumann, Christoph. 2014. "Heritage Agnosticism: A Third Path for the Study of Cultural Heritage." *Social Anthropology*. https://doi.org/10.1111/1469-8676.12068.

Fabre, Daniel, ed. 2013. *Les Émotions Patrimoniales*. Paris: Éditions de la Maison des Sciences de l'Homme.

Giono, Jean. 1931. *Le Serpent d'Étoiles*. Paris: Grasset.

———. 1933. *Le Grand Troupeau*. Paris: Gallimard.

Herzfeld, Michael. 1997. *Cultural Intimacy. Social Poetics in the Nation-States*. New York: Routledge.

Isnart, Cyril. 2020. "Materiality, Morality and (Un)Easiness. Association(s), Anthropology and Music Heritage-Making in Portugal." *International Journal of Heritage Studies*. https://doi.org/10.1080/13527258.2020.1719537.

Isnart, Cyril, and Alexandro Testa. 2020. "Reconfiguring Tradition(s) in Europe: An Introduction to the Special Issue." *Ethnologia Europaea* 50(1). https://doi.org/10.16995/ee.1917.

Smith, Laurajane. 2006. *Uses of Heritage*. London: Routledge.

Tornatore, Jean-Louis. 2014. "Words for Expressing What We Care About. The Continuity and the Exteriority of the Heritage Experience." In *Les vocabulaires locaux du "patrimoine." Traductions, négociations, transformations*, ed. Julien Bondaz, Florence Graezer Bideau, Cyril Isnart, and Anaïs Leblon, 31–54. Berlin: Lit Verlag.

Index

A
Abruzzo, 124, 149, 151
Adriatic Sea, 102
Agency for Payments and Interventions in Agriculture (APIA), 217
Agrigento, 269, 276n4
Albania, 6, 8, 102–18
Alto Adige. *See* South Tyrol
Almwirtschaft, 124, 126, 140n5, 222
Alós d'Isil, 87–90, 98n3
Alpilles, 66, 68, 71–72, 75
Alps, 5, 12, 44, 46, 55, 64–65, 69, 94–95, 103, 105, 114, 117, 121, 126–28, 222–33, 234n1, 235n3, 235nn6-7, 300
Alternative Food Networks (AFN), 122–24, 139nn1–2
Amatrice, 150, 158–64
Apennines, 149, 161
APPIA Network for Pastoralism, 153
Apulia, 149, 151, 153, 155
Ariège, 82–83, 86–87, 91, 93, 98n2
Arzana, 268, 273
Assmann, Jan, 186
Association of Transhumant Farmers in Epirus, 36
Associazione Allevatori Friuli Venezia Giulia, 235n5
Attica, 27
Austrian Kingdom of Lombardy-Venetia, 226

B
Bajo, Pietro, 47
Balkans, 117, 119, 177
Barbagia di Seulo, 272
Bare, 285
Baronie, 268, 272
Barthes, Roland, 186
Basilicata, 149, 151
bear, bears, 9, 52, 81–97, 98, 98n2, 114, 118, 209, 244
Beskid Mountains, 11. *See also* Silesian Beskids
Bjelašnica, 13, 280–94, 296nn3–4, 296n6, 296nn8–9
Biella, 46
biocultural heritage, 1, 5–8, 10, 12, 15, 21, 149, 161, 165, 167, 169–70
biodiversity, 1–2, 4–5, 7, 9–10, 23, 61, 65–67, 71, 74, 77, 83, 85, 103, 106, 109, 116, 137, 150, 162, 166–67, 169, 177, 205, 219, 232. *See also* biodiversity conservation
biodiversity conservation, 2, 9–10, 83, 85, 169
Biros, 83, 86, 91–93, 95, 98n3
Bistriţa-Năsăud, 217
Blanc, Jean, 70
Bonabé, 83, 86–90, 95, 98n3
Bonorva, 264
Borşa, 212
Bosnia and Herzegovina, 8, 13, 104, 280, 282, 284, 286. *See also* Federation of Bosnia and Herzegovina
Botiza, 206, 209–12, 215–16
Bozen (Bolzano), 128
Braşov, 207
Brumann, Christoph, 306
Bucovina, 217

Bulgaria, 104
Burns, Robert, 223
Busachi, 265

C
Caltavuturo, 269
Camineras de tramuda, 272
Campania, 151
Campidano, 265, 267–68
Campitello Matese, 153
Canale Valley, 224, 227, 229
Candela, 153
Carinthia, 224, 229
Carnia, 227
Carpathian Convention, 182
Carpathian Trailing of the Sheep, 185
Carpathians, 176–77, 180, 182, 184–85, 189, 192, 204
Casanova, Eugeni, 89
Castel del Giudice, 156
Catalonia, 89, 161
Catania, 269
cattle, 2, 25, 44–46, 124, 127–29, 131, 149, 153, 155–56, 203, 206, 212–14, 216–18, 228–32, 248, 274, 301, 303–4
Cavnic, 215
Cea, 268
Cemi, 113
Center for Pastoral Studies and Implementations of the Alpes-Méditerranée (CERPAM), 66, 72, 74–75, 77–78
Cévennes National Park, 65
Chilivani, 264
Cieszyn, 174, 180–81
Coghinas, 264
Colantuono family, 153, 155–56
collectivization, 11, 104–5, 206, 214–15, 284
Collina Po Biosphere Reserve, 44, 47, 49
Common Agricultural Policy (CAP), 6, 8, 11, 24–26, 31–32, 36–38, 151, 217, 227, 234

Communism, 11, 105, 113, 175, 181, 187, 215–16
Čomor, Ismet and Tidža, 280, 296n4
Connerton, Paul, 187
Conservatory of Natural Areas of Provence (CEN PACA), 67, 73
Consortium for the Protection of Montasio Cheese, 228
Conti, Alberto, 51
Convention on Biological Diversity (CBD), 61–62, 66
Costa, Roberta, 51
Council of Europe's Landscape Convention, 151
Covasna, 207
cow, cows, 92, 127–36, 140n6, 140nn8–9, 208, 210, 212–13, 216–17, 234
Crau, 9, 62, 66–68, 72–74, 77. *See also* National Natural Reserve of the Coussouls of Crau
Crete, 26–28, 31, 33, 37, 305
Croatia, 6, 104
Čuhoviči, 286
cultural heritage, 5–6, 11, 16, 24, 78, 85, 96, 105, 108–10, 117, 125, 150–51, 153, 155, 158, 162, 174, 176, 178–79, 182–85, 187, 192–93, 272, 275, 300–5
Czechoslovakia, 181, 206

D
Danube, 207
Dechavanne, Vincent, 68
Defense of the Forest against Wildfires (DFCI), 66, 72
De Gasperi, Giovanni Battista, 223
De Giorgi, Annalisa, 52
demonticazione, 163
Desulo, 266
Dinarides, 282
DIVERS—Biodiversity of the Mountain Taste, 235n3
Dobrogea, 207
duodji, 250–51
Durance, 65

Dvorsky, Viktor, 223

E
Early Modern Period, 224
earthquake, earthquakes, 150, 160–63, 225, 227
Ecrins national park, 64–65, 68, 70
Enna, 270, 276n4
Epirus, 26, 28, 33, 36
Etna, 269
Etoloakarnania, 27
Europe, x, 2–5, 7–8, 24, 34–35, 46, 67, 126, 138, 155, 157, 161, 175, 180, 189, 204, 216, 219, 289, 296n2, 300, 302, 304
European network of "Green and Blue infrastructure" (TVB), 65, 71
European Union (EU), 7, 11, 24, 36–37, 48, 84, 140n3, 151, 180, 182, 191, 199, 217, 227, 233, 243, 262, 303
extensive pastoralism, 3–4, 10, 12, 14–15, 152–53, 162, 164, 167, 169

F
Federation of Bosnia and Herzegovina, 13, 280, 282
Finland, 242–43, 245–46, 250–55. *See also* Finnish Arctic region
Finnish Arctic region, 8
Fordongianus, 265
Fonni, 264–66
forestry, 64, 66, 244, 249, 255
France, 6, 8, 46, 62, 67, 73, 84, 105, 116–18, 124–25, 153, 301
Friuli-Venezia Giulia (FVG), 12, 224, 228, 235nn4–5
Frosolone, 153, 155

G
gazdas, 174, 177, 183, 191, 197
Gela, 269
Gemona del Friuli, 225
Gennargentu, 264, 268
Geographic Information Systems (GIS), 227

Geraci Siculo, 274–75
Gheg language, 106
Gibson-Graham, J. K., 121, 135
Giono, Jean, 301
Giraldi, Nicola, 228
goat, goats, 8, 23–29, 32–35, 64, 71, 76, 89, 104, 106, 124, 127, 130–34, 136, 141n11, 176, 182, 188, 203, 205, 208, 210, 212–13, 216, 226, 273
Gradina, 286
grassland, grasslands, 1–2, 11, 48–49, 63–64, 73, 106, 108, 127, 205, 207, 209, 212, 217, 219, 222, 267
Greece, 8, 23–37, 155, 300
Greek Agricultural Organization "ELGO-DIMITRA," 38
Greek Payment Authority of CAP Aid Schemes (OPEKEPE), 26
Gutâi, 212, 216–17

H
Hadžići, 284–85, 293
Hala Barania, 176
Hala Ochodzita, 176
Haute-Garonne, 84
herder, herders, 1–2, 4, 9, 45, 48, 104, 131, 151–53, 157, 159, 164, 166, 168, 212, 217–18, 224–25, 228–30, 232, 242–55
heritage turn, 5–6, 10, 162, 167, 169
heritagization, 5–6, 9, 13, 105, 109, 113, 118–19, 149–50, 152–53, 156, 169, 176, 301
Herzfeld, Michael, 305
Hoffman, Luc, 62
House of Shepherds and Transhumance, 117
husbandry, 1, 7–8, 11–13, 116, 124–25, 137, 175, 203, 205, 217–18, 222–23, 227–29, 232–33, 235n7, 250, 283

I
Iceland, 200n2
Ieud, 205–6, 208–16, 218
Igman, 284
Iliđa, 284

inclusion of transhumance in the UNESCO ICH list. *See* UNESCO ICH List
Ingold, Tim, 2, 106, 110, 160, 247–48, 254
Integrated Rangeland Management Plans (IGMPs), 34
Ionian Sea, 102
Isil, 87–90, 98n3
Istebna, 179, 189, 192
Italian Alpine Club (CAI), 162
Italian Ministry of Agricultural, Food and Forestry Policies (MiPAAF), 232
Italy, 8, 10, 24, 27, 32, 44–46, 121, 124, 126, 128, 140n5, 149, 157, 225–27, 300
itinerant shepherd, itinerant shepherds, 45–50, 52–53, 55–57

J
Jacoby, Karl, 82, 98n1
Jehlička, Petr, 138
Jones, Schuyler, 174

K
Kamesznica, 176
Kanun, 111
Kędzior, Jan, 179
Kelmend, 103–18
Koniaków, 10–11, 174, 176–79, 186, 188, 190, 192–93, 196–97
Kohut family, 174, 176–79, 183–85, 189–90, 192–97, 199
Kohut, Maria. *See* Kohut family
Kohut, Piotr. *See* Kohut family
Koniaków, 10–11, 174, 176–79, 186, 188, 190, 192, 196–97. *See also* Shepherd's Centre (Koniaków)
Kosovo, 102, 104
Križnar, Naško, 281
Kuwait, 287, 294

L
Laga and Gran Sasso mountains, 150
lamb, lambs, 29, 45, 159, 176, 197, 212, 213, 265

landscape, landscapes, ix, 1–2, 4–7, 9–12, 14–15, 24, 33, 44–46, 48–51, 55, 57, 63, 66, 69–70, 81–85, 88, 93, 95, 97–98, 103, 105–6, 108–13, 116, 137–38, 149–53, 155–57, 162, 165–69, 170n1, 175, 182, 194, 205–6, 209, 213, 217, 219, 232, 244–46, 249, 252, 255, 260, 264, 271–72, 276n9, 291–93, 303–5
Lapland, 12, 241, 243–44, 246–47, 249–54
Laslaz, Lionel, 68, 73n9
Lazio, 149, 151, 261
Leone, Fabrizio, 52
Licata, 269
Liechtenstein, 200n2
LIFE project (EU), 81, 161
LIFE program (EU). *See* LIFE project (EU)
livestock guardian dogs (LGDs), 81, 83, 85–86, 91, 93–95, 97
Local Action Groups (LAGs), 13, 150, 166, 271, 276n9
Lombardy, 50
Lora Moretto, Albino, 57
Luberon Regional Natural Park, 74
Lukomir, 13, 280–94
Lula, 268

M
Macedonia, 26, 33–34
Macomer, 264
Madonie, 262, 269
Magurka Radziechowska, 176
Maison of Transhumance, 8–9
Malësi e Madhe, 103
Maramureș, 11, 203–8, 212, 214–19
Marx, Karl, ix
Mauss, Marcel, 118
Mediterranean, 5, 23, 31, 37, 67, 72–73, 102, 140n4, 150, 261–62, 276n1, 282, 300
Mercantour National Park, 64–65, 69, 76
Michałek, Józef, 179
Middle Ages, 65, 223–25
Mientjes, Antoon Cornelis, 264

milking, 29, 31–33, 132–33, 185, 194, 210–12, 229, 249, 251, 262, 265–67, 269
Molise, 149, 151, 153, 155–57, 164
Montalbo, 264, 272
Montenegro, 102, 104, 113
Moore, Jason, 82
mountain pasture, mountain pastures, 12, 33, 44, 63, 86–87, 90–94, 97–98, 126–27, 174, 177, 179, 183, 190, 200, 204–7, 210, 212, 222–34, 235n7, 265, 269–85
Multidimensional Tourism Institute (MTI), 241

N
National Council for Research on Agriculture and analysis of the Agrarian Economy (CREA), 153
National Institute of Agricultural Research (INRA), 74
National Institute of Research for Agriculture, Food and the Environment (IRSTEA), 75
National Natural Reserve of the Coussouls of Crau (RNNCC), 66–67, 72, 77
National Natural Reserve of the Coussouls of Crau (RNNCC), 66, 72, 77
National School of Pastoralism (SNAP), 153
National Strategy on Inner Areas, 151
Natisone, 229
Nebrodi, 269, 274
Nenets, 247
Neoneli, 265
Netting, Robert, 223
NGOs, 112, 156, 286–87, 303
Norway, 200n1, 245–46
Nurra, 264, 267

O
Ogliastra, 271–72
Olbia, 264
Orgosolo, 265–66

Orland, Barbara, 127
Orosei, 264
Ozieri, 264

P
Palermo, 270, 274, 276n4
Pallars Sobirà, 83, 86–91
Park Authority of the Po River, 49, 57
Pascolini, Mauro, 224
pastoral routes, 4, 8, 10, 15, 156, 166, 168, 272
Pazarić, 284
Peloponnese, 26–28, 31, 33–34
Peloritani, 269
Perestroika, 105
Pescasseroli, 153
Piedmont, 44–49
Pietrosul Rodnei, 204
Pine, Frances, 175
Po River, 44, 49, 57. *See also* Park Authority of the Po River
Podgrapy, 176
Poland, 8, 174–76, 182–83, 185, 191, 196
Polizzi Generosa, 269–70
Porto Empedocle, 269
Prealps, 224, 231, 235n7
Protected Designation of Origin (PDO), 3, 228, 232, 263
Protected Geographical Indication (PGI), 3, 183
Provence. *See* Region Provence-Alpes-Côte d'Azur
Pyrenees, 8, 9, 81–87, 90–91, 95, 97–98

Q
Quirra, 268

R
Raccolana Valley, 229
Riabitare l'Italia, 153
redyk, 200n1
regeneration, 1, 6, 9, 14, 110, 116, 149–50, 156, 158, 163
Regional Agency for Rural Development (ERSA), 227, 232

Region Provence-Alpes-Côte d'Azur, 46, 61–62, 65, 67, 74
return, 81–84, 87, 95–97
Ristolas, 77
Rodna Mountains, 206, 212, 216
Rodnei mountains. *See* Rodna Mountains
Romania, 6, 8, 11, 105, 125, 185, 190, 203–7, 215–16

S
sałasz, 174–76, 179–81, 185–86, 191, 197–98, 200n1
Sámi, 8, 12, 241–55
San Marco in Lamis, 155
Sarajevo, 281, 284–87, 289–90, 293
Sarcidano, 272
Sardinia, 8, 12, 149, 259–64, 266–67, 270–73, 275, 276n1, 276nn4–5
Saudi Arabia, 287, 294
Sciacca, 269
Scialanga family, 162–64
Scialanga, Silvestro. *See* Scialanga family
Scialanga, Vittoria. *See* Scialanga family
Sclafani, 269
Sennerei, 132–33. *See also Sennerin*
Sennerin, 133–34, 138
Sentein, 91, 98n3
sheep, 1–3, 8, 10, 23, 24–30, 32–35, 44–48, 50, 53, 64–78, 81, 83, 85–98, 104, 106, 124, 127, 149, 157, 161–63, 165–67, 170n1, 174, 176–99, 200n1, 203–8, 210–14, 216–19, 226, 248, 259, 261–74, 282–83, 285, 287, 290, 292–94, 296n2, 296n5, 296n9, 301
Shepherd's Centre (Koniaków), 176–79, 186, 192, 195
Sibiu, 207
Sicily, 52, 149, 259–77
siida, siidat, 245–46, 253–54
Silesian Beskids, 174–75, 179–81, 185, 190192, 199
Siracusa, 269
Slovenia, 8, 81, 93, 223–25, 229, 234
Slow Food, 156, 232, 235n4

Society for the Study and Valorization of Alpine Animal Husbandry Systems (SoZooAlp), 228
South Tyrol, 10, 121–39
Soviet Union, 206
Spain, 6, 84, 86, 90
Stammler, Florian, 247
Stilfs, 121, 128–31, 136–37
Supramonte, 264
Șurdești, 203–4, 209–10, 212, 215–16
sustainable development, 5, 6, 9, 14–15, 23, 102, 114, 117, 119, 161, 182, 287, 303
Sweden, 245–46
Switzerland, 10, 127
Szacka, Barbara, 184

T
Tarčin, 284
Tatra Mountains, 176
Tessarin, Nicoletta, 224
Thessaly, 26–28, 32–33
Thrace, 26
Tirana, 117
Tirta, Mark, 103
Tisa River, 204
Tortoli, 268
traditional pastoralism, 3, 9–10, 13–14, 103, 105, 113, 150, 164, 177, 178, 181, 190, 260
transhumant shepherd, transhumant shepherds, 14, 70, 103, 105, 112, 117, 165, 207, 264, 267
Transylvania, 204, 206–7
Trapani, 270, 276n4
tratturo, tratturi, 150–51, 153, 155, 274
tseligato, tseligata, 29
Turkey, 287, 294
Tuscany, 149

U
Ukraine, 185, 189–90, 196, 204
Umuljani, 286–87
UNESCO, 6, 10, 24, 105, 113, 117–18, 125, 152, 155, 156, 167–68, 232, 271, 300, 306

UNESCO ICH convention, 304
UNESCO ICH List, 5–6, 10, 15, 50, 55, 156, 165
UNESCO MAB (Man and Biosphere), 49
UNESCO World Heritage Sites list, 155
United Arab Emirates, 294

V
Vajont, 225
Val d'Aran, 86, 91–92
Valgaudemar, 64
Ventimiglia di Sicilia, 271
Verdon, 65, 68, 71–72
Vergunst, John L., 160
Viazzo, Pier Paolo, 223
Villassalto, 268
Vinschgau, 126, 128, 131, 136
Vișeu, 212
Vlorë, 102
Vuia, Romulus, 204

W
wildlife, 3, 49, 52, 56, 84, 95–96, 110
Wisła, 179, 186
wolf, wolves, 4, 65–66, 70–71, 76–78, 94–95, 244
World War I, 92, 181, 212, 227, 247, 260, 301
World War II, 92, 122, 175, 181, 189, 203, 212, 227, 244–45, 259–63, 270
World Wildlife Fund (WWF), 52, 62

Y
Yugoslavia, 225, 284, 286

Z
Zakarpatia, 204
Zakopane, 192
Żywiec Beskids, 197

The analytical index was developed by Dr. Omerita Ranalli who we take the opportunity to thank for her careful analysis and composition.